Eulerian theory of Newtonian deformable lattices

Dislocation and disclination charges in solids

Illustration of book cover:

Jenny Gremaud, illustrator

http://wuika-jennyburgy.blogspot.ch/?view=classic

burgy.jenny@gmail.com

Eulerian theory of Newtonian deformable lattices

Dislocation and disclination charges in solids

Gérard Gremaud

Published by Gérard Gremaud
Rte du Pâquier 1, 1723 Marly, Switzerland

ISBN: 978-2-8399-1943-2

Translated from French by Marc Fleury

This book is a short summary
of the original French book:

***Théorie eulérienne
des milieux déformables
Charges de dislocation
et désinclinaison dans les solides***

2013
ISBN 978-2-88074-964-4
http://gerardgremaud.ch/

Introduction

In this book, one summarizes autonomously a first book[1] published during year 2013, which lays methodically the foundations of an original approach of the solid lattices deformation using the Euler coordinates, and which introduces in details the concept of tensor dislocation charges and tensor disclination charges within a lattice. This new concept allows one to quantify the topological singularities which can appear at the microscopic scale of a solid lattice. On the basis of this original approach of the solid lattices and their topological singularities, one can deduce a set of fundamental and phenomenological equations allowing to treat rigorously the macroscopic spatiotemporal evolution of a newtonian solid lattice which deforms in the absolute space of an external observer laboratory.

Motivation and genesis of this book

When one desires to study the solid deformation, one generally uses lagrangian coordinates to describe the evolution of the deformations, and diverse differential geometries to describe the topological defects contained in the solid.

The use of lagrangian coordinates presents a number of inherent difficulties. From the mathematical point of view, the tensors describing the continuous solid deformation are always of order higher than one concerning the spatial derivatives of the displacement field components, which leads to a very complicated mathematical formalism when the solid presents strong distortions (deformations and rotations). To these mathematical difficulties are added physical difficulties when one has to introduce some known properties of solids. Indeed, the lagrangian coordinates become practically unusable, for example when one has to describe the temporal evolution of the microscopic structure of a solid lattice (phase transitions) and of its structural defects (point defects, dislocations, disclinations, boundaries, etc.), or when it is necessary to introduce some physical properties of the medium (thermal, electrical, magnetic or chemical properties) leading to scalar, vectorial or tensorial fields in the real space.

The use of differential geometries in order to introduce topological defects as dislocations in a deformable continuous medium has been initiated by the work of Nye[2] (1953), who showed for the first time the link between the dislocation density tensor and the lattice curvature. On the other hand, Kondo[3] (1952) and Bilby[4] (1954) showed independently that the dislocations can be

---

[1] *Théorie eulérienne des milieux déformables, charges de dislocation et de désinclinaison dans les solides, G. Gremaud, Presses polytechniques et universitaires romandes, Lausanne, Suisse, 2013, 750 pages (ISBN 978-2-88074-964-4)*

[2] *J.F. Nye, Acta Metall.,vol. 1, p.153, 1953*

[3] *K. Kondo, RAAG Memoirs of the unifying study of the basic problems in physics and engineering science by means of geometry, volume 1. Gakujutsu Bunken Fukyu- Kay, Tokyo, 1952*

[4] *B. A. Bilby , R. Bullough and E. Smith, «Continuous distributions of dislocations: a new application of the methods of non-riemannian geometry», Proc. Roy. Soc. London, Ser. A 231, p. 263–273, 1955*

identified as a crystalline version of the Cartan's concept[5] of torsion of a continuum. This approach was generalized in details by Kröner[6] (1960). However, the use of differential geometries in order to describe the deformable media leads very quickly to difficulties similar to those of the lagrangian coordinates system. A first difficulty arises from the complexity of the mathematical formalism which is similar to the formalism of general relativity, what makes very difficult to handle and to interpret the obtained general field equations. A second difficulty arises with the differential geometries when one has to introduce topological defects other than dislocations. For example, Kröner[7] (1980) has proposed that the existence of extrinsic point defects could be considered as extra-matter and introduced in the same manner that matter in general relativity under the form of Einstein equations, which would lead to a pure riemannian differential geometry in the absence of dislocations. He has also proposed that the intrinsic point defects (vacancies and interstitials) could be approached as a non-metric part of an affine connection. Finally, he has also envisaged introducing other topological defects, as disclinations for example, by using higher order geometries much more complex, as Finsler or Kawaguchi geometries. In fact, the introduction of differential geometries implies generally a heavy mathematical artillery (metric tensor and Christoffel symbols) in order to describe the spatiotemporal evolution in infinitesimal local referentials, as shown for example in the mathematical theory of dislocations of Zorawski[8] (1967).

In view of the complexity of calculations in the case of lagrangian coordinates as well as in the case of differential geometries, it seemed to me that it would be better to develop a much simpler approach of deformable solids, but at least equally rigorous, which has been finally published in a first book[1] during year 2013: *la théorie eulérienne des milieux déformables*.

Regarding *the description of defects (topological singularities)* which can appear within a solid, as dislocations and disclinations, it is a domain of physics initiated principally by the idea of macroscopic defects of Volterra[9] (1907). This domain experienced a fulgurant development during the twentieth century, as well illustrated by Hirth[10] (1985). The lattice dislocation theory started up in 1934, when Orowan[11], Polanyi[12] and Taylor[13] published independently papers describing the edge dislocation. In 1939, Burgers[14] described the screw and mixed dislocations. And finally in 1956, Hirsch, Horne et Whelan[15] and Bollmann[16] observed independently dislocations in me-

---

[5] *E. Cartan, C.R. Akad. Sci., 174, p. 593, 1922 & C.R. Akad. Sci., 174, p.734, 1922*

[6] *E. Kröner, «Allgemeine Kontinuumstheorie der Versetzungen und Eigenspannungen», Arch. Rat. Mech. Anal., 4, p. 273-313, 1960*

[7] *E. Kröner, «Continuum theory of defects», in «physics of defects», ed. by R. Balian et al., Les Houches, Session 35, p. 215–315. North Holland, Amsterdam, 1980.*

[8] *M. Zorawski, «Théorie mathématique des dislocations», Dunod, Paris, 1967.*

[9] *V. Volterra, «L'équilibre des corps élastiques», Ann. Ec. Norm. (3), XXIV, Paris, 1907*

[10] *J.-P. Hirth, «A Brief History of Dislocation Theory», Metallurgical Transactions A, vol. 16A, p. 2085, 1985*

[11] *E. Orowan, Z. Phys., vol. 89, p. 605,614 et 634, 1934*

[12] *M. Polanyi, Z. Phys., vol.89, p. 660, 1934*

[13] *G. I. Taylor, Proc. Roy. Soc. London, vol. A145, p. 362, 1934*

[14] *J. M. Burgers, Proc. Kon. Ned. Akad. Weten schap., vol.42, p. 293, 378, 1939*

[15] *P. B. Hirsch, R. W. Horne, M. J. Whelan, Phil. Mag., vol. 1, p. 667, 1956*

[16] *W. Bollmann, Phys. Rev., vol. 103, p. 1588, 1956*

tals by using electronic microscopes. Concerning the disclinations, it is in 1904 that Lehmann[17] observed them in molecular crystals, and in 1922 that Friedel[18] gave them a physical explanation. From the second part of the century, the physics of lattice defects has grown considerably.

In the first part of this essay, the dislocations and the disclinations are approached by introducing intuitively the concept of dislocation charges by using the famous Volterra pipes[19] (1907) and an analogy with the electrical charges. With Euler coordinates, the concept of dislocation charge density appears then in an equation of geometro-compatibility of the solid, when the concept of flux of charges is introduced in an equation of geometro-kinetics of the solid.

The *rigorous formulation of the charge concept* in the solids makes the essential originality of this approach of the topological singularities. The detailed development of this concept leads to the appearance of tensorial charges of first order, *the dislocation charges*, associated with *the plastic distortions* of the solid (plastic deformations and rotations), and of tensorial charges of second order, *the disclination charges*, associated with *the plastic contortions* of the solid (plastic flexions and torsions). It appears that these topological singularities are quantified in a solid lattice and that they have to appear as *strings (thin tubes)* which can be modelized as unidimensional lines of dislocation or disclination, or as *membranes (thin sheets)* which can be modelized as two-dimensional boundaries of flexion, torsion or accommodation.

The concept of dislocation and disclination charges allows one to find rigorously the main results obtained by the classical dislocation theory. But it allows above all to define a tensor $\vec{\Lambda}_i$ of *linear dislocation charge*, from which one deduces a scalar Λ of *linear rotation charge*, which is associated with the screw part of the dislocation, and a vector $\vec{\Lambda}$ of *linear flexion charge*, which is associated with the edge part of the dislocation. For a given dislocation, both charges Λ and $\vec{\Lambda}$ are perfectly defined without needing a convention at the contrary of the classical definition of a dislocation with its Burger vector! On the other hand, the description of the dislocations in the eulerian coordinate system by the concept of dislocation charges allows one to treat exactly the evolution of the charges and the deformations during very strong volumetric contractions and expansions of a solid medium.

Content of the book

This book is composed of four sections which summarize autonomously the eulerian theory of deformation introduced in my first book[1] published in 2013.

The first section (A) introduces *the eulerian deformation theory of newtonian lattices.*

The deformation of a lattice is characterized by *distortions and contortions (chap. 1 to 3)*. A vectorial representation of the tensors, presenting undeniable advantages over purely tensorial representation thanks the possibility to use the powerful formalism of the vectorial analysis, allows to obtain *the geometro-compatibility equations* of the lattice which insure its solidity, and *the geometro-kinetics equations* of the lattice, which allow one to describe the deformation kinetics.

---

[17] *O. Lehmann, «Flussige Kristalle», Engelman, Leibzig, 1904*

[18] *G. Friedel, Ann. Physique, vol. 18, p. 273, 1922*

[19] *V. Volterra, «L'équilibre des corps élastiques», Ann. Ec. Norm. (3), XXIV, Paris, 1907*

One introduces then the physics in this topological context *(chap. 4)*, namely *the newtonian dynamics* and *the eulerian thermo-kinetics* (based on the first and second principles of thermodynamics). With all these ingredients, it becomes possible to describe the particular behaviors of a solid lattice *(chap. 5)*, as *the elasticity, the anelasticity, the plasticity* and *the self-diffusion*. This first section ends with the establishment of the complete set of evolution equations of a lattice in the Euler coordinate system *(chap. 6)*.

The second section (B) is dedicated to *the applications of the eulerian theory (chap. 7)*.

It presents very succinctly some examples of *phenomenologies of everyday solids*. One shows how to obtain the functions and equations of state of an isotropic solid, what are the elastic and thermal properties which can appear, how waves propagate and why there exist thermoelastic relaxations, what are the mass transport phenomena and why it could appear inertial relaxations, what are the common phenomenologies of anelasticity and plasticity, and finally how it can appear structural transitions of first and second order in a solid lattice.

The third section (C) is dedicated to the introduction of *dislocation charges* and *disclination charges* in the eulerian lattices.

After the analytical introduction of the concepts of density and flux of dislocation and disclination charges in the lattices *(chap. 8)*, one presents a detailed review of *the lattice macroscopic and microscopic topological singularities* which can be associated to the dislocation and disclination charges *(chap. 9)*.

Then one discusses the motion of dislocation charges within the lattice by introducing *the dislocation charges flux of the dislocation charges* and *the Orowan relations (chap. 10)*. Finally, one deduces the Peach and Koehler force which acts on the dislocations, and one establishes the new set of evolution equations of a lattice in the Euler coordinate system *(chap. 11)*, which takes into account the existence of topological singularities within the lattice.

The fourth section (C) is dedicated to *the applications of the charge concept* within the eulerian solid lattice *(chap. 12)*.

It shows the elements of the dislocation theory in the everyday solids. One begins to show that, in the particular case of the deformation of isotropic lattices by pure shears, one can replace the shear strain tensor by the rotation vector, which allows one to find a set of equations which corresponds strictly *to all the Maxwell equations of electromagnetism*! One shows then how to calculate the fields and energies of the screw and edge dislocations in an isotropic lattice, just as the interactions which can occur between dislocations. One finishes this section of applications by presenting *the string model* of dislocations, which is the fundamental model allowing one to explain most of the macroscopic behaviors of anelasticity and plasticity of crystalline solids.

An other application of the theory

In this book, the reader will see numerous analogies between the eulerian theory of deformable media and the theories of electromagnetism, gravitation, special relativity and general relativity. These analogies are sufficiently surprising and remarkable to alert any open and curious scientific spirit! But it is also clear that these analogies are, by far, not perfect in this book. It is then tantalizing to analyze much more carefully these analogies and to try to find how to perfect them. That is done in an other book[20],[21], which is entirely allotted to the deepening, the improvement and the understanding of these analogies.

[20] *Gérard Gremaud, "Universe and Matter conjectured as a 3-dimensional Lattice with Topological Singularities", Amazon company, Lausanne, 2016, 646 pages, ISBN 978-2-8399-1934-0*

[21] *Gérard Gremaud, "Univers et Matière conjecturés comme un Réseau Tridimensionnel avec des Singularités Topologiques", Amazon company, Lausanne, 2016, 660 pages, ISBN 978-2-8399-1940-1*

Table of contents

Part B - Application: examples of phenomenologies of usual solids

Appendix

Mathematical formulas

PART A

Eulerian theory of newtonian deformable lattices

Distortions and contortions

Geometro-compatibility and geometro-kinetic

Newtonian dynamics and eulerian thermo-kinetics

Elasticity, anelasticity, plasticity and self-diffusion

Evolution equations of a newtonian lattice

Chapter 1

Distortions of a lattice

In this chapter, one introduces a description of the solid deformable lattices in *the Euler coordinate system*. From the definition of the field of average velocity $\vec{\phi}$ of the sites of a lattice in the absolute frame of an observer ***GO***, one deduces *a volumetric expansion scalar* τ and *a rotation vector* $\vec{\omega}$, which satisfy *geometro-kinetic equations* of spatiotemporal evolution.
One shows then these two entities are in fact the trace and the antisymmetric part of a more general tensor, called *the distortion tensor*, which will be represented *in vectorial notation* by $\vec{\beta}_i$, and which satisfies also a *geometro-kinetic equations* in the Euler coordinates. From this tensor, one deduces also *a strain tensor* $\vec{\varepsilon}_i$ and *a shear tensor* $\vec{\alpha}_i$ of the solid lattice.
Later, with some simple examples of velocity fields $\vec{\phi}(\vec{r},t)$ associated with diverse known movements of the medium, one verifies that the topological tensors $\vec{\beta}_i$, $\vec{\varepsilon}_i$, $\vec{\alpha}_i$, $\vec{\omega}$ and τ describe perfectly the distortions, the deformations, the shear strains, the rotations and the volumetric expansions which can appear within a solid lattice.

1.1 - Spatiotemporal evolution of a deformable medium

If an observer, who will be called *the Great Observer* ***GO***, wants to describe in his laboratory the spatiotemporal evolution of a given continuous medium which moves in the space by translation and rotation, and which can moreover deform in the course of time (figure 1.1), he has first to define the type of kinetic which he has to use. Using as basic axiom that the spatiotemporal evolution of the medium is described by *a galilean kinetic*, and as a consequence by a principle of velocity additivity, the observer ***GO*** can describe this evolution on the basis of the absolute frame of his laboratory. This reference frame is composed of an orthonormal basic frame $Q\xi_1\xi_2\xi_3$, that is *three orthogonal measuring rods of unit length* $(\mathbf{\vec{e}}_1,\mathbf{\vec{e}}_2,\mathbf{\vec{e}}_3)$, and *one universal clock*, insuring that the time t is measured in identical way everywhere in his laboratory (figure 1.1).
In order to describe simply and completely the spatiotemporal evolution of the continuous medium, the observer can use *a lagrangian coordinate system*. First of all, he carries out a marking of the material medium at the initial time $t=0$ by means of a grid of points $\boldsymbol{P}_0$. He can then define a moveless local frame $Ox_1x_2x_3$, situated at the point identified by $\vec{\xi}_O$ in the frame of his laboratory. By equipping this fixed frame $Ox_1x_2x_3$ with measuring rods of unit length $(\vec{e}_1,\vec{e}_2,\vec{e}_3)$, and by positioning it judiciously with regard to the initial position of the medium at time $t=0$, he can measure the positions of all the points $\boldsymbol{P}_0$ of the medium at the initial time $t=0$ by means of vectors $\vec{r}$. At a time $t>0$, a point $\boldsymbol{P}_0$ of the medium will move at $\boldsymbol{P}$, and

the observer can then link up the point P_0 to the point P by using a vector $\vec{u}$, which is called *the displacement vector* of the point P_0. As this vector is dependent on the initial position $\vec{r}$ of the point P_0 and on the time t, the whole of vectors $\vec{u}(\vec{r},t)$ locating all the points of the medium is called *the displacement field* of the medium in lagrangian coordinates.

The concept of continuous medium at the macroscopic scale

The expression of continuous medium is an intuitive concept meaning that the medium does not present during its spatiotemporal evolution and at the macroscopic scale where it is observed, neither the appearance of discontinuous structures, nor the forming of discontinuities such as tears or local breakings, nor the forming of cavities.

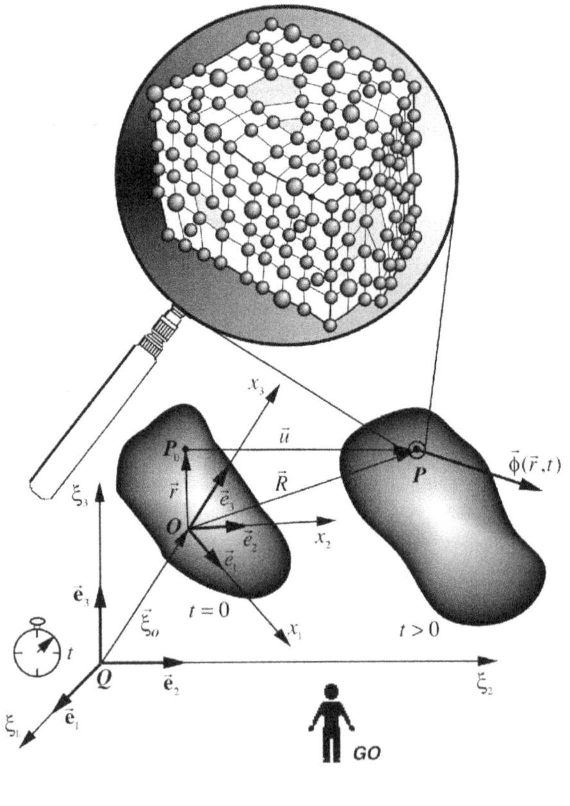

Figure 1.1 - *spatiotemporal evolution of a medium in the absolute frame*

From the macroscopic observation of the behavior of the medium, and in particular of *the continuity of the displacement field* $\vec{u}$, it is possible to attribute some appellations to the observed medium.

If the medium presents a displacement field $\vec{u}$ perfectly continuous during its spatiotemporal evolution, it is qualified of *solid medium.* It owns then a macroscopic shape which is difficult to modify.
On the other hand, if the medium presents a discontinuous displacement field $\vec{u}$ which forms an inextricable tangle in the course of time, it is qualified of *fluid medium.* It owns then the macroscopic property to flow, and has consequently to be maintained in a container from which it takes the shape. In this case, the displacement field $\vec{u}$ lost its original physical sense and meaning, and only *the local velocity* $\vec{\phi}(\vec{\xi},t)$ of the fluid at time t and at the space coordinates $\vec{\xi}$ of the absolute frame maintains a physical sense. In this book, one will be essentially concerned by the case of solid medium.

The concept of continuous medium at the microscopic scale

The concept of continuous solid medium is applicable only when the medium is observed at *a macroscopic scale.* Indeed, an enlargement of the same medium *at a sufficiently microscopic scale* will show a discontinuous collection of objects (fig. 1.1), at which one will afterwards attribute *the generic name of particles* (for example corpuscles, atoms, molecules, etc.). One logically concludes that the global phenomenological properties observed at the macroscopic scale where the medium seems continuous are in fact statistical effects resulting from the great number of particles which interact between themselves at the microscopic scale.
The enlargement of the medium allows also to define some of its important microscopic characteristics, as *its structure,* that is the manner in which its particles are assembled together, and *its chemical composition,* that is the nature of its particles.
A continuous medium will be called *solid* when, at the microscopic scale, it corresponds to a collection of particles such as *the identity of the nearest neighbour particles of a given particle does not change in the course of time.* In other words, each particle is bounded to its nearest neighbour particles by stiff bond which prevent long distance motions. As a consequence, only short distance motions from the nearest neighbour particles are permitted, due to the elasticity of the bonds. By the action of these bonds, the particles of the medium form *a solid lattice.*

The different classes of solid lattices

It is possible to define diverse classes of solid lattices, depending on the arrangement of the particles relative to each other. If the particles presents an arrangement well established, which repeats at long range by translation of an elementary unit cell, one speaks about *a lattice with a crystalline structure.* For example, the two-dimensional lattice reported in figure 1.2(a) and obtained by the translation of a hexagonal unit cell presents a perfect order at long range as well as at short range.
Some solid lattices can present arrangements of their particles without long range order, but only a certain order at short range. One speaks in this case about a lattice with *an amorphous structure.* The two-dimensional example reported in figure 1.2(b) represents an amorphous lattice of particles, obtained by paving the surface with irregular pentagons, hexagons and heptagons with fixed length sides. The short range order of the lattice is reflected by the fact that each particle possesses exactly three closest neighbours.

Solid lattices can also exist with particle arrangement which does not present long range translation order, but a certain order by rotation. One speaks then about *a lattice with a quasicrystalline structure*, represented by the example in figure 1.2(c) which shoes clearly the absence of long range order by translation.

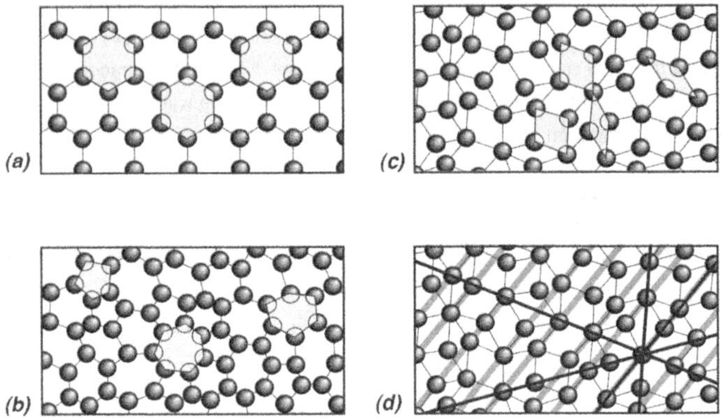

Figure 1.2 *- two-dimensional lattices with crystalline (a), amorphous (b) and quasicrystalline (c-d) arrangements of their particles*

This lattice is obtained by paving the surface with two types of diamond shaped unit cells with different apex angles (Penrose paving). At first sight, this lattice seems amorphous. But a more detailed analysis reported in figure 1.2(d) shows that the particles aligned with parallel straight lines. The distances between these parallel lines are not regular, and there exists in fact five different privileged directions for the orientation of these alignments. This means that this two-dimensional quasicrystalline structure presents a kind of fifth order rotation symmetry, which is not allowed in the case of the crystalline structures obtained by translation of base patterns.

The examples shown in figure 1.3b are two-dimensional representations. It is then necessary to generalize these notions to the three-dimensional space. In three dimensions, the crystalline lattices are constituted by the translation of a three-dimensional elementary volume, called *the unit cell of the lattice* (figure 1.3). The crystalline lattices can be defined by the lengths of the edges of the unit cell and the angles between them, which are called *the lattice parameters*. The crystal structures can be grouped in *seven lattice systems* according to the axial system used to describe their lattice. Considering the possible arrangement of the atoms relative to each other in the unit cell *(lattice centerings P, C, I or F)*, one arrives at *the fourteen Bravais lattices* represented in figure 1.3. Using the symmetry operations of rotation, reflection and inversion that leave at least one point unmoved and the appearance of the crystal structure unchanged, one can still define *32 possible crystallographic point groups*. Using the symmetry operations associated to the translation, one can finally define *230 distinct space groups*.

It is interesting to note that the lattice centerings leads to different values of the number of substitutional sites of the unit cell which can contain a bounded particle. For example in the case of the cubic lattice system, the number of substitutional sites of the unit cell is one for *the simple cubic lattice*, 2 for *the body-centered cubic lattice* and 4 for *the face-centered cubic lattice.*

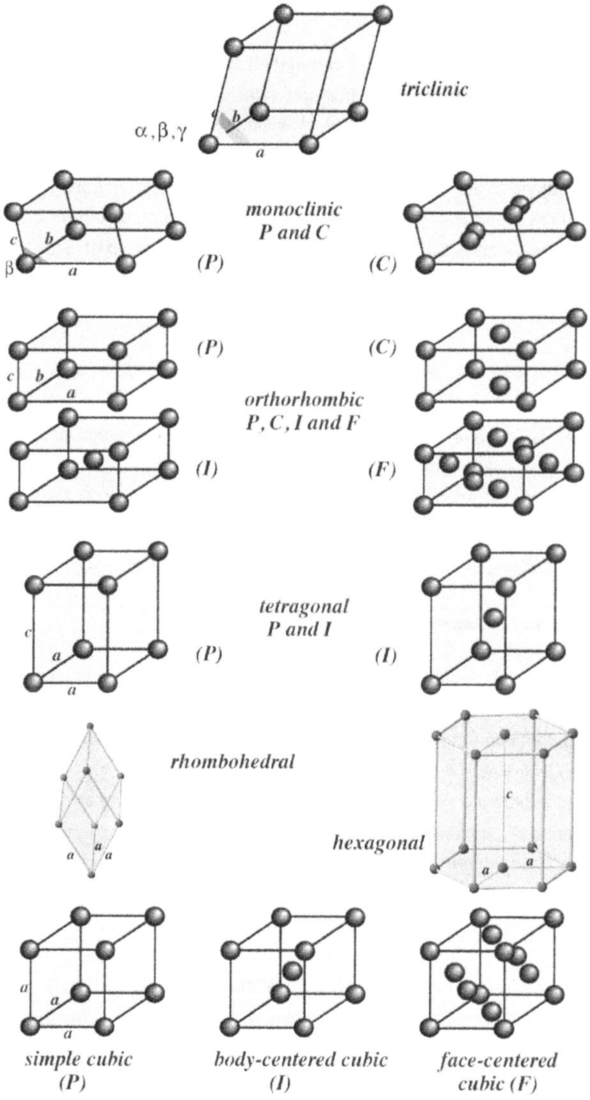

Figure 1.3 *- the seven crystalline systems and the fourteen Bravais lattices*

In the case of unordered solid medium as *the amorphous solids, the quasicrystalline solids* or *the polycrystalline solids with very fine grains*, the notion of lattice unit cell has no more meaning. But the concept of lattice site keeps a physical meaning, in the sense that it corresponds to a substitutional site of the unordered solid lattice.

The concept of topological singularities in a solid lattice

In the case of an ordered solid lattice, it can appear structural defects inside the regular arrangement of the lattice particles. These structural defects have diverse origins as irregularities of the chemical species of the lattice, or topological singularities, that is irregularities in the topological structure of the lattice which can be punctual, linear or planar following their topology. It is also by observations of the dynamics at the microscopic scale, during the macroscopic spatiotemporal evolution of the lattice, that it will be possible to understand the objective reasons of some macroscopic behaviors. For example, one will see that it can exist close links between the macroscopic properties of deformation of ordered lattices and the distorsions of the lattice unit cell induced by the presence of mobile topological singularities within the lattice, as dislocations or disclinations.
As a conclusion, a complete description of the spatiotemporal evolution of a lattice which can be considered as continuous at the macroscopic scale *cannot be obtained without a description of the phenomena which take place at the microscopic scale*. The search of a theory describing the macroscopic spatiotemporal evolution of a continuous deformable lattice has to be based on the definition of average macroscopic fields (scalar, vectorial and tensorial) deduced from a statistical description of the dynamics at the microscopic scale of a multitude of objects interacting with each other.

The lagrangian and the eulerian coordinate systems

In order to describe the spatiotemporal evolution of a deformable continuous medium, there exist two well known coordinate systems: *the lagrangian coordinate system*, used generally to describe the deformation of solids, and *the eulerian coordinate system*, used generally to describe the hydrodynamics of fluids.
The lagrangian coordinate system is based on *the description of the temporal evolution of the previously defined displacement field* $\vec{u}(\vec{r},t)$, knowing the initial coordinates $\vec{r}$ of all the points of the solid in a fixed reference frame $Ox_1x_2x_3$ of the observer laboratory, as illustrated in figure 1.1.
Concerning the eulerian coordinate system, it is based on *the description of the temporal evolution of the velocity field* $\vec{\phi}(\vec{\xi},t)$ of the medium points situated at the space coordinates $\vec{\xi}$ at the time t in the absolute reference frame $Q\xi_1\xi_2\xi_3$ of the observer laboratory.
The lagrangian coordinate system is well adapted to the description of the evolution of solid lattices which deform very weakly, but becomes perfectly unusable to describe strong deformations of a lattice, or to describe a lattice containing topological singularities. The eulerian coordinate system is much more general, because it allows one to describe not only the fluids, but also the solid lattices presenting strong deformations or containing topological singularities. It is the reason why we develop in this essay *an eulerian description of the spatiotemporal evolution of deformable solid lattices*.

1.2 - Definition of local quantities in Euler coordinates

The average local velocity $\vec{\phi}(\vec{r},t)$

In the case of a collection of solid state particles in the space, each particle i has its own velocity $\vec{v}_i$. In order to determine the average local velocity of the particles, one has to fix a small element of volume V_f centered on the space coordinates $\vec{r}$ (figure 1.4), then to measure the velocities $\vec{v}_i$ of all the particles contained in this fixed volume V_f. If the instantaneous number of particles contained in this volume V_f is equal to N, and if N is sufficiently large, *the average velocity* $\vec{\phi}$ at the place $\vec{r}$ and the instant t can be defined by the following expression

$$\vec{\phi} = \vec{\phi}(\vec{r},t) = \frac{1}{N}\sum_{i \in V_f} \vec{v}_i(t) \tag{1.1}$$

If an average velocity $\vec{\phi}$ different of zero is measured, this means also that it is possible to find for each particle a fluctuation $\Delta\vec{v}_i$ to the average velocity $\vec{\phi}$ by the relation

$$\vec{v}_i = \vec{\phi} + \Delta\vec{v}_i \quad \Rightarrow \quad \frac{1}{N}\sum_{i \in V_f} \Delta\vec{v}_i(t) = 0 \tag{1.2}$$

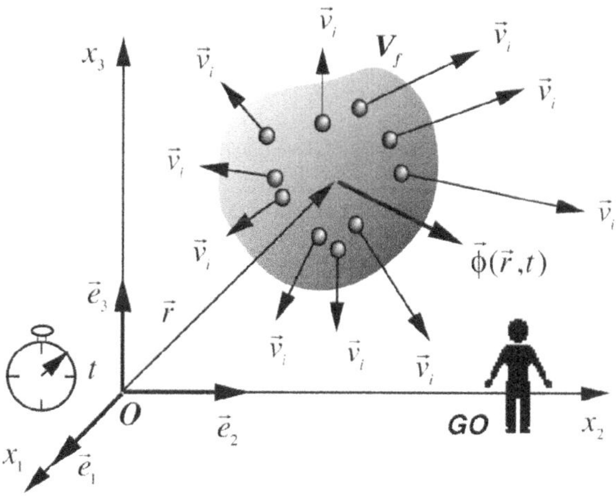

Figure 1.4 *- search of the average local velocity of a lattice in the volume* V_f

In a solid lattice, the existence of an average velocity $\vec{\phi}(\vec{r},t)$ different from zero involves that the solid lattice of particles be submitted to a collective movement. The velocity $\vec{\phi}(\vec{r},t)$ represents then the average local velocity of displacement of the particles bound to the lattice sites, which corresponds to *the average velocity of the lattice sites*, when the $\Delta\vec{v}_i$ are the fluctuations of velocity of the particles bounded to the lattice around each of these sites. For example, in a real solid, such fluctuations are due to the disordered motions of the thermal agitation of the particles associated directly to the solid temperature.

The local density $n(\vec{r},t)$ of elementary substitutional sites of the lattice

Besides the average local velocity $\vec{\phi}(\vec{r},t)$ of the sites of a macroscopically continuous lattice, there is also an other quantity which will play a fundamental role in Euler coordinates: it is *the volume density of elementary substitutional sites of the lattice* which will be written n. This choice will involve to define all the physical quantities characterizing the solid lattice as *average values taken on each site of the lattice.* It is clear also that, in disordered lattice, the quantity n can be related to the volume density of elementary sites of the disordered lattice.

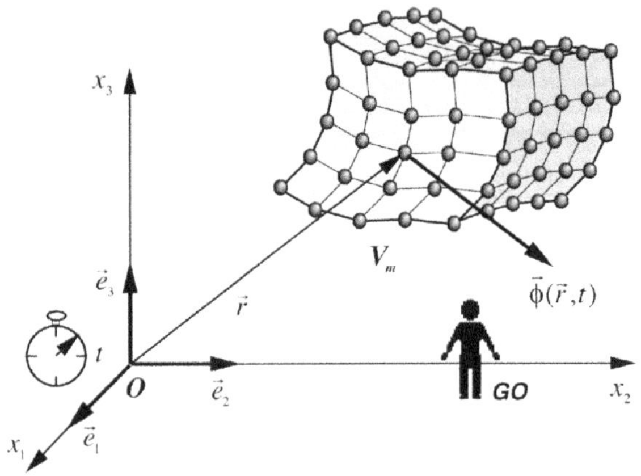

Figure 1.5 - *average velocity of the collective motion of the sites of a solid lattice*

This density is a local average value which can be expressed as a function of $\vec{r}$ and t

$$n = n(\vec{r},t) \tag{1.3}$$

The quantity n has to satisfy a *continuity equation.* In order to express this equation, one has to consider a volume V_m moving with the lattice at the velocity $\vec{\phi}$. The total number N of sites of the lattice contained in the volume V_m is given by

$$N = \iiint_{V_m} n\,dV \tag{1.4}$$

Over times, this total number N can vary only if it exists sources of lattice sites inside the volume V_m, since the volume V_m follows the medium at the velocity $\vec{\phi}$. The continuity equation can be written by equating the temporal variation of N with *a term of source* S_n

$$\frac{dN}{dt} = \frac{d}{dt}\iiint_{V_m} n\,dV = \iiint_{V_m} S_n\,dV \tag{1.5}$$

In this relation S_n is the term of *volume source of sites of lattice,* which is the number of sites of the lattice created of annihilated by unit of volume and unit of time within the lattice under consideration.

With *(A.58)* applied to a mobile volume, expression *(1.5)* becomes

$$\iiint_{V_m} \left[\frac{\partial n}{\partial t} + \operatorname{div}\left(n\vec{\phi} \right) \right] dV = \iiint_{V_m} S_n \, dV \qquad (1.6)$$

which holds for any mobile volume V_m we choose, and the equation of local continuity for n can be read off

$$\frac{\partial n}{\partial t} = S_n - \operatorname{div}\left(n\vec{\phi} \right) \qquad (1.7)$$

This equation links the local density $n(\vec{r},t)$ of sites of lattice and the average velocity $\vec{\phi}(\vec{r},t)$ of these sites. We will discuss later the possibility of existence of sources of sites S_n within the lattice.

The material derivative in eulerian coordinates

Relation *(1.7)* can be transformed by expanding the term in the divergence

$$\frac{\partial n}{\partial t} = S_n - n \operatorname{div}\vec{\phi} - \vec{\phi}\, \overrightarrow{\operatorname{grad}}\, n \qquad (1.8)$$

and then using operator $\vec{\nabla}$, defined in section A.3

$$\frac{\partial n}{\partial t} + (\vec{\phi}\vec{\nabla})n = S_n - n \operatorname{div}\vec{\phi} \qquad (1.9)$$

We introduce here an important operator in eulerian coordinates called *the material derivative*, and represented by the symbol (straight derivative instead of curled) d / dt and defined as

$$\frac{d}{dt} = \frac{\partial}{\partial t} + (\vec{\phi}\vec{\nabla}) \qquad (1.10)$$

This operator corresponds to the temporal derivative of a quantity observed along the trajectory of the sites of the lattice. It consists in fact in calculating the total derivative of $n(\vec{r},t)$, with respect to time, as shown in the following equalities

$$\frac{dn}{dt} = \frac{\partial n}{\partial t} + \sum_i \frac{\partial n}{\partial x_i}\frac{\partial x_i}{\partial t} = \frac{\partial n}{\partial t} + \sum_i \frac{\partial n}{\partial x_i}\phi_i = \frac{\partial n}{\partial t} + (\vec{\phi}\vec{\nabla})n \qquad (1.11)$$

With this operator, the continuity equation n can be presented as a path derivative

$$\frac{dn}{dt} = S_n - n \operatorname{div}\vec{\phi} \qquad (1.12)$$

The equation of continuity for density n of sites of lattice is fundamental, because we are going to base all solid lattices in eulerian coordinates.

Non-commutativity of spatial and temporal operators in eulerian coordinates

In eulerian coordinates, the principal operator of time, the material derivative d / dt along the path, does not, in general commute with the operators of space $\overrightarrow{\operatorname{rot}}$, $\overrightarrow{\operatorname{grad}}$ and div. This important property of operators can be verified. Indeed for any vector $\vec{A}$ defined in the medium we have

$$\begin{cases} \overrightarrow{\mathrm{rot}}\left(\frac{d\vec{A}}{dt}\right) = \overrightarrow{\mathrm{rot}}\left[\frac{\partial \vec{A}}{\partial t} + (\vec{\phi}\vec{\nabla})\vec{A}\right] = \frac{\partial}{\partial t}\overrightarrow{\mathrm{rot}}\,\vec{A} + (\vec{\phi}\vec{\nabla})\overrightarrow{\mathrm{rot}}\,\vec{A} + \overrightarrow{\mathrm{grad}}(\vec{\phi}\vec{\nabla}) \wedge \vec{A} \\ \overrightarrow{\mathrm{rot}}\left(\frac{d\vec{A}}{dt}\right) = \overrightarrow{\mathrm{rot}}\left[\frac{\partial \vec{A}}{\partial t} + (\vec{\phi}\vec{\nabla})\vec{A}\right] = \frac{d}{dt}\left(\overrightarrow{\mathrm{rot}}\,\vec{A}\right) + \sum_k \overrightarrow{\mathrm{grad}}\,\phi_k \wedge \frac{\partial \vec{A}}{\partial x_k} \end{cases} \tag{1.13}$$

We deduce the relation of commutation between the rotational and the material derivative which can be written in the form

$$\overrightarrow{\mathrm{rot}}\left(\frac{d\vec{A}}{dt}\right) - \frac{d}{dt}\left(\overrightarrow{\mathrm{rot}}\,\vec{A}\right) = \sum_k \overrightarrow{\mathrm{grad}}\,\phi_k \wedge \frac{\partial \vec{A}}{\partial x_k} \tag{1.14}$$

Similar expressions can be found for the gradient of a scalar B or the divergence of a vector $\vec{A}$

$$\overrightarrow{\mathrm{grad}}\left(\frac{dB}{dt}\right) - \frac{d}{dt}\left(\overrightarrow{\mathrm{grad}}\,B\right) = \sum_k \overrightarrow{\mathrm{grad}}\,\phi_k \frac{\partial B}{\partial x_k} \tag{1.15}$$

$$\mathrm{div}\left(\frac{d\vec{A}}{dt}\right) - \frac{d}{dt}\left(\mathrm{div}\,\vec{A}\right) = \sum_k \overrightarrow{\mathrm{grad}}\,\phi_k \frac{\partial \vec{A}}{\partial x_k} \tag{1.16}$$

1.3 - Eulerian geometro-kinetic equations

A definition of the volume expansion of a lattice

The volume density n of sites of the solid lattice is directly linked to the volume expansion of the medium. In fact $n \to 0$ for large expansions and $n \to \infty$ for large contractions. This notion of volume expansion of the lattice can be better captured with a quantity v defined as the inverse of n

$$v = 1/n \tag{1.17}$$

Indeed this quantity v has the dimension of a volume. It represents the average volume occupied by an elementary site of the lattice. This volume v translates the intuitive notion of volume expansion of the medium, as $v \to \infty$ for large expansions and $v \to 0$ for intense contractions. But relationship *(1.12)* transforms also in the following way

$$\frac{1}{n}\frac{dn}{dt} = \frac{d}{dt}\left(\ln n + cste\right) = \frac{S_n}{n} - \mathrm{div}\,\vec{\phi} \tag{1.18}$$

In this expression, we can choose the constant to be $-\ln n_0$, and we can then define the dimension-less scalar τ with relation

$$\tau = -\ln\frac{n}{n_0} = \ln\frac{v}{v_0} \tag{1.19}$$

We will call it the *scalar of volume expansion*, which satisfies the following equation of geometro-kinetic

$$\frac{d\tau}{dt} = -\frac{S_n}{n} + \mathrm{div}\,\vec{\phi} \tag{1.20}$$

and which measures perfectly the notion of volume expansion of the lattice as this time $\tau \to \infty$ for large expansions (when $v \to \infty$), and $\tau \to -\infty$ for intense contractions (when $v \to 0$) and

$\tau \to 0$ when $v \to v_0$. Given the construction of the scalar τ, the constants v_0 et n_0 introduced here can be freely adjusted to signify a null expansion of the lattice ($\tau = 0$).

A definition of the global rotation and the contorsions of a lattice

If the field $\vec{\phi}(\vec{r},t)$ is not homogenous in space, then the lattice can have, aside from volume expansions, movements that correspond to a *global rotation* and *local rotations,* which imply contorsions (*flexions and torsions)* of the medium in space. To explain these rotational movements, we must determine the average local angular speed of the lattice. To that end, one must imagine within the lattice a circle of diameter R centered on O' with coordinates $\vec{r}_{O'}$ and oriented perpendicular to axis Ox_k as shown in figure 1.6. The average tangential velocity $\langle\phi\rangle$ along the circle can be written

$$\langle\phi\rangle = \frac{1}{2\pi R}\oint_C \vec{\phi}\,d\vec{r} \tag{1.21}$$

If the medium is rotating locally, or globally, about O' around axis Ox_k, the average tangential velocity $\langle\phi\rangle$ thus calculated will be different than zero. The average angular velocity $\langle\dot{\theta}_k\rangle$ of the lattice at O', around the axis Ox_k, is obtained simply by dividing the average tangential velocity $\langle\phi\rangle$ by the circle radius R

$$\langle\dot{\theta}_k\rangle = \frac{\langle\phi\rangle}{R} = \frac{1}{2\pi R^2}\oint_C \vec{\phi}\,d\vec{r} \tag{1.22}$$

With this expression of $\langle\dot{\theta}_k\rangle$, the theorem of the rotational *(A.38)* allows one to transform the integral on a circular boundary C to an integral on the surface S enclosed within the boundary

$$\langle\dot{\theta}_k\rangle = \frac{1}{2\pi R^2}\iint_S \overrightarrow{\mathrm{rot}}\,\vec{\phi}\,d\vec{S}_k = \frac{1}{2}\left(\frac{1}{\pi R^2}\iint_S \overrightarrow{\mathrm{rot}}\,\vec{\phi}\,dS_k\right)\vec{e}_k \tag{1.23}$$

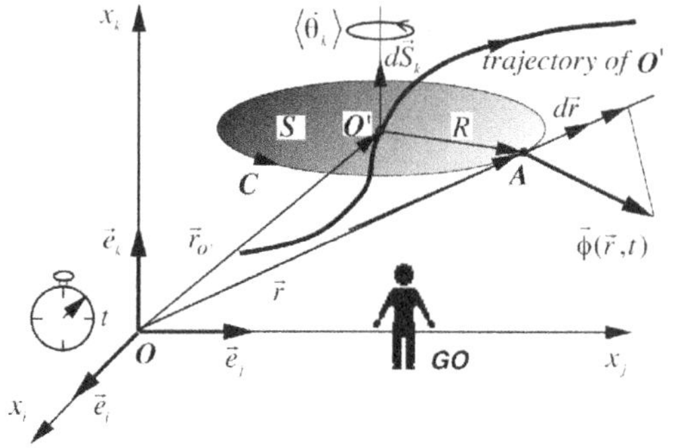

Figure 1.6 *- measure of local angular velocity of rotation of the lattice*

The expression between parenthesis is the average value of $\overrightarrow{\mathrm{rot}}\,\vec{\phi}$ on the disk of radius R, so that when $R \to 0$, there appears the *local angular velocity* $\dot{\theta}_k$ of the medium around the axis Ox_k at the coordinate $\vec{r}$ of space

$$\dot{\theta}_k = \frac{1}{2}\vec{e}_k \lim_{R\to 0}\left(\frac{1}{\pi R^2} \iint_S \overrightarrow{\mathrm{rot}}\,\vec{\phi}\, dS_k \right) = \frac{1}{2}\vec{e}_k\, \overrightarrow{\mathrm{rot}}\,\vec{\phi} \qquad (1.24)$$

The vector of local angular velocity $\dot{\vec{\theta}}$ of the medium is easily deduced

$$\dot{\vec{\theta}} = \sum_k \dot{\theta}_k \vec{e}_k = \frac{1}{2}\overrightarrow{\mathrm{rot}}\,\vec{\phi} \qquad (1.25)$$

This vector $\dot{\vec{\theta}}$ allows us to define the *vector of rotation* $\vec{\omega}$. Along the path in the medium, the local variation $d\vec{\omega}$ of the angle of rotation is given by $d\vec{\omega} = \dot{\vec{\theta}}dt$, so that it is possible to write a geometro-kinetic equation similar to expression *(1.20)* of the scalar of volume expansion, which involves the material derivative of the vector of rotation $\vec{\omega}$

$$\frac{d\vec{\omega}}{dt} = \dot{\vec{\theta}} = \frac{1}{2}\overrightarrow{\mathrm{rot}}\,\vec{\phi} \qquad (1.26)$$

It is interesting to note here that the variation of the scalar of volume expansion τ is due to a divergent part in the field of velocity $\vec{\phi}$ *(1.20)*, whereas the variations in the local rotation vector $\vec{\omega}$ is a consequence of the existence of a rotational part of the same field $\vec{\phi}$ *(1.26)*.

1.4 - Eulerian distortion tensors

In the presence of a velocity field $\vec{\phi}(\vec{r},t)$ which is not homogenous in space, a lattice can exhibit movements, besides the global translation and rotation, corresponding to local rotations (torsions) as well as various types of deformations, such as volume expansion and shear.

To describe these deformations with rotations, which we will generically call *distortions* of the lattice, we have to precisely describe the spatial variations of each of the components of the field of velocities, and we will link them to the temporal variation of a topological measure characterizing the distortions of the lattice. An elegant way of proceeding is to note that the volume expansion τ and the rotation vector $\vec{\omega}$ which we have just defined correspond to the trace and the anti-symmetrical part of a general geometro-kinetic equation based on a *tensor of second order,* β_{ij}, *the gradient of the vectorial velocity field* $\vec{\phi}$.

The vectorial notation and the decomposition properties of second order tensors

For convenience, second order tensor, such as the tensor gradient β_{ij}, can be represented as a *field of 3 vectors.* To that end, we write

$$\begin{pmatrix} \vec{\beta}_1 \\ \vec{\beta}_2 \\ \vec{\beta}_3 \end{pmatrix} = \begin{pmatrix} \beta_{11} & \beta_{12} & \beta_{13} \\ \beta_{21} & \beta_{22} & \beta_{23} \\ \beta_{31} & \beta_{32} & \beta_{33} \end{pmatrix} \begin{pmatrix} \vec{e}_1 \\ \vec{e}_2 \\ \vec{e}_3 \end{pmatrix} \qquad (1.27)$$

We will see that this *vectorial representation of tensorial fields* is very powerful and simplifies considerably the physical interpretation of tensorial fields.

From the general tensor $\vec{\beta}_i$ of distortions, we will vectorially describe the deformations of the lattice and we will extract information concerning the different types of deformations we can encounter. The way to proceed is to use the mathematical properties of a tensor $\vec{\beta}_i$, such as separating its symmetrical part and calculating its trace. We will do so now in the next section, in all generality of a second order tensor, without regards to the fact that said tensor $\vec{\beta}_i$ is derived from the field of velocities by a geometro-kinetic equation.

The generic operation of *transposition of a second order tensor* is defined by exchanging the terms of the matrix representation symmetrically to the diagonal. In component view this is achieved by inverting indices. In the vectorial notation of tensors, this operation is simply transcribed as

$$\vec{\beta}_i = \sum_k \beta_{ik} \vec{e}_k \quad \Rightarrow \quad \left[\vec{\beta}_i\right]^{\mathrm{T}} = \sum_k \beta_{ki} \vec{e}_k = \sum_k \vec{e}_k \left(\vec{\beta}_k \vec{e}_i\right) \tag{1.28}$$

By using properties *(A.24)*, the operation of transposition can also be written

$$\left[\vec{\beta}_i\right]^{\mathrm{T}} = \sum_k \vec{e}_k \left(\vec{e}_i \vec{\beta}_k\right) = \vec{\beta}_i + \vec{e}_i \wedge \left(\sum_k \vec{e}_k \wedge \vec{\beta}_k\right) \tag{1.29}$$

This formulation of transposition corresponds to subtracting from the tensor to be transposed the double of its antisymmetric part. In the vectorial notation, *the anti-symmetric part of tensor* β_{ij} can be build as a vector $\vec{\omega}$ by writing

$$\vec{\omega} = -\frac{1}{2} \sum_k \vec{e}_k \wedge \vec{\beta}_k \tag{1.30}$$

To verify that this vector $\vec{\omega}$ indeed represents the anti-symmetric part of tensor β_{ij} as a vector, let's look at the component view where $\vec{\omega}$ is written using the circular permutation on indices ijk

$$\vec{\omega} = -\frac{1}{2} \sum_k \vec{e}_k \wedge \vec{\beta}_k = \sum_k \frac{1}{2}\left(\beta_{ij} - \beta_{ji}\right) \vec{e}_k \tag{1.31}$$

Similarly one can build the *symmetrical part of tensor* $\vec{\beta}_i$, with

$$\left[\vec{\beta}_i\right]^{\mathrm{S}} = \frac{1}{2}\left\{\vec{\beta}_i + \left[\vec{\beta}_i\right]^{\mathrm{T}}\right\} = \vec{\beta}_i + \frac{1}{2} \vec{e}_i \wedge \left(\sum_k \vec{e}_k \wedge \vec{\beta}_k\right) \tag{1.32}$$

Let's call $\vec{\varepsilon}_i$ the symmetrical part of $\vec{\beta}_i$, by using *(1.31)*, we have

$$\vec{\varepsilon}_i = \vec{\beta}_i - \vec{e}_i \wedge \vec{\omega} \tag{1.33}$$

The cyclical permutation on indices ijk to obtain the component view, shows that the tensor $\vec{\varepsilon}_i$ indeed represents the symmetrical part of β_{ij}

$$\vec{\varepsilon}_i = \vec{\beta}_i - \vec{e}_i \wedge \vec{\omega} = \beta_{ii} \vec{e}_i + \frac{1}{2}\left(\beta_{ij} + \beta_{ji}\right) \vec{e}_j + \frac{1}{2}\left(\beta_{ik} + \beta_{ki}\right) \vec{e}_k \tag{1.34}$$

The trace τ *of tensors* $\vec{\beta}_i$ *et* $\vec{\varepsilon}_i$ is defined as the sum of the diagonal elements

$$\tau = \sum_k \vec{e}_k \vec{\beta}_k = \sum_k \beta_{kk} = \sum_k \varepsilon_{kk} = \sum_k \vec{e}_k \vec{\varepsilon}_k \tag{1.35}$$

The tensor $\vec{\alpha}_i$ obtained by subtracting the trace from the symmetrical part $\vec{\varepsilon}_i$ is called *the transverse symmetrical part of tensor* β_{ij}

$$\vec{\alpha}_i = \vec{\varepsilon}_i - \frac{1}{3} \tau \vec{e}_i \tag{1.36}$$

This tensor has a null trace, by construction

$$\operatorname{trace}(\vec{\alpha}_i) = \sum_k \vec{e}_k \vec{\alpha}_k = \sum_k \alpha_{kk} \equiv 0 \tag{1.37}$$

Given the definition of the transpose operation, it is clear that the transposition of a *symmetrical* tensor does not change the tensor. This is the case of the symmetrical tensor $\vec{\varepsilon}_i$,

$$\vec{\varepsilon}_i = [\vec{\varepsilon}_i]^{\mathrm{T}} \quad \Rightarrow \quad \vec{\varepsilon}_i = \sum_k \vec{e}_k(\vec{e}_i\vec{\varepsilon}_k) = \vec{\varepsilon}_i + \vec{e}_i \wedge \left(\sum_k \vec{e}_k \wedge \vec{\varepsilon}_k\right) \tag{1.38}$$

from which we deduce that the symmetry of tensors $\vec{\varepsilon}_i$ and $\vec{\alpha}_i$ implies the following properties

$$\vec{\varepsilon}_i = \sum_k \vec{e}_k(\vec{e}_i\vec{\varepsilon}_k) \quad \text{and} \quad \sum_k \vec{e}_k \wedge \vec{\varepsilon}_k \equiv 0 \tag{1.39}$$

$$\vec{\alpha}_i = \sum_k \vec{e}_k(\vec{e}_i\vec{\alpha}_k) \quad \text{and} \quad \sum_k \vec{e}_k \wedge \vec{\alpha}_k \equiv 0 \tag{1.40}$$

The eulerian tensors of distortion, deformation and shear

By using the vectorial notation for second order tensors, we define the *distortion tensor* $\vec{\beta}_i$ with the following geometro-kinetic equation

$$\frac{d\vec{\beta}_i}{dt} = -\frac{S_n}{3n}\vec{e}_i + \overrightarrow{\operatorname{grad}}\,\phi_i \tag{1.41}$$

whose trace gives us the geometro-kinetic equation for the *volume expansion* τ (fig. 1.7)

$$\frac{d\tau}{dt} = -\frac{S_n}{n} + \operatorname{div}\vec{\phi} \tag{1.42}$$

and its antisymmetric part represents the geometro-kinetic equation for the *vector of rotation* $\vec{\omega}$ (fig. 1.7)

$$\frac{d\vec{\omega}}{dt} = \frac{1}{2}\overrightarrow{\operatorname{rot}}\,\vec{\phi} \tag{1.43}$$

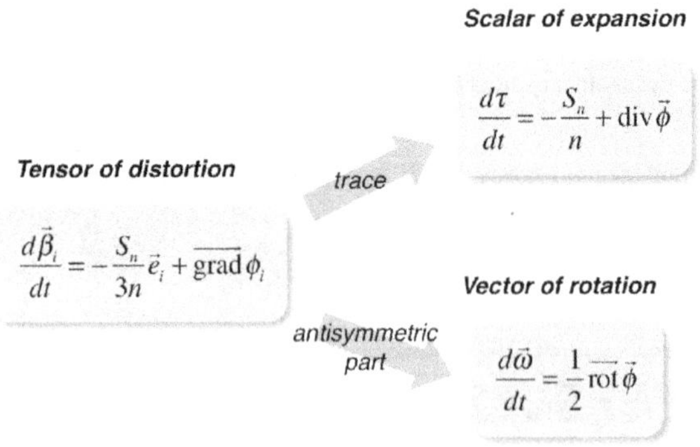

Figure 1.7 - *trace and antisymmetric part of the equation of geometro-kinetic*

Based on the decomposition of tensors of second order, as we have just seen, we can now define generally speaking, the tensor $\vec{\varepsilon}_i$ representing the symmetrical part of tensor $\vec{\beta}_i$. It is called the *tensor of deformation* $\vec{\varepsilon}_i$, since this tensor is equal to the tensor of distortion from which the rotation part (global and local) is subtracted. It satisfies the following geometro-kinetic equation

$$\frac{d\vec{\varepsilon}_i}{dt} = -\frac{S_n}{3n}\vec{e}_i + \overrightarrow{\text{grad}}\,\phi_i - \frac{1}{2}\vec{e}_i \wedge \overrightarrow{\text{rot}}\,\vec{\phi} \qquad (1.44)$$

Finally, it is possible to define a tensor $\vec{\alpha}_i$ corresponding to the *transverse symmetrical part* of tensor $\vec{\beta}_i$. As this tensor is obtained from that of deformations $\vec{\varepsilon}_i$ from which the trace representing volume expansion τ is subtracted, it will be called the *shear tensor* $\vec{\alpha}_i$. It satisfies the following equation of geometro-kinetic

$$\frac{d\vec{\alpha}_i}{dt} = \overrightarrow{\text{grad}}\,\phi_i - \frac{1}{2}\vec{e}_i \wedge \overrightarrow{\text{rot}}\,\vec{\phi} - \frac{1}{3}\vec{e}_i\,\text{div}\,\vec{\phi} \qquad (1.45)$$

The complete decomposition of the tensor of distortion follows the schema presented at figure 1.8, it shows the symmetric and anti-symmetric parts, the trace and the transverse symmetric part of the tensor.

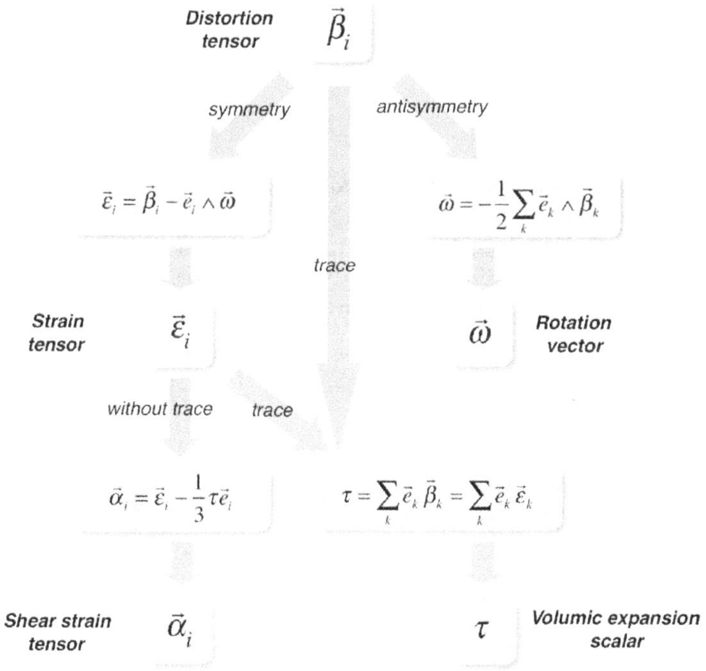

Figure 1.8 *- Decomposition of the eulerian tensor of distortions*

The full set of equations we have obtained is transcribed in table 1.1. With examples we will see later, we will show the adequacy of this system of 'galilean' equations to describe the geometro-kinetic of the topological distortions of any solid lattice in eulerian coordinates.

Table 1.1 - Geometro-kinetics of the distortions of a lattice in eulerian coordinates

Definition of volume expansion

$$\frac{\partial n}{\partial t} = S_n - \operatorname{div}\left(n\vec{\phi}\right) \quad (1) \qquad \Rightarrow \qquad \tau = -\ln\frac{n}{n_0} = \ln\frac{v}{v_0} \quad (2) \qquad \Rightarrow$$

Geometro-kinetics of the distortions

$$\begin{cases} \dfrac{d\vec{\beta}_i}{dt} = -\dfrac{S_n}{3n}\vec{e}_i + \overrightarrow{\operatorname{grad}}\,\phi_i & (3) \\ \dfrac{d\vec{\omega}}{dt} = \dfrac{1}{2}\overrightarrow{\operatorname{rot}}\,\vec{\phi} & (4) \\ \dfrac{d\tau}{dt} = -\dfrac{S_n}{n} + \operatorname{div}\vec{\phi} & (5) \end{cases} \qquad \begin{cases} \vec{\omega} = -\dfrac{1}{2}\sum_k \vec{e}_k \wedge \vec{\beta}_k & (6) \\ \tau = \sum_k \vec{e}_k\,\vec{\beta}_k = \sum_k \vec{e}_k\,\vec{\varepsilon}_k & (7) \\ \vec{\varepsilon}_i = \vec{\beta}_i - \vec{e}_i \wedge \vec{\omega} & (8) \\ \vec{\alpha}_i = \vec{\varepsilon}_i - \dfrac{1}{3}\tau\vec{e}_i & (9) \end{cases}$$

1.5 - Examples of velocity and distortion fields

With simple examples of the velocity field $\vec{\phi}(\vec{r},t)$ belonging to known movements of the lattice, this paragraph will show that the topological tensors β_{ij}, ε_{ij}, α_{ij}, $\vec{\omega}$ and τ completely capture the distortions, deformations, shear, rotations and volume expansions that can appear in a lattice, including in the case of large distortions. The values of these tensors simply go to infinity when distortions become very large!

Global translation

Consider the velocity field $\vec{\phi}(\vec{r},t)$ describing a global translation of the medium in the direction $\vec{a}$, where $\vec{a}$ is a constant vector

$$\vec{\phi}(\vec{r},t) = \vec{a}\,g(t) \qquad (1.46)$$

With the geometro-kinetic equations, such a field gives us the following particulate derivatives

$$\frac{dn}{dt} = \frac{dv}{dt} = \frac{d\tau}{dt} = \frac{d\vec{\omega}}{dt} = \frac{d\vec{\beta}_i}{dt} = \frac{d\vec{\varepsilon}_i}{dt} = \frac{d\vec{\alpha}_i}{dt} = 0 \qquad (1.47)$$

which shows that all topological tensors remain constant along the translation path of the medium.

Figure 1.9(a) represents a velocity field with a global translation of the medium, in the case where the direction $\vec{a}$ of translation, remains in the basis plane defined by $\vec{e}_1$ and $\vec{e}_2$. Figure 1.9(b) shows the temporal evolution between 2 different instants ($t = 0$ and $t > 0$) of a square portion of the medium with this velocity field.

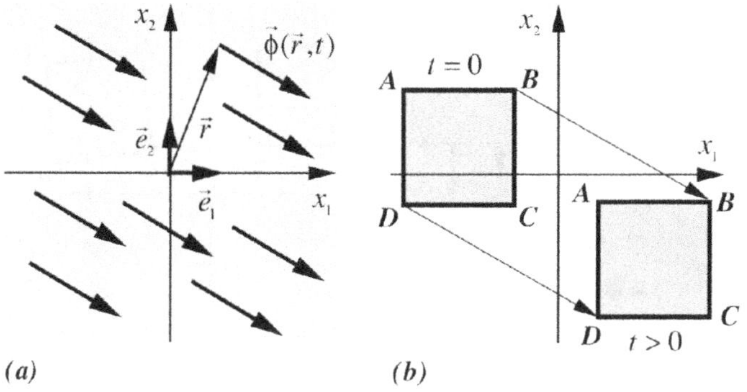

***Figure 1.9** - global translation*

Global rotation

Consider the following velocity field

$$\vec{\phi}(\vec{r},t) = \dot{\vec{\theta}}(t) \wedge \vec{r} \quad \text{with} \quad \dot{\vec{\theta}}(t) = \vec{a}\, g(t) \tag{1.48}$$

where $\vec{a}$ is a constant vector. The component view of this field reads

$$\vec{\phi}(\vec{r},t) = \left[\left(a_2 x_3 - a_3 x_2\right)\vec{e}_1 + \left(a_3 x_1 - a_1 x_3\right)\vec{e}_2 + \left(a_1 x_2 - a_2 x_1\right)\vec{e}_3 \right] g(t) \tag{1.49}$$

Such a velocity field represents a global rotation of the medium around a vector $\vec{a}$. With the equations of geometro-kinetic, we have

$$\frac{dn}{dt} = \frac{dv}{dt} = \frac{d\tau}{dt} = 0 \quad \text{and} \quad \frac{d\vec{\varepsilon}_i}{dt} = \frac{d\vec{\alpha}_i}{dt} = 0 \tag{1.50}$$

As a consequence, the tensor of deformation $\vec{\varepsilon}_i$, the shear tensor $\vec{\alpha}_i$ and the scalar of volume expansion τ do not change during the evolution of the medium about an axis. The temporal evolution of $\vec{\omega}$ reads

$$\frac{d\vec{\omega}}{dt} = \frac{1}{2}\overrightarrow{\mathrm{rot}}\ \vec{\phi} = \frac{1}{2}\overrightarrow{\mathrm{rot}}\ \left(\dot{\vec{\theta}}(t) \wedge \vec{r}\right) = \dot{\vec{\theta}}(t) \tag{1.51}$$

Its temporal evolution along the path is equal to the angular velocity $\dot{\vec{\theta}}(t)$, as we deduced in section 1.3.

We can compute the evolution of tensor β_{ij} along the path, and it appears that in the case of a global rotation, the tensor β_{ij} is purely anti-symmetric

$$\frac{d\vec{\beta}_i}{dt} = \vec{e}_i \wedge \frac{d\vec{\omega}}{dt} = \vec{e}_i \wedge \dot{\vec{\theta}}(t) \tag{1.52}$$

Figure 1.10(a) shows the velocity field corresponding to a global rotation of the medium, in the case where direction $\vec{a}$ of the axis of rotation is parallel to the base vector $\vec{e}_3$. Figure 1.10(b) shows the evolution at 2 different times ($t=0$ and $t>0$) of a square portion of the medium.

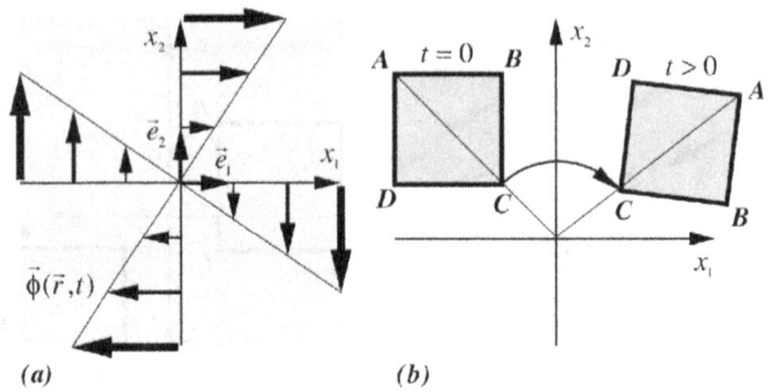

***Figure 1.10** - global rotation*

Elongations and volume expansion

The following velocity field is associated with homogeneous expansions and contractions along the 3 different axis of coordinates

$$\vec{\phi}(\vec{r},t)=\left[a_1x_1\vec{e}_1+a_2x_2\vec{e}_2+a_3x_3\vec{e}_3\right]g(t) \tag{1.53}$$

where the a_i are constants.

With such a field, the medium is elongating or retracting along the 3 coordinate axis depending on the sign of the constants a_i. The equations of geometro-kinetic allow us to deduce β_{ij} and ε_{ij} along the path

$$\frac{d\vec{\beta}_i}{dt}=\frac{d\vec{\alpha}_i}{dt}=a_i g(t)\vec{e}_i \tag{1.54}$$

Tensors $\vec{\beta}_i$ and $\vec{\varepsilon}_i$ are symmetrical and only have trace components. It is rather easy to compute the evolution of τ, which corresponds to an expansion or a contraction of the medium along the trajectory of points

$$\frac{d\tau}{dt}=(a_1+a_2+a_3)g(t) \tag{1.55}$$

Furthermore it is clear that this type of deformation by dilation is not associated with any local rotation in the medium as we can see from the evolution of vector $\vec{\omega}$

$$\frac{d\vec{\omega}}{dt}=0 \tag{1.56}$$

With regards to shear tensor $\vec{\alpha}_i$, it only contains diagonal elements, which take non-null values in this case since, along a give path of a particle of the lattice, we have

$$\frac{d\vec{\alpha}_i}{dt}=\left[a_i-\frac{1}{3}(a_1+a_2+a_3)\right]g(t)\vec{e}_i \tag{1.57}$$

From $\tau(t)$, values for $n(t)$ and $v(t)$ are computed with

$$n(t)=n_0\,\mathrm{e}^{-\tau(t)} \quad \text{and} \quad v(t)=v_0\,\mathrm{e}^{\tau(t)} \tag{1.58}$$

The scalar τ can be expressed in terms of trajectories $X_i(t)$ of 3 points initially situated at X_i^0 (figure 1.11) on the 3 axis of coordinates.

$$\tau = -\ln\frac{n(t)}{n_0} = \ln\frac{v(t)}{v_0} = \ln\left(1+\frac{\Delta v(t)}{v_0}\right) \qquad (1.59)$$

And since the volumes $v(t)$ and v_0 are directly proportional to $\prod X_i(t)$ and $\prod X_i^0$, it follows

$$\tau = \ln\left(\prod_i \frac{X_i(t)}{X_i^0}\right) = \sum_i \ln\frac{X_i(t)}{X_i^0} = \sum_i \ln\left(1+\frac{\Delta X_i(t)}{X_i^0}\right) \qquad (1.60)$$

By using $\tau = \sum \beta_{ii} = \sum \varepsilon_{ii}$, we can directly compute β_{ii} et ε_{ii}

$$\beta_{ii} = \varepsilon_{ii} = \ln\frac{X_i(t)}{X_i^0} = \ln\left(1+\frac{\Delta X_i(t)}{X_i^0}\right) \qquad (1.61)$$

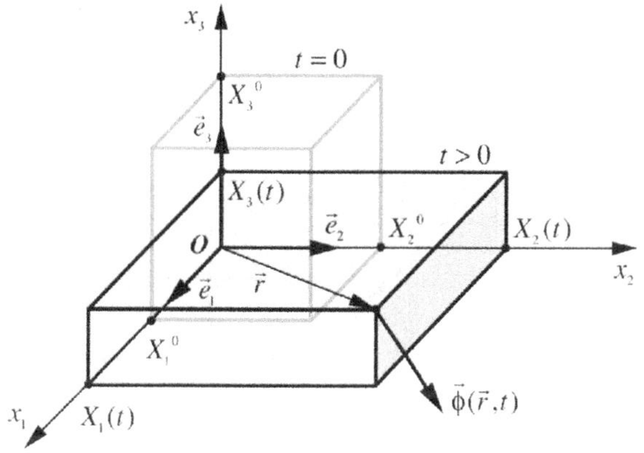

Figure 1.11 *- elongations along the axis of coordinates*

The diagonal elements β_{ii} and ε_{ii} measure the individual expansions and contractions in the 3 directions of space respectively, while the trace τ measures the variation of the volume said expansions and contractions.

Finally it should be noted that in the limit where elongations are small, the scalar τ can be expanded as

$$\tau \cong \frac{\Delta v(t)}{v_0} \quad \text{if} \quad |\Delta v(t)| << v_0 \qquad (1.62)$$

The diagonal elements of the tensors of distortion and deformation can also be expanded as such

$$\beta_{ii} = \varepsilon_{ii} \cong \frac{\Delta X_i(t)}{X_i^0} \quad \text{if} \quad |\Delta X_i(t)| << X_i^0 \qquad (1.63)$$

Another important remark can be made about the asymptotical behavior of the topological tensors τ, β_{ii} and ε_{ii}. They all go to $\pm\infty$ when the elongations become large

$$\tau, \beta_{ii}, \varepsilon_{ii} \to +\infty \quad \text{if} \quad \Delta X_i \to +\infty \qquad (1.64)$$

$$\tau, \beta_{ii}, \varepsilon_{ii} \to -\infty \quad \text{if} \quad \Delta X_i \to -X_i^0 \qquad (1.65)$$

2 extreme cases of evolution of elongation along coordinate axis can now be studied:

- *The case where the volume expansion/contraction is isotropic in space*, meaning where the 3 constants a_i are equal ($a_1 = a_2 = a_3$). We then have

$$\frac{d\tau}{dt} = 3a_i g(t) \quad ; \quad \frac{d\vec{\alpha}_i}{dt} = 0 \quad ; \quad \frac{d\vec{\omega}}{dt} = 0 \qquad (1.66)$$

The figure 1.12(a) shows a 2D cut of such a velocity field in the plane defined by vectors $\vec{e}_1$ and $\vec{e}_2$. Figure 1.12(b) shows the evolution between 2 instants ($t = 0$ and $t > 0$) of a square portion of medium under such field. What we see is a volume expansion without shear.

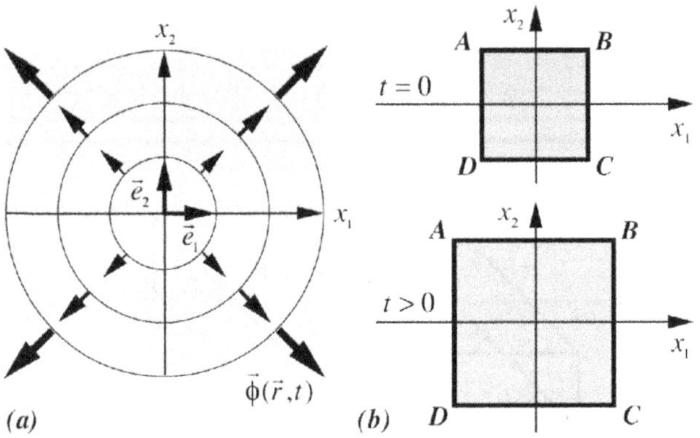

Figure 1.12 *- elongations with volume expansion and no shear*

- *The case where expansions and contractions in the 3 directions of space are such that volume expansion τ of the medium is null*, meaning where the sum of the a_i is null ($a_1 + a_2 + a_3 = 0$). It then follows that

$$\frac{d\tau}{dt} = 0 \quad \text{and} \quad \frac{d\vec{\beta}_i}{dt} = \frac{d\vec{\varepsilon}_i}{dt} = \frac{d\vec{\alpha}_i}{dt} = a_i g(t)\vec{e}_i \qquad (1.67)$$

Figure 1.13(a) shows the cut of such a field with $a_1 = -a_2 = a$ and $a_3 = 0$.

Figure 1.13(b) shows the spatiotemporal evolution of a square portion of the medium with such a field at 2 different times ($t = 0$ and $t > 0$). By visual inspection we can see there is no volume expansion, the shape changes, but not the volume (as captured by the null trace). We can also see the presence of strong shear in the medium, which can also be seen in the vectorial equations

$$\frac{d\vec{\omega}}{dt} = 0 \quad ; \quad \frac{d\vec{\alpha}_1}{dt} = ag(t)\vec{e}_1 \quad ; \quad \frac{d\vec{\alpha}_2}{dt} = -ag(t)\vec{e}_2 \quad ; \quad \frac{d\vec{\alpha}_3}{dt} = 0 \qquad (1.68)$$

Indeed the main difference between the 2 cases we just studied, lies in the isotropic versus anisotropic nature of the elongations along the 3 axis. The different values of expansion gives rise to shear, which is absent in the isotropic case. We conclude that the diagonal components of tensor $\vec{\alpha}_i$ measure shear associated with non-isotropic homogeneous volume expansion of the medium.

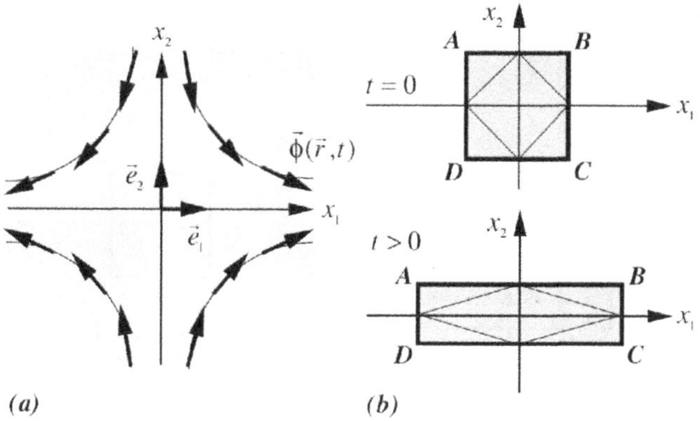

Figure 1.13 - *elongations with null volume expansion (constant volume)*

Shear and local rotations, or torsions

Consider the following velocity field

$$\vec{\phi}(\vec{r},t) = \left[\left(ax_2 + bx_3\right)\vec{e}_1 + \left(cx_1 + dx_3\right)\vec{e}_2 + \left(ex_1 + fx_2\right)\vec{e}_3\right]g(t) \tag{1.69}$$

where a,b,c,d,e,f are constants.

Given such a field, the equations of geometro-kinetics for tensor β_{ij} gives us the temporal evolution of the β_{ij} components along the path

$$\left(\frac{d\vec{\beta}_i}{dt}\right) = \begin{pmatrix} 0 & a & b \\ c & 0 & d \\ e & f & 0 \end{pmatrix} g(t) \tag{1.70}$$

It clearly shows that the trace of the tensor of distortion cannot change, and volume expansion τ is a constant

$$\frac{d\tau}{dt} = 0 \tag{1.71}$$

We can compute the geometro-kinetic equations satisfied by the rotation vector and the tensors of deformation and shear, those are

$$\begin{cases} \dfrac{d\vec{\omega}}{dt} = \dfrac{1}{2}\left[(f-d)\vec{e}_1 + (b-e)\vec{e}_2 + (c-a)\vec{e}_3\right]g(t) \\ \left(\dfrac{d\varepsilon_{ij}}{dt}\right) = \left(\dfrac{d\alpha_{ij}}{dt}\right) = \dfrac{1}{2}\begin{pmatrix} 0 & a+c & b+e \\ a+c & 0 & d+f \\ b+e & d+f & 0 \end{pmatrix} g(t) \end{cases} \tag{1.72}$$

Figure 1.14(a) represents such a velocity field in the case where only the a constant is different from zero ($a \neq 0$ and $b=c=d=e=f=0$).

The spatio-temporal evolution between 2 instants ($t=0$ and $t>0$) of a square portion of medium is shown in figure 1.14(b). The presence of shear in the medium is observed, and there also is a *local rotation* which we will call *torsion* of the medium and we have

$$\frac{d\vec{\omega}}{dt}=-\frac{a}{2}g(t)\vec{e}_3 \quad ; \quad \frac{d\vec{\alpha}_1}{dt}=\frac{a}{2}g(t)\vec{e}_2 \quad ; \quad \frac{d\vec{\alpha}_2}{dt}=\frac{a}{2}g(t)\vec{e}_1 \quad ; \quad \frac{d\vec{\alpha}_3}{dt}=0 \tag{1.73}$$

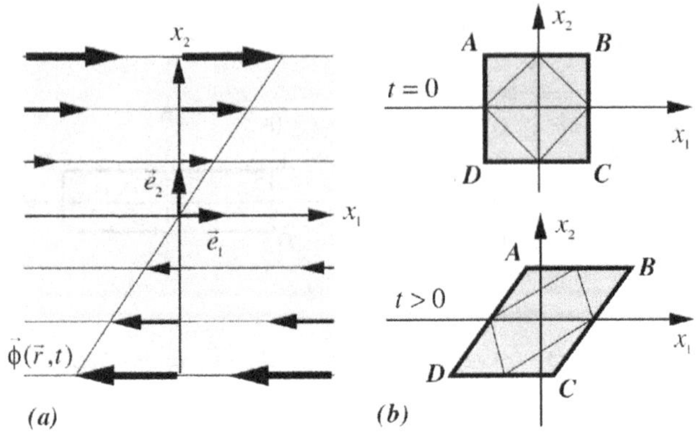

Figure 1.14 *- shear*

In the same way we did for elongations, it is possible to give a topological interpretation to the components of the shear tensor by using the example of distortion presented in figure 1.14.
In that case, the velocity field is parallel to $\vec{e_1}$ and is computed as

$$\vec{\phi}(\vec{r},t)=ax_2g(t)\vec{e}_1 \tag{1.74}$$

The displacement $X_1(x_2,t)$ along the $\vec{e}_1$ at a given point A found at coordinate $x_2\neq 0$ (figure 1.15) can be calculated. We have

$$X_1(x_2,t)=ax_2\int_0^t g(t)dt \tag{1.75}$$

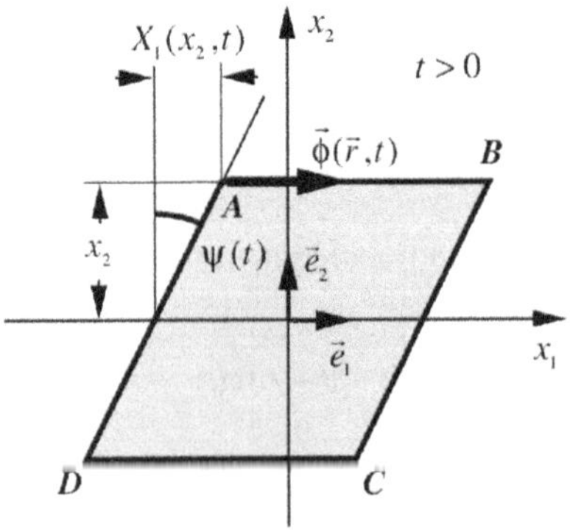

Figure 1.15 *- relation with the tangent of the angle of shear*

We can deduce the temporal evolution of the tangent of shear $\psi(t)$ shown in figure 1.15

$$\operatorname{tg}\psi(t)=\frac{X_1(x_2,t)}{x_2}=a\int_0^t g(t)dt \qquad (1.76)$$

The evolution of the vector of rotation and evolution of the tensor of shear along the trajectory can be linked to the value of $\operatorname{tg}\psi(t)$

$$\vec{\omega}=-\frac{a}{2}\vec{e}_3\int_0^t g(t)dt=-\frac{1}{2}\vec{e}_3\operatorname{tg}\psi(t) \qquad (1.77)$$

$$\vec{\alpha}_1=\frac{a}{2}\vec{e}_2\int_0^t g(t)dt=\frac{1}{2}\vec{e}_2\operatorname{tg}\psi(t) \qquad (1.78)$$

$$\vec{\alpha}_2=\frac{a}{2}\vec{e}_1\int_0^t g(t)dt=\frac{1}{2}\vec{e}_1\operatorname{tg}\psi(t) \qquad (1.79)$$

The vector of rotation $\vec{\omega}$ and shear tensor $\vec{\alpha}_i$ are linked to half the angle of shear $\psi(t)$.

It is possible to consider now 2 extreme cases of evolution for shear and rotations:

- *The case where tensor* β_{ij} *is perfectly anti-symmetric,* meaning when $c=-a$, $e=-b$ and $f=-d$

$$\left(\frac{d\beta_{ij}}{dt}\right)=\begin{pmatrix}0 & a & b\\ -a & 0 & d\\ -b & -d & 0\end{pmatrix}g(t) \qquad (1.80)$$

In this case, the tensor β_{ij} is entirely described by a rotation vector $\vec{\omega}$ and the global rotation of the medium is the one we have already seen with

$$\frac{d\vec{\omega}}{dt}=\dot{\vec{\theta}}(t)=\left(-d\vec{e}_1+b\vec{e}_2-a\vec{e}_3\right)g(t) \qquad (1.81)$$

- *In the case where* β_{ij} *is perfectly symmetric,* meaning when $c=a$, $e=b$ and $f=d$. In this case the 3 tensors β_{ij}, ε_{ij} et α_{ij} are equal and the variation of their components along the path is written

$$\left(\frac{d\beta_{ij}}{dt}\right)=\left(\frac{d\varepsilon_{ij}}{dt}\right)=\left(\frac{d\alpha_{ij}}{dt}\right)=\begin{pmatrix}0 & a & b\\ a & 0 & d\\ b & d & 0\end{pmatrix}g(t) \qquad (1.82)$$

The scalar of volume expansion τ and the vector of rotation $\vec{\omega}$ do not change along the path

$$\frac{d\tau}{dt}=0 \quad \text{and} \quad \frac{d\vec{\omega}}{dt}=0 \qquad (1.83)$$

The medium does not have volume expansion, nor local or global rotation. As a consequence, only pure shear can appear.

We visually show it in figure 1.16(a), in which we have represented, in the plane of base vectors $\vec{e}_1$ and $\vec{e}_2$, the cut of such a velocity field with $a\neq b$ and $b=c=0$.

Figure 1.16(b) shows the evolution at 2 different times ($t=0$ and $t>0$) of a square portion of medium under such a velocity field. The presence of a head of the medium, without rotation, is seen as

$$\frac{d\vec{\omega}}{dt}=0 \quad ; \quad \frac{d\vec{\alpha}_1}{dt}=ag(t)\vec{e}_2 \quad ; \quad \frac{d\vec{\alpha}_2}{dt}=ag(t)\vec{e}_1 \quad ; \quad \frac{d\vec{\alpha}_3}{dt}=0 \qquad (1.84)$$

The shear is identical to that reported in figure 1.14(b), if a rotation of 45° of the system of coordinates is done, which leads to very similar expressions for the shear tensor in both cases:

$$\left(\frac{d\alpha_{ij}}{dt}\right)=\begin{pmatrix} a & 0 & 0 \\ 0 & -a & 0 \\ 0 & 0 & 0 \end{pmatrix} g(t) \ \textit{(fig. 1.14)} \quad \text{and} \quad \left(\frac{d\alpha_{ij}}{dt}\right)=\begin{pmatrix} 0 & a & 0 \\ a & 0 & 0 \\ 0 & 0 & 0 \end{pmatrix} g(t) \ \textit{(fig. 1.16)} \qquad (1.85)$$

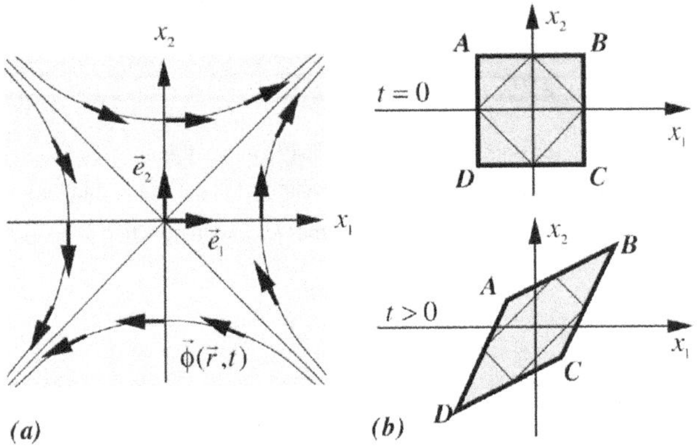

Figure 1.16 *- shear*

Examples of non-uniform distortions

As an example, it is possible to calculate the tensors of distortions with velocity fields that lead to non-uniform distortions of the medium. Consider the following field in figure 1.17(a), *with a cylindrical (axial) symmetry*

$$\vec{\phi}(\vec{r},t)=\left[r\vec{e}_3\wedge\left(x_1\vec{e}_1+x_2\vec{e}_2\right)\right]g(t)=r\left(x_1\vec{e}_2-x_2\vec{e}_1\right)g(t) \quad \text{with} \quad r=\sqrt{x_1^{\,2}+x_2^{\,2}} \qquad (1.86)$$

This velocity field describes a rotation of the medium about the origin of our system of coordinates. The tensor β_{ij} is deduced from the geometro-kinetic equations as

$$\left(\frac{d\beta_{ij}}{dt}\right)=\frac{1}{r}\begin{pmatrix} -x_1x_2 & -\left(r^2+x_2^{\,2}\right) & 0 \\ +\left(r^2+x_1^{\,2}\right) & x_1x_2 & 0 \\ 0 & 0 & 0 \end{pmatrix} g(t) \qquad (1.87)$$

This is a non-uniform rotation, and therefore local, as $\vec{\omega}$ depends on radius r. There is also a volume expansion τ along the path

$$\frac{d\vec{\omega}}{dt}=\frac{3r}{2}g(t)\vec{e}_3 \quad \text{and} \quad \frac{d\tau}{dt}=0 \qquad (1.88)$$

The tensor of deformation ε_{ij} and of shear α_{ij} are identical and show the existence of non-uniform shear

$$\left(\frac{d\varepsilon_{ij}}{dt}\right)=\left(\frac{d\alpha_{ij}}{dt}\right)=\frac{1}{2r}\begin{pmatrix}-2x_1x_2 & x_1^2-x_2^2 & 0\\ x_1^2-x_2^2 & 2x_1x_2 & 0\\ 0 & 0 & 0\end{pmatrix}g(t) \qquad (1.89)$$

The second velocity field, represented in figure 1.17(b), has a spherical symmetry

$$\vec{\phi}(\vec{r},t)=r\left[x_1\vec{e}_1+x_2\vec{e}_2+x_3\vec{e}_3\right]g(t) \quad \text{with} \quad r=\sqrt{x_1^2+x_2^2+x_3^2} \qquad (1.90)$$

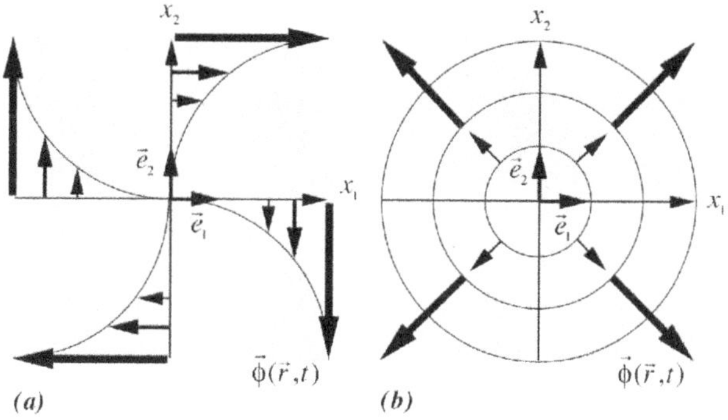

Figure 1.17 *- examples of non-uniform distortions*

This velocity field implies a volume expansion of the medium about the center of system of coordinates. The evolution of tensors β_{ij} and ε_{ij} is the same

$$\left(\frac{d\beta_{ij}}{dt}\right)=\left(\frac{d\varepsilon_{ij}}{dt}\right)=\frac{1}{r}\begin{pmatrix}r^2+x_1^2 & x_1x_2 & x_1x_3\\ x_1x_2 & r^2+x_2^2 & x_2x_3\\ x_1x_3 & x_2x_3 & r^2+x_3^2\end{pmatrix}g(t) \qquad (1.91)$$

This means it is an evolution without rotation along the path, but with a non-uniform volume expansion τ which depends on radius r

$$\frac{d\vec{\omega}}{dt}=0 \quad \text{and} \quad \frac{d\tau}{dt}=4r\,g(t) \qquad (1.92)$$

Finally, the evolution of shear tensor α_{ij} shows the existence of non-uniform shear as a function of radius r

$$\left(\frac{d\alpha_{ij}}{dt}\right)=\frac{1}{r}\begin{pmatrix}x_1^2-r^2/3 & x_1x_2 & x_1x_3\\ x_1x_2 & x_2^2-r^2/3 & x_2x_3\\ x_1x_3 & x_2x_3 & x_3^2-r^2/3\end{pmatrix}g(t) \qquad (1.93)$$

Chapter 2

Local frames and geometro-compatibility of a lattice

In this chapter, we show that eulerian coordinates are well suited to the description of solid lattices. They allow for the description of lattices with large deformations, something that is hard to achieve with lagrangian coordinates. However, we will show that this description of the solid lattice with Euler coordinates requires us to introduce local referential frames. This introduces its own complexity, inherent to the differential geometry description, which makes use of microscopic local frames. To bypass such complexity we will define the concept of *macroscopic local frame*. Finally, in the eulerian coordinate description, we derive the conditions of compatibility that define the lattice as a solid: one where we can define a *continuous* displacement field for the lattice.

2.1 - Definition of local referential frames in solid lattices

The eulerian description of the spatiotemporal evolution of a solid lattice of particles is defined by the *velocity field* $\vec{\phi}(\vec{r},t)$ which describes the average movement of the individual nodes of the lattice relative to an absolute fixed observer (Grand Observer: ***GO***) with a fixed, cartesian, coordinate system $Q\xi_1\xi_2\xi_3$.

From this velocity field we will derive and calculate the *local angular velocity* $d\vec{\omega}/dt$ of the solid rotation as well as the *scalar of local volume expansion* τ of the solid. With these we will describe the time dependent distortions of the solid lattice with the equations of geometro-kinetics we introduced in the first chapter (table 1.1). These equations were based on the 3 following principal equations:

$$\begin{cases} \dfrac{d\vec{\beta}_i}{dt} = -\dfrac{S_n}{3n}\vec{e}_i + \overrightarrow{\operatorname{grad}}\,\phi_i \\ \dfrac{d\vec{\omega}}{dt} = \dfrac{1}{2}\overrightarrow{\operatorname{rot}}\,\vec{\phi} \\ \dfrac{d\tau}{dt} = -\dfrac{S_n}{n} + \operatorname{div}\vec{\phi} \end{cases} \qquad (2.1)$$

This method is valid in the case of a perfectly isotropic lattice. However, in the majority of the cases it is not well suited to describe the distortions of a lattice in eulerian coordinates for the following reasons:

- A solid may present an *anisotropic elasticity,* which depends on the direction considered with respect to the *local crystalline directions of the lattice itself.* While the distortion tensors *(2.1)* correctly measure the distortions of a solid, whether it be isotropic or not, the tensors measure and describe the distortions *in the absolute global frame* $Q\xi_1\xi_2\xi_3$ and not along the *local crystalline directions of the lattice*. This complicates the calculation of the free elastic energy of anisotropic solid lattice from the distortion tensors *(2.1)*, specifically if the local rotation is important because the free energy of an anisotropic solid is not built from the invariants of the distortion tensors.

- In the case of a perfect solid lattice with isotropic elasticity (without structural singularities), the free elastic energy can indeed be expressed in terms of the distortion tensors *(2.1)* because it only depends on the invariant quantities of the distortion tensors expressed in said absolute global frame. On the other hand, in the case of an *imperfect* solid lattice with isotropic elasticity, meaning an isotropic lattice which contains anisotropic structural defects and singularities (such as *disclinations* and *dislocations, joints* and other usual crystal defects) the presence of such anisotropic defects requires a description which *uses the local crystalline directions and associated local frame.*

It is imperative to describe the distortions of a lattice with anisotropic elasticity, or an isotropic lattice with defects (non perfect isotropic lattice) with *local frames and local coordinates.* This system of local frames reflects the same average local translation and rotation of the lattice so that the directions of the local system of coordinates maps, on average, to the directions of the local crystalline lattice.

Definition of an orthonormal macroscopic local frame.

In the absolute frame $Q\xi_1\xi_2\xi_3$ of the ***GO*** observer, we can choose to follow a point O of the solid by knowing its instantaneous velocity $\vec{\phi}_O(t)$ expressed in said referential. To this point O we also associate the instantaneous vector of rotation $\vec{\omega}_O(t)$. We can then define at this point O a *macroscopic orthonormal local frame,* $Ox_1x_2x_3$, which follows the solid with a *drag-along velocity* $\vec{\phi}_O(t)$ and which rotates in space with the *drag-along rotational angular velocity* $\vec{\omega}_O(t)$ (figure 2.1).

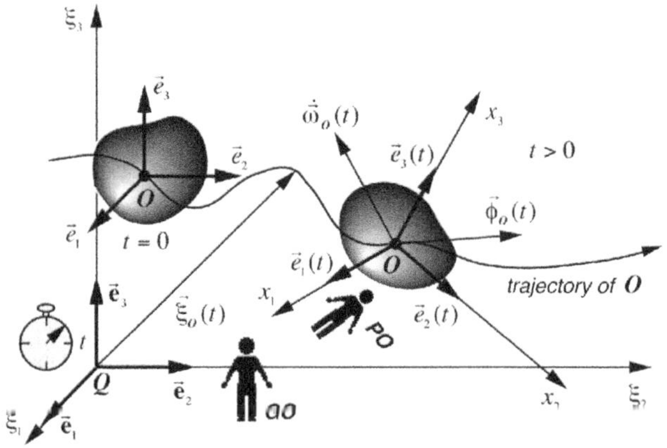

Figure 2.1 *- the orthonormal macroscopic local frame*

The origin O of this new coordinate system can be found with a *location vector* $\vec{\xi}_O(t)$ whose equation of motion, written in the absolute frame $Q\xi_1\xi_2\xi_3$, is simply stated as

$$\frac{d\vec{\xi}_O(t)}{dt} = \vec{\phi}_O(t) \tag{2.2}$$

To this local frame $Ox_1x_2x_3$, we can associate the 3 unitary orthogonal vectors $\vec{e}_i(t)$, the frame. These vectors turn in space with the drag-along rotation speed $\vec{\omega}_O(t)$. We can also define the Small Observer ***(PO)*** ("Petit Observateur") attached to this referential and who uses rulers $\vec{e}_i$, *which have the same length as those of* ***GO***, as well as a clock perfectly synchronized with that of the ***GO*** of the absolute referential.

On the matrix of rotation of the coordinate change

In the absolute referential $Q\xi_1\xi_2\xi_3$ of the ***GO*** observer, the vectors $\vec{e}_i(t)$ of the local coordinate system satisfy the following equation of change, which capture the fact that the vectors $\vec{e}_i(t)$ are rotating with an angular velocity $\vec{\omega}_O(t)$ along the trajectory of the O point

$$\frac{d\vec{e}_i(t)}{dt} = \vec{\omega}_O(t) \wedge \vec{e}_i(t) \tag{2.3}$$

The local frame vectors $\vec{e}_i(t)$ can be decomposed on the global frame vectors $\vec{\mathbf{e}}_j$ with the matrix of rotation $g_{ij}(t)$, which is then defined as

$$\vec{e}_i(t) = \sum_j g_{ij}(t)\vec{\mathbf{e}}_j \tag{2.4}$$

This relation can be inverted as

$$\vec{\mathbf{e}}_j = \sum_i \left[\vec{e}_i(t)\vec{\mathbf{e}}_j\right]\vec{e}_i(t) = \sum_i g_{ij}(t)\vec{e}_i(t) \tag{2.5}$$

The matrix $g_{ij}(t)$ is in fact the matrix describing the coordinate transformation (map) resulting from the *change of frame* between the two frames, from the global coordinate $Q\xi_1\xi_2\xi_3$ to the local one $Ox_1x_2x_3$. As a rotation, it is an *orthogonal matrix*, and its inverse is its transpose.

$$\left(g_{ij}\right)^{-1} = \left(g_{ij}\right)^{\mathrm{T}} \;\Rightarrow\; \left(g_{ij}\right)\left(g_{ij}\right)^{\mathrm{T}} = \left(g_{ij}\right)\left(g_{ij}\right)^{-1} = \begin{pmatrix} 1 & 0 & 0 \\ 0 & 1 & 0 \\ 0 & 0 & 1 \end{pmatrix} \tag{2.6}$$

As a rotation, its determinant is also equal to 1

$$\left|\left(g_{ij}\right)\left(g_{ij}\right)^{\mathrm{T}}\right| = \left|\left(g_{ij}\right)\left(g_{ij}\right)^{-1}\right| = 1 \;\Rightarrow\; \left|\left(g_{ij}\right)\right| = 1 \tag{2.7}$$

Such a matrix also satisfies the following relations, translating the fact that both sets of vectors $\vec{e}_i$ et $\vec{e}_i$ are indeed unitary

$$\begin{cases} \left|\vec{e}_i\right| = 1 \;\Rightarrow\; \sum_j \left[g_{ij}(t)\right]^2 = 1 \\ \left|\vec{\mathbf{e}}_j\right| = 1 \;\Rightarrow\; \sum_i \left[g_{ij}(t)\right]^2 = 1 \end{cases} \tag{2.8}$$

Given relation *(2.3)*, we deduce the differential equations describing the evolution of the matrix $g_{ij}(t)$ of frame change

$$\sum_j \frac{dg_{ij}(t)}{dt}\vec{\mathbf{e}}_j = \sum_j g_{ij}(t)\left(\vec{\omega}_O(t) \wedge \vec{\mathbf{e}}_j\right) \qquad (2.9)$$

In the case of a rigid solid (non deformable), the solid only displaces in space by translation or global rotation, i.e. without deformation, and the directions represented by the unit frame vectors $\vec{e}_i$ always correspond exactly to the directions of the *crystalline lattice of the solid.* In the case where the solid under consideration moves and deforms at the same time, as the vectors $\vec{e}_i(t)$ move with a velocity $\vec{\phi}_O(t)$ and rotate with an angular velocity $\dot{\vec{\omega}}_O(t)$, the directions represented by the local unit vectors $\vec{e}_i$ still correspond to the *average directions of the crystal lattice* at the point O.

As a matter of fact, in a vicinity of point O, the directions of the crystalline lattice correspond more or less to the directions of the unit vectors $\vec{e}_i$. The size of this neighborhood depends on how deformed the solid around the point O. We can empirically define this neighborhood as the volume V_m around the point O which satisfies the following criteria: *the volume V_m of lattice around O such that the components of the rotational vector $\vec{\omega}$ and the scalar of volume expansion τ do not vary more than 1% over said volume.*
In figure *2.2*, we represent the crystalline lattice around a point O in the presence of a deformation of the lattice.

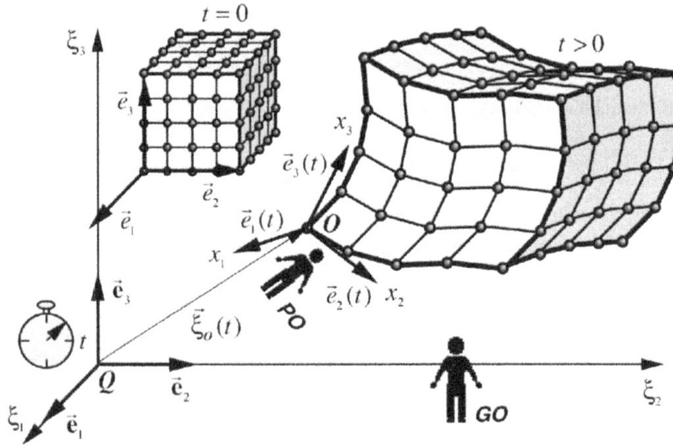

Figure 2.2 *- the solid lattice around the neighborhood of point O with local frame $Ox_1x_2x_3$*

Projection of the fields on the local frame

Every scalar field or vectorial field defined in the absolute space can be projected and decomposed in the global frame $Q\xi_1\xi_2\xi_3$ or locally in a neighborhood of the point O, in the mobile local frame $Ox_1x_2x_3$ (figure 2.5).
For example, the velocity field $\vec{\phi}$ of the solid lattice in the absolute frame is defined at point $\vec{\xi}$ and instant t by its 3 projections ϕ_i on the 3 axes $Q\xi_1$, $Q\xi_2$ and $Q\xi_3$

$$\vec{\phi}(\vec{\xi},t)=\sum_{i=1}^{3}\phi_i(\xi_k,t)\vec{\mathbf{e}}_i \qquad \text{with} \qquad \vec{\xi}=\sum_{k=1}^{3}\xi_k\vec{\mathbf{e}}_k \tag{2.10}$$

In the neighborhood of point $\boldsymbol{O}$ *of the local frame*, it can also be projected on the 3 axes $\boldsymbol{O}x_1$, $\boldsymbol{O}x_2$ and $\boldsymbol{O}x_3$ (figure 2.3). At point $\boldsymbol{B}$ of the local referential, marked by $\vec{r}$, and at instant t, the velocity field $\vec{\phi}$ can we written in the local frame projected as such

$$\vec{\phi}(\vec{r},t)=\sum_{i=1}^{3}\phi_i(x_k,t)\vec{e}_i \qquad \text{with} \qquad \vec{r}=\sum_{k=1}^{3}x_k\vec{e}_k \tag{2.11}$$

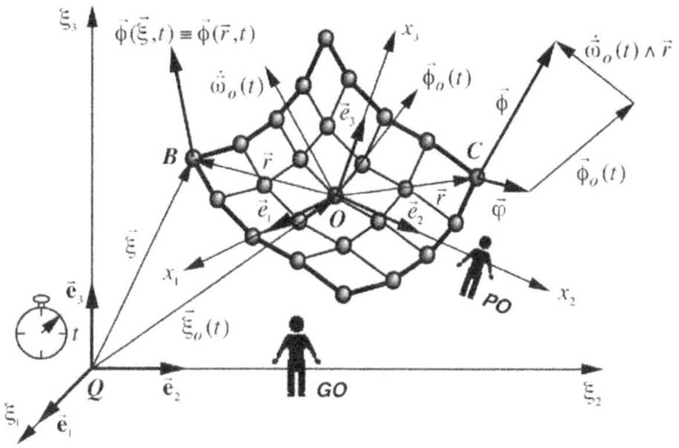

Figure 2.3 - *projection of velocity field* $\vec{\phi}$ *in the macroscopic local frame*

The expressions *(2.4)* and *(2.5)* giving the local frame vectors $\vec{e}_i$ as functions of the global frame vectors $\vec{\mathbf{e}}_i$ and vice-versa allow us to find a mapping between the ϕ_i components of the velocity field $\vec{\phi}$ in the absolute and local frame.

$$\vec{\phi}=\sum_i\phi_i(\xi_k,t)\vec{\mathbf{e}}_i=\sum_{ij}\phi_i(\xi_k,t)g_{ji}\vec{e}_j=\sum_j\phi_j(x_k,t)\vec{e}_j=\sum_{ij}\phi_j(x_k,t)g_{ji}\vec{\mathbf{e}}_i \;\Rightarrow$$

$$\phi_i(x_k,t)=\sum_j g_{ij}\,\phi_j(\xi_k,t) \qquad \text{and} \qquad \phi_j(\xi_k,t)=\sum_i g_{ij}\,\phi_i(x_k,t) \tag{2.12}$$

Furthermore, since the position vectors $\vec{r}$ and $\vec{\xi}$ identify respectively the nodes of the lattice in the local and global frame, they are related to each other by the simple relation

$$\vec{r}=\vec{\xi}-\vec{\xi}_O(t) \tag{2.13}$$

The mapping between the components ξ_k et x_k, the position vectors in the absolute frame and the local frame (which identify the same point) are given by

$$\vec{r}=\sum_i x_i\vec{e}_i=\sum_{ij}x_i\,g_{ij}\,\vec{\mathbf{e}}_i=\sum_j\xi_j\,\vec{\mathbf{e}}_j-\vec{\xi}_O(t)=\sum_{ij}\xi_j\,g_{ij}\,\vec{e}_i-\vec{\xi}_O(t) \;\Rightarrow$$

$$x_i=\sum_j g_{ij}\xi_j-\vec{\xi}_O(t)\vec{e}_i \qquad \text{and} \qquad \xi_j=\sum_i g_{ij}x_i+\vec{\xi}_O(t)\vec{\mathbf{e}}_j \tag{2.14}$$

Table 2.1 - Local macroscopic frame $Ox_1x_2x_3$ in a solid lattice

Rotation matrix of frame change

$$\begin{cases} \dfrac{d\vec{e}_i(t)}{dt} = \vec{\omega}_O(t) \wedge \vec{e}_i(t) & (1) \\ \displaystyle\sum_j \frac{dg_{ij}(t)}{dt}\vec{\mathbf{e}}_j = \sum_j g_{ij}(t)\left(\vec{\omega}_O(t) \wedge \vec{\mathbf{e}}_j\right) & (2) \end{cases}$$

$$\begin{cases} \left|\left(g_{ij}\right)\right| = 1 & (3) \\ \displaystyle\sum_j \left[g_{ij}(t)\right]^2 = 1 & (4) \\ \displaystyle\sum_i \left[g_{ij}(t)\right]^2 = 1 & (5) \end{cases}$$

Relations of frame change

$$\begin{cases} \vec{e}_i(t) = \displaystyle\sum_j g_{ij}(t)\vec{\mathbf{e}}_j & (6) \\ \phi_i(x_k,t) = \displaystyle\sum_j g_{ij}\,\phi_j(\xi_k,t) & (7) \\ x_i = \displaystyle\sum_j g_{ij}\xi_j - \vec{\xi}_O(t)\vec{e}_i & (8) \\ \dfrac{\partial}{\partial x_i} = \displaystyle\sum_j g_{ij}\frac{\partial}{\partial \xi_j} & (9) \end{cases}$$

$$\begin{cases} \vec{\mathbf{e}}_j = \displaystyle\sum_i g_{ij}(t)\vec{e}_i(t) & (10) \\ \phi_j(\xi_k,t) = \displaystyle\sum_i g_{ij}\,\phi_i(x_k,t) & (11) \\ \xi_j = \displaystyle\sum_i g_{ij}x_i + \vec{\xi}_O(t)\vec{\mathbf{e}}_j & (12) \\ \dfrac{\partial}{\partial \xi_j} = \displaystyle\sum_i g_{ij}\frac{\partial}{\partial x_i} & (13) \end{cases}$$

Local field of relative velocity

$$\begin{cases} \vec{\varphi}(\vec{r},t) = \vec{\phi}(\vec{r},t) - \vec{\phi}_O(t) - \vec{\omega}_O(t) \wedge \vec{r} & (1) \\ \vec{\varphi}(\vec{\xi},t) = \vec{\phi}(\vec{\xi},t) - \vec{\phi}_O(t) - \vec{\omega}_O(t) \wedge \left(\vec{\xi} - \vec{\xi}_O(t)\right) & (2) \end{cases}$$

Finally, the relations that exist between the partial derivatives in the absolute frame and the local frame are easily deducted

$$\frac{\partial}{\partial x_i} = \sum_j \frac{\partial}{\partial \xi_j}\frac{\partial \xi_j(y_k)}{\partial x_i} \quad \text{and} \quad \frac{\partial}{\partial \xi_j} = \sum_i \frac{\partial}{\partial x_i}\frac{\partial x_i(x_k)}{\partial \xi_j} \quad \Rightarrow$$

$$\frac{\partial}{\partial x_i} = \sum_j g_{ij}\frac{\partial}{\partial \xi_j} \quad \text{and} \quad \frac{\partial}{\partial \xi_j} = \sum_i g_{ij}\frac{\partial}{\partial x_i} \tag{2.15}$$

The local field of relative velocity

It is possible to define the *local field of relative velocity* $\vec{\varphi}$ of the points of the solid in the neighborhood of point O, in relation to the local mobile frame $Ox_1x_2x_3$. The velocity field is in fact the one measured at point C for example (figure 2.3), by the *Small Observer **(PO)*** linked to the local frame $Ox_1x_2x_3$, who would use rulers $\vec{e}_i$ and a clock synchronized to the clock of the absolute frame to measure the velocities $\vec{\varphi}$ of the nodes of the lattice, while the *Great Observer **(GO)*** is linked to the absolute frame and he measures the absolute velocities $\vec{\phi}$ of the nodes of the lattice.

The local field of relative velocity $\vec{\varphi}$ is obtained by subtracting the drag-along velocity $\vec{\phi}_O(t)$ of the local frame $Ox_1x_2x_3$, from the absolute field velocity $\vec{\phi}$ as well as the field of rotational velocity $\dot{\vec{\omega}}_O(t) \wedge \vec{r}$ associated with the rotation of the local frame $Ox_1x_2x_3$ around point O characterized by the angular drag-along velocity $\dot{\vec{\omega}}_O(t)$. We obtain the following equations

$$\vec{\varphi}(\vec{r},t) = \vec{\phi}(\vec{r},t) - \vec{\phi}_O(t) - \dot{\vec{\omega}}_O(t) \wedge \vec{r} \tag{2.16}$$

$$\vec{\varphi}(\vec{\xi},t) = \vec{\phi}(\vec{\xi},t) - \vec{\phi}_O(t) - \dot{\vec{\omega}}_O(t) \wedge \left(\vec{\xi} - \vec{\xi}_O(t)\right) \tag{2.17}$$

On the projection of the invariant operators of space and time in the local frame

The space operators $\overrightarrow{\mathrm{grad}}$, $\overrightarrow{\mathrm{rot}}$ and div applied to scalar or vectorial fields are *invariants*, meaning that they provide a result, scalar or vectorial, that does not depend on the choice of frame in which they are calculated (section A.2.). The same goes for the time operator of *material derivative* d/dt as this measures the temporal variation of quantities along the path. This means that any mathematical relations which invokes the operators $\overrightarrow{\mathrm{grad}}$, $\overrightarrow{\mathrm{rot}}$ and div, as well as d/dt, can be computed in either the absolute frame $Q\xi_1\xi_2\xi_3$ or the local frame $Ox_1x_2x_3$.

In the case of the temporal operator d/dt, the velocity one must introduce in the expression of material derivative is the absolute velocity $\vec{\phi}$ when d/dt is calculated in the absolute frame $Q\xi_1\xi_2\xi_3$ and the relative velocity $\vec{\varphi}$ when d/dt is calculated in the local frame $Ox_1x_2x_3$.

Regarding the formal operator $\vec{\nabla}$, we show, by using relations *(2.4)* et *(2.5)*, that there exists the following link between its formulation in the absolute frame $Q\xi_1\xi_2\xi_3$ and its formulation in the local frame $Ox_1x_2x_3$

$$\vec{\nabla} = \sum_i \vec{e}_i \frac{\partial}{\partial x_i} = \sum_{ij} \vec{e}_i\, g_{ij} \frac{\partial}{\partial \xi_j} = \sum_j \left(\sum_i \vec{e}_i\, g_{ij} \right) \frac{\partial}{\partial \xi_j} = \sum_j \vec{\mathbf{e}}_j \frac{\partial}{\partial \xi_j} \tag{2.18}$$

We can restate the various space and time operators, both in local and absolute frame, in terms of the $\vec{\nabla}$ operator. The following relations allow us to recover the expressions $\overrightarrow{\mathrm{grad}}$, $\overrightarrow{\mathrm{rot}}$ and div, expressions which we show in table 2.2

$$\overrightarrow{\mathrm{grad}}\, f = \vec{\nabla} f \quad \text{and} \quad \overrightarrow{\mathrm{rot}}\, \vec{u} = \vec{\nabla} \wedge \vec{u} \quad \text{and} \quad \mathrm{div}\, \vec{u} = \vec{\nabla} \vec{u} \tag{2.19}$$

We can also define the derivative of a quantity, scalar or vectorial, along a given direction $\vec{n}$, the directional derivative. It is written $(\vec{n}\vec{\nabla})$ and can be expanded in terms of the $\vec{\nabla}$ operator.

An important remark can be made here. In the case of a *perfectly non-deformable* solid lattice, the field of relative velocity $\vec{\varphi}$ defined by the relations *(2.16)* et *(2.17)* is everywhere null in the local frame $Ox_1x_2x_3$. On the other hand, in the case of a *deformable* solid lattice the field of relative velocity $\vec{\varphi}$ only contains velocities associated with the deformations of the lattice as measured in the local frame $Ox_1x_2x_3$, as well as the local rotations associated with these deformations, since the drag-along velocities of the lattice, both of the translational and rotational kind have been subtracted. Notably on the point O relative velocity $\vec{\varphi}$ is null by definition.

Table 2.2 - Expressions of the invariant operators in $Q\xi_1\xi_2\xi_3$ and $Ox_1x_2x_3$

	In absolute frame $Q\xi_1\xi_2\xi_3$ $f = f(\xi_k,t)$ *or* $\vec{u} = \sum u_i(\xi_k,t)\vec{\mathbf{e}}_i$	*In local frame* $Ox_1x_2x_3$ $f = f(x_k,t)$ *or* $\vec{u} = \sum u_i(x_k,t)\,\vec{e}_i$
$\vec{\nabla}$	$\sum_i \vec{\mathbf{e}}_i \dfrac{\partial}{\partial \xi_i}$ (1)	$\sum_i \vec{e}_i \dfrac{\partial}{\partial x_i}$ (2)
$\overrightarrow{\mathrm{grad}}\, f$	$\sum_i \vec{\mathbf{e}}_i \dfrac{\partial f(\xi_k,t)}{\partial \xi_i}$ (3)	$\sum_i \vec{e}_i \dfrac{\partial f(x_k,t)}{\partial x_i}$ (4)
$\overrightarrow{\mathrm{rot}}\,\vec{u}$	$\begin{vmatrix} \vec{\mathbf{e}}_1 & \vec{\mathbf{e}}_2 & \vec{\mathbf{e}}_3 \\ \partial/\partial\xi_1 & \partial/\partial\xi_2 & \partial/\partial\xi_3 \\ u_1(\xi_k,t) & u_2(\xi_k,t) & u_3(\xi_k,t) \end{vmatrix}$ (5)	$\begin{vmatrix} \vec{e}_1 & \vec{e}_2 & \vec{e}_3 \\ \partial/\partial x_1 & \partial/\partial x_2 & \partial/\partial x_3 \\ u_1(x_k,t) & u_2(x_k,t) & u_3(x_k,t) \end{vmatrix}$ (6)
$\mathrm{div}\,\vec{u}$	$\sum_i \dfrac{\partial u_i(\xi_k,t)}{\partial \xi_i}$ (7)	$\sum_i \dfrac{\partial u_i(x_k,t)}{\partial x_i}$ (8)
$(\vec{n}\vec{\nabla})$	$\sum_i (\vec{n}\vec{\mathbf{e}}_i)\dfrac{\partial}{\partial \xi_i}$ (9)	$\sum_i (\vec{n}\vec{e}_i)\dfrac{\partial}{\partial x_i}$ (10)
$\dfrac{d}{dt}$	$\dfrac{\partial}{\partial t} + (\vec{\phi}\vec{\nabla}) = \dfrac{\partial}{\partial t} + \sum_i \phi_i(\vec{\xi},t)\dfrac{\partial}{\partial \xi_i}$ (11)	$\dfrac{\partial}{\partial t} + (\vec{\varphi}\vec{\nabla}) = \dfrac{\partial}{\partial t} + \sum_i \varphi_i(\vec{r},t)\dfrac{\partial}{\partial x_i}$ (12)

We can conclude that in the case of *weakly deformable* solid, meaning when the displacements of the nodes of the lattice as seen in the local frame $Ox_1x_2x_3$ stay small in a neighborhood of point O, it is possible to replace the material derivative d/dt, corresponding normally to the temporal variation along the trajectories of the nodes in the local frame $Ox_1x_2x_3$, by the temporal derivative at a fixed point of the local frame $Ox_1x_2x_3$, meaning the partial derivative with respect to time.

The same goes for a solid which is *deforming very slowly,* since in that case the field of relative velocity $\vec{\varphi}$ is negligible. We derive a remark which will be *very important* for the rest of the treatise:

- *In the vicinity of a point O of a local frame $Ox_1x_2x_3$, if the deformations are small or happen slowly over time, we can use the following approximation for the temporal derivative:*

$$\boxed{\frac{d}{dt} = \frac{\partial}{\partial t} + (\vec{\varphi}\vec{\nabla}) \cong \frac{\partial}{\partial t}} \quad \textbf{\textit{in the vicinity of } } O \tag{2.20}$$

2.2 - Projection of the geometro-kinetic equations in the local frame

It is now possible to redefine the distortion tensors of a solid by using a formalism similar to that introduced in the first chapter, but in the local frame $Ox_1x_2x_3$ of the small observer ***PO*** (figure 2.3).

Geometro-kinetic equations of distortion in the local frame

In the local frame $Ox_1x_2x_3$, the angular velocity of rotation $d\vec{\omega}/dt$ and the material derivative of the scalar τ of volume expansion of the solid can be computed in a way perfectly similar to that used in the second chapter already. Therefore, the equations of geometro-kinetic introduced above remain valid in the local mobile frame.

$$\frac{d\vec{\omega}}{dt} = \frac{1}{2}\overrightarrow{\mathrm{rot}}\,\vec{\phi} \quad \text{and} \quad \frac{d\tau}{dt} = -\frac{S_n}{n} + \mathrm{div}\,\vec{\phi} \tag{2.21}$$

We can now define in the local frame a new *tensor of global distortion* $\vec{\beta}_i$ such that its trace is equal to $d\tau/dt$ and that half of its anti-symmetric part, with an inverted sign, gives us $d\vec{\omega}/dt$. The geometro-kinetic equations which we defined in the second chapter in the following way

$$\frac{d\vec{\beta}_i}{dt} = -\frac{S_n}{3n}\vec{e}_i + \overrightarrow{\mathrm{grad}}\,\phi_i = -\frac{S_n}{3n}\vec{e}_i + \overrightarrow{\mathrm{grad}}\left(\vec{\phi}\vec{e}_i\right) = -\frac{S_n}{3n}\vec{e}_i + \left(\vec{e}_i\vec{\nabla}\right)\vec{\phi} + \vec{e}_i \wedge \overrightarrow{\mathrm{rot}}\,\vec{\phi} \tag{2.22}$$

yield a tensor which satisfy these requirements. Indeed, it follows for the trace of $d\vec{\beta}_i/dt$ and for the half of its anti-symmetric part with inverted sign, the expressions of $d\tau/dt$ and of $d\vec{\omega}/dt$

$$\sum_k \vec{e}_k \frac{d\vec{\beta}_k}{dt} = -\frac{S_n}{n} + \vec{\nabla}\vec{\phi} + \sum_k \underbrace{\overrightarrow{\mathrm{rot}}\,\vec{\phi}\left(\vec{e}_k \wedge \vec{e}_k\right)}_{=0} = -\frac{S_n}{n} + \mathrm{div}\,\vec{\phi} = \frac{d\tau}{dt} \tag{2.23}$$

$$-\frac{1}{2}\sum_k \vec{e}_k \wedge \frac{d\vec{\beta}_k}{dt} = -\frac{1}{2}\left[\sum_k \vec{e}_k \wedge \left(\vec{e}_k\vec{\nabla}\right)\vec{\phi} + \sum_k \vec{e}_k \wedge \left(\vec{e}_k \wedge \overrightarrow{\mathrm{rot}}\,\vec{\phi}\right)\right] = \frac{1}{2}\overrightarrow{\mathrm{rot}}\,\vec{\phi} = \frac{d\vec{\omega}}{dt} \tag{2.24}$$

The new *deformation tensor* $\vec{\varepsilon}_i$ is obtained by symmetrization of the geometro-kinetic equation $d\vec{\beta}_i/dt$ in the local frame $Ox_1x_2x_3$

$$\frac{d\vec{\varepsilon}_i}{dt} = \frac{d\vec{\beta}_i}{dt} - \vec{e}_i \wedge \frac{d\vec{\omega}}{dt} \quad \Rightarrow \quad \frac{d\vec{\varepsilon}_i}{dt} = -\frac{S_n}{3n}\vec{e}_i + \overrightarrow{\mathrm{grad}}\,\phi_i - \frac{1}{2}\vec{e}_i \wedge \overrightarrow{\mathrm{rot}}\,\vec{\phi} \tag{2.25}$$

The new *shear tensor* $\vec{\alpha}_i$, is obtained from the transversal symmetric part of $d\vec{\beta}_i/dt$ in the local frame $Ox_1x_2x_3$

$$\frac{d\vec{\alpha}_i}{dt} = \frac{d\vec{\varepsilon}_i}{dt} - \frac{1}{3}\vec{e}_i\frac{d\tau}{dt} \quad \Rightarrow \quad \frac{d\vec{\alpha}_i}{dt} = \overrightarrow{\mathrm{grad}}\,\phi_i - \frac{1}{2}\vec{e}_i \wedge \overrightarrow{\mathrm{rot}}\,\vec{\phi} - \frac{1}{3}\vec{e}_i\,\mathrm{div}\,\vec{\phi} \tag{2.26}$$

The tensors of global distortion $\vec{\beta}_i, \vec{\varepsilon}_i, \vec{\alpha}_i$ thus defined depend on the choice of the local frame $Ox_1x_2x_3$, via the presence of the frame vectors $\vec{e}_i$ in their definitions. They are computed from the components of the absolute velocity field $\vec{\phi}$ of the solid projected in the mobile local frame $Ox_1x_2x_3$, in the fashion that the distortion tensors $\vec{\beta}_i, \vec{\varepsilon}_i, \vec{\alpha}_i$ were computed in the second chapter from the projections of the $\vec{\phi}$ field in a fixed frame linked to $Q\xi_1\xi_2\xi_3$.

For the Small Observer ***PO*** in the local mobile frame $Ox_1x_2x_3$, the frame vectors $\vec{e}_i$ appear as constants independent of time, and therefore in $Ox_1x_2x_3$ we have

$$\begin{cases} \dfrac{d\tau}{dt} = \sum_k \vec{e}_k \dfrac{d\vec{\beta}_k}{dt} \;\Rightarrow\; \tau = \sum_k \vec{e}_k \vec{\beta}_k \\ \dfrac{d\vec{\omega}}{dt} = -\dfrac{1}{2}\sum_k \vec{e}_k \wedge \dfrac{d\vec{\beta}_k}{dt} \;\Rightarrow\; \vec{\omega} = -\dfrac{1}{2}\sum_k \vec{e}_k \wedge \vec{\beta}_k \\ \dfrac{d\vec{\varepsilon}_i}{dt} = \dfrac{d\vec{\beta}_i}{dt} - \vec{e}_i \wedge \dfrac{d\vec{\omega}}{dt} \;\Rightarrow\; \vec{\varepsilon}_i = \vec{\beta}_i - \vec{e}_i \wedge \vec{\omega} \\ \dfrac{d\vec{\alpha}_i}{dt} = \dfrac{d\vec{\varepsilon}_i}{dt} - \dfrac{1}{3}\vec{e}_i \dfrac{d\tau}{dt} \;\Rightarrow\; \vec{\alpha}_i = \vec{\varepsilon}_i - \dfrac{1}{3}\tau\vec{e}_i \end{cases} \tag{2.27}$$

As a consequence the decompositions of the distortion tensors we saw in figures 1.7 and 1.8 remain valid when we replace the tensors $\vec{\beta}_i, \vec{\varepsilon}_i, \vec{\alpha}_i$ defined in the fixed absolute frame $Q\xi_3\xi_2\xi_3$ by the same tensors $\vec{\beta}_i, \vec{\varepsilon}_i, \vec{\alpha}_i$, this time defined in the mobile local frame $Ox_1x_2x_3$.

Table 2.3 - Geometro-kinetic equations of the distortions of a solid lattice in the local frame $Ox_1x_2x_3$

$$\begin{cases} \dfrac{d\vec{\beta}_i}{dt} = -\dfrac{S_n}{3n}\vec{e}_i + \overrightarrow{\text{grad}}\,\phi_i & (1) \\ \dfrac{d\vec{\omega}}{dt} = \dfrac{1}{2}\overrightarrow{\text{rot}}\,\vec{\phi} & (2) \\ \dfrac{d\tau}{dt} = -\dfrac{S_n}{n} + \text{div}\,\vec{\phi} & (3) \end{cases}$$

$$avec \quad \frac{d}{dt} = \frac{\partial}{\partial t} + \left(\vec{\varphi}\vec{\nabla}\right) \quad (4)$$

$$\begin{cases} \vec{\omega} = -\dfrac{1}{2}\sum_k \vec{e}_k \wedge \vec{\beta}_k & (5) \\ \tau = \sum_k \vec{e}_k \vec{\beta}_k = \sum_k \vec{e}_k \vec{\varepsilon}_k & (6) \\ \vec{\varepsilon}_i = \vec{\beta}_i - \vec{e}_i \wedge \vec{\omega} & (7) \\ \vec{\alpha}_i = \vec{\varepsilon}_i - \dfrac{1}{3}\tau\vec{e}_i & (8) \end{cases}$$

One should note here that only the vectorial field $\vec{\omega}$ and the scalar field τ (representing respectively the anti-symmetric part and the trace of the distortion tensor $\vec{\beta}_i$) are truly *invariant* quantities, in the sense that we are dealing with a vector and a scalar that do not depend on the choice of frame we use to compute them. The geometro-kinetic equations in the local frame are shown in table 2.2. One should compare them to the geometro-kinetic equations of table 1.1, expressed in the absolute frame: we can observe that they have exactly the same form, but that in these expressions it is *essentially the material derivative that is different* since it depends on $\vec{\phi}$ in the absolute frame $Q\xi_3\xi_2\xi_3$ while it depends on $\vec{\varphi}$ in the mobile frame $Ox_1x_2x_3$.

Local distortions $\vec{\beta}_i^{(\delta)}$ and local rotations $\vec{\omega}^{(\delta)}$ due to pure deformations

Using the rotational part of the local field of relative velocity $\vec{\varphi}$ expressed in the frame $Ox_1x_2x_3$

$$\overrightarrow{\mathrm{rot}}\,\vec{\varphi} = \overrightarrow{\mathrm{rot}}\left(\vec{\phi} - \vec{\phi}_O(t) - \dot{\vec{\omega}}_O(t) \wedge \vec{r}\right) = \overrightarrow{\mathrm{rot}}\,\vec{\phi} - \dot{\vec{\omega}}_O \,\mathrm{div}\,\vec{r} + \left(\dot{\vec{\omega}}_O \vec{\nabla}\right)\vec{r} = \overrightarrow{\mathrm{rot}}\,\vec{\phi} - 2\dot{\vec{\omega}}_O \tag{2.28}$$

we can define a local rotation vector $\vec{\omega}^{(\delta)}$ *due to pure deformation* which measures the rotations associated only with the deformations of the solid about the O point, since its material derivative is equal to the global angular velocity $d\vec{\omega}\,/\,dt$ from which we subtract the drag-along angular velocity $\dot{\vec{\omega}}_O$ of the frame $Ox_1x_2x_3$

$$\frac{d\vec{\omega}^{(\delta)}}{dt} = \frac{1}{2}\overrightarrow{\mathrm{rot}}\,\vec{\varphi} = \frac{1}{2}\overrightarrow{\mathrm{rot}}\,\vec{\phi} - \dot{\vec{\omega}}_O = \frac{d\vec{\omega}}{dt} - \dot{\vec{\omega}}_O \tag{2.29}$$

It is also possible to build a *local distortion tensor* $\vec{\beta}_i^{(\delta)}$ *due to pure deformation* such that the trace is equal to the scalar of volume expansion τ and that the half of its anti-symmetric part with inverted sign gives us the vector of local rotation $\vec{\omega}^{(\delta)}$. The following geometro-kinetic equation will do

$$\frac{d\vec{\beta}_i^{(\delta)}}{dt} = -\frac{S_n}{3n}\vec{e}_i + \overrightarrow{\mathrm{grad}}\,\varphi_i = -\frac{S_n}{3n}\vec{e}_i + \overrightarrow{\mathrm{grad}}\left(\vec{\varphi}\vec{e}_i\right) \tag{2.30}$$

Indeed, by rewriting the expression for $d\vec{\beta}_i^{(\delta)}\,/\,dt$ as

$$\frac{d\vec{\beta}_i^{(\delta)}}{dt} = -\frac{S_n}{3n}\vec{e}_i + \left(\vec{e}_i\vec{\nabla}\right)\vec{\varphi} + \vec{e}_i \wedge \overrightarrow{\mathrm{rot}}\,\vec{\varphi} \tag{2.31}$$

We can verify the 2 following expressions

$$\sum_k \vec{e}_k \frac{d\vec{\beta}_k^{(\delta)}}{dt} = -\frac{S_n}{n} + \mathrm{div}\,\vec{\varphi} = \frac{d\tau}{dt} \tag{2.32}$$

$$-\frac{1}{2}\sum_k \vec{e}_k \wedge \frac{d\vec{\beta}_k^{(\delta)}}{dt} = \frac{1}{2}\overrightarrow{\mathrm{rot}}\,\vec{\varphi} = \frac{d\vec{\omega}^{(\delta)}}{dt} \tag{2.33}$$

The tensor of local distortion $\vec{\beta}_i^{(\delta)}$ thus defined can be directly linked to the field of absolute velocity . Indeed, the following

$$\begin{aligned}\overrightarrow{\mathrm{grad}}\left(\vec{\phi}\vec{e}_i\right) &= \vec{e}_i \wedge \overrightarrow{\mathrm{rot}}\,\vec{\phi} + \left(\vec{e}_i\vec{\nabla}\right)\vec{\phi} = \vec{e}_i \wedge \overrightarrow{\mathrm{rot}}\,\vec{\varphi} + \vec{e}_i \wedge 2\dot{\vec{\omega}}_O \\ &+ \left(\vec{e}_i\vec{\nabla}\right)\vec{\varphi} + \left(\vec{e}_i\vec{\nabla}\right)\left(\dot{\vec{\omega}}_O(t) \wedge \vec{r}\right) = \vec{e}_i \wedge \overrightarrow{\mathrm{rot}}\,\vec{\varphi} + \left(\vec{e}_i\vec{\nabla}\right)\vec{\varphi} - \dot{\vec{\omega}}_O \wedge \vec{e}_i\end{aligned} \tag{2.34}$$

shows that

$$\frac{d\vec{\beta}_i^{(\delta)}}{dt} = -\frac{S_n}{3n}\vec{e}_i + \overrightarrow{\mathrm{grad}}\left(\vec{\phi}\vec{e}_i\right) - \vec{e}_i \wedge \dot{\vec{\omega}}_O = \frac{d\vec{\beta}_i}{dt} - \vec{e}_i \wedge \dot{\vec{\omega}}_O \tag{2.35}$$

It is easy to show that the expressions for the deformation field $\vec{\varepsilon}_i$ and shear field $\vec{\alpha}_i$ as a function of the field $\vec{\varphi}$ of relative velocities or as a function of the field $\vec{\phi}$ of absolute velocities are the same, which seems logical since $\vec{\varepsilon}_i$ and $\vec{\alpha}_i$ are only dependent on the pure deformations.

$$\frac{d\vec{\varepsilon}_i}{dt} = -\frac{S_n}{3n}\vec{e}_i + \overrightarrow{\mathrm{grad}}\,\varphi_i - \frac{1}{2}\vec{e}_i \wedge \overrightarrow{\mathrm{rot}}\,\vec{\varphi} = -\frac{S_n}{3n}\vec{e}_i + \left(\vec{e}_i\vec{\nabla}\right)\vec{\varphi} + \frac{1}{2}\vec{e}_i \wedge \overrightarrow{\mathrm{rot}}\,\vec{\varphi} \tag{2.36}$$

$$\frac{d\vec{\alpha}_i}{dt} = \overrightarrow{\mathrm{grad}}\,\varphi_i - \frac{1}{2}\vec{e}_i \wedge \overrightarrow{\mathrm{rot}}\,\vec{\varphi} - \frac{1}{3}\vec{e}_i\,\mathrm{div}\,\vec{\varphi} = \left(\vec{e}_i\vec{\nabla}\right)\vec{\varphi} + \frac{1}{2}\vec{e}_i \wedge \overrightarrow{\mathrm{rot}}\,\vec{\varphi} - \frac{1}{3}\vec{e}_i\,\mathrm{div}\,\vec{\varphi} \tag{2.37}$$

Since the frame vectors $\vec{e}_i$ are constants independent of time for the Small Observer ***PO*** in the local mobile frame $Ox_1x_2x_3$, the following relations of decomposition of the tensors of local distortion hold in $Ox_1x_2x_3$

$$\begin{cases} \dfrac{d\tau}{dt} = \sum_k \vec{e}_k \dfrac{d\vec{\beta}_k^{(\delta)}}{dt} \Rightarrow \tau = \sum_k \vec{e}_k \vec{\beta}_k^{(\delta)} \\ \dfrac{d\vec{\omega}^{(\delta)}}{dt} = -\dfrac{1}{2}\sum_k \vec{e}_k \wedge \dfrac{d\vec{\beta}_k^{(\delta)}}{dt} \Rightarrow \vec{\omega}^{(\delta)} = -\dfrac{1}{2}\sum_k \vec{e}_k \wedge \vec{\beta}_k^{(\delta)} \\ \dfrac{d\vec{\varepsilon}_i}{dt} = \dfrac{d\vec{\beta}_i^{(\delta)}}{dt} - \vec{e}_i \wedge \dfrac{d\vec{\omega}^{(\delta)}}{dt} \Rightarrow \vec{\varepsilon}_i = \vec{\beta}_i^{(\delta)} - \vec{e}_i \wedge \vec{\omega}^{(\delta)} \\ \dfrac{d\vec{\alpha}_i}{dt} = \dfrac{d\vec{\varepsilon}_i}{dt} - \dfrac{1}{3}\vec{e}_i \dfrac{d\tau}{dt} \Rightarrow \vec{\alpha}_i = \vec{\varepsilon}_i - \dfrac{1}{3}\tau \vec{e}_i \end{cases} \tag{2.38}$$

Table 2.4 - Geometro-kinetic equations of local distortions and rotations of a solid lattice in the local frame $Ox_1x_2x_3$

$$\begin{cases} \dfrac{d\vec{\beta}_i^{(\delta)}}{dt} = -\dfrac{S_n}{3n}\vec{e}_i + \overrightarrow{\operatorname{grad}}\,\varphi_i = \dfrac{d\vec{\beta}_i}{dt} - \vec{e}_i \wedge \dot{\vec{\omega}}_O(t) & (1) \\ \dfrac{d\vec{\omega}^{(\delta)}}{dt} = \dfrac{1}{2}\overrightarrow{\operatorname{rot}}\,\vec{\varphi} = \dfrac{d\vec{\omega}}{dt} - \dot{\vec{\omega}}_O(t) & (2) \\ \dfrac{d\tau}{dt} = -\dfrac{S_n}{n} + \operatorname{div}\vec{\varphi} & (3) \end{cases}$$

$$\begin{cases} \vec{\beta}_i^{(\delta)} = \vec{\beta}_i - \vec{e}_i \wedge \vec{\omega}_O(t) & (4) \\ \vec{\omega}^{(\delta)} = -\dfrac{1}{2}\sum_k \vec{e}_k \wedge \vec{\beta}_k^{(\delta)} = \vec{\omega} - \vec{\omega}_O(t) & (5) \\ \tau = \sum_k \vec{e}_k \vec{\beta}_k^{(\delta)} = \sum_k \vec{e}_k \vec{\beta}_k = \sum_k \vec{e}_k \vec{\varepsilon}_k & (6) \\ \vec{\varepsilon}_i = \vec{\beta}_i^{(\delta)} - \vec{e}_i \wedge \vec{\omega}^{(\delta)} = \vec{\beta}_i - \vec{e}_i \wedge \vec{\omega} & (7) \\ \vec{\alpha}_i = \vec{\varepsilon}_i - \dfrac{1}{3}\tau \vec{e}_i & (8) \end{cases}$$

For the same reason as above, the tensors of local distortion from deformation can be directly linked to the tensors of global distortion in $Ox_1x_2x_3$ by the following relations, in which the rotational vector $\vec{\omega}_O$ only depends on time

$$\dfrac{d\vec{\beta}_i^{(\delta)}}{dt} = \dfrac{d\vec{\beta}_i}{dt} - \vec{e}_i \wedge \dot{\vec{\omega}}_O \Rightarrow \vec{\beta}_i^{(\delta)} = \vec{\beta}_i - \vec{e}_i \wedge \vec{\omega}_O(t) \tag{2.39}$$

$$\dfrac{d\vec{\omega}^{(\delta)}}{dt} = \dfrac{d\vec{\omega}}{dt} - \dot{\vec{\omega}}_O \Rightarrow \vec{\omega}^{(\delta)} = \vec{\omega} - \vec{\omega}_O(t) \tag{2.40}$$

The local tensors $\vec{\beta}_i^{(\delta)}$ and $\vec{\omega}^{(\delta)}$, associated with local deformations of the lattice in the neighborhood of point O, differ from the global tensors $\vec{\beta}_i$ and $\vec{\omega}$ obtained in the previous paragraph only due to the existence of a drag-along rotation $\vec{\omega}_O(t)$ of the local frame $Ox_1x_2x_3$. In short, the tensors τ, $\vec{\varepsilon}_i$ et $\vec{\alpha}_i$, can be computed indifferently from the field of absolute speed $\vec{\phi}$ or the field of relative speed $\vec{\varphi}$. The equations of geometro-kinetic for the tensors of local distortion due to deformation, deduced from the field $\vec{\varphi}$ in the local mobile frame $Ox_1x_2x_3$, are given in table 2.4.

2.3 - Geometro-compatibility in eulerian coordinates

We have shown section 1.1 that the description of the deformation of compatible solids in lagrangian coordinates is characterized by the existence of a field of displacement. Indeed, in lagrangian coordinates (figure 2.4), the solid is described by a position vector $\vec{r}$ which marks all the points in a frame $Ox_1x_2x_3$ fixed in the absolute referential $Q\xi_1\xi_2\xi_3$ of the **GO** observer. The lagrangian vector displacement field $\vec{u}_L(\vec{r},t)$ allows us to find the position of all points of the solid at time t, which were originally at coordinate $\vec{r}$ of the frame $Ox_1x_2x_3$.

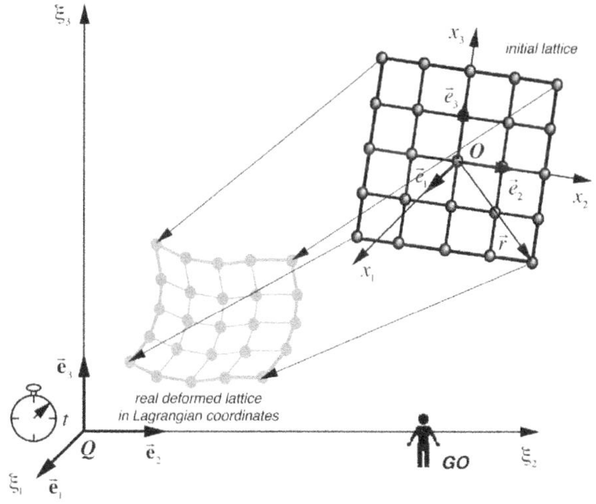

Figure 2.4 *- the displacement field $\vec{u}_L(\vec{r},t)$ in lagrangian coordinates*

It is intuitively clear that the description of the distortions of a solid in eulerian coordinates should allow us to recover the displacement field. Indeed in eulerian coordinates (figure 2.5), the deformed solid is described at a time t in the absolute frame of the observer **GO**. For a point A of the solid of coordinate $\vec{\xi}$ in this frame we define the eulerian displacement vector $\vec{u}_E(\vec{\xi},t)$ as linking point A to point A' where the A point of the solid was at initial time $t=0$. The same type of construction can be applied in the mobile frame $Ox_1x_2x_3$ of the **PO** observer: at a given time t, for each point $\boldsymbol{B}$ of the solide at coordinate $\vec{r}$ in that frame, we define the displacement vector $\vec{u}_E^{(\delta)}(\vec{r},t)$ which links point $\boldsymbol{B}$ to the location $\boldsymbol{B}''$ of frame $Ox_1x_2x_3$ where point $\boldsymbol{B}$ was at time $t=0$.

The displacement fields $\vec{u}_E(\vec{\xi},t)$ and $\vec{u}_E^{(\delta)}(\vec{r},t)$ thus defined allow us to rebuild the solid as it was at time $t=0$, before the various translations, rotations and deformations it underwent, both in the absolute frame of ***GO*** and in the mobile frame $Ox_1x_2x_3$ of the ***PO*** observer.

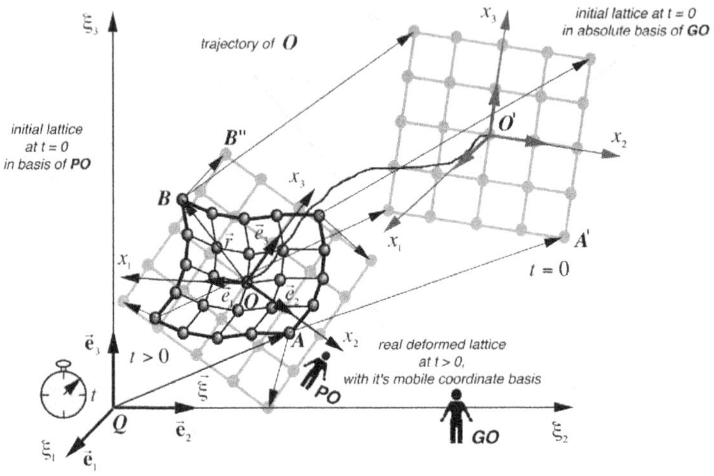

Figure 2.5 - *displacement fields* $\vec{u}_E(\vec{\xi},t)$ *and* $\vec{u}_E^{(\delta)}(\vec{r},t)$ *in eulerian coordinates*

Condition of geometro-compatibility in eulerian coordinates

According to the form of the geometro-kinetic equation for the distortion tensor $\vec{\beta}_i$ we have

$$\frac{d\vec{\beta}_i}{dt} = -\frac{S_n}{3n}\vec{e}_i + \overrightarrow{\text{grad}}\,\phi_i \qquad (2.41)$$

It is clear that there is a link between the temporal derivative of the tensor $\vec{\beta}_i$ and the gradient of the field of velocity $\vec{\phi}(\vec{\xi},t)$. Also the velocity field $\vec{\phi}(\vec{\xi},t)$ is related to the time derivative of the displacement field $\vec{u}_E(\vec{\xi},t)$, thus there is link between the tensor $\vec{\beta}_i$ and the gradient of the displacement field $\vec{u}_E(\vec{\xi},t)$. The distortion tensor has to be linked to the gradient of the displacement field.

By decomposing the global distortion $\vec{\beta}_i$ in a local distortion tensor $\vec{\beta}_i^{(\delta)}$, associated to the deformations of the solid in the $Ox_1x_2x_3$ frame, and a second component associated to the drag-along rotation $\vec{\omega}_O(t)$ of said frame we write

$$\vec{\beta}_i = \vec{\beta}_i^{(\delta)} + \vec{e}_i \wedge \vec{\omega}_O(t) \quad \Rightarrow \quad \overrightarrow{\text{rot}}\,\vec{\beta}_i = \overrightarrow{\text{rot}}\,\vec{\beta}_i^{(\delta)} \qquad (2.42)$$

Supposing that the tensor of distortion is indeed the gradient of the displacement field we can write *a priori* the following condition of compatibility in eulerian coordinates

Condition of geometro-compatibility: $\overrightarrow{\mathrm{rot}}\,\vec{\beta}_i = \overrightarrow{\mathrm{rot}}\,\vec{\beta}_i^{(\delta)} = 0$ *(2.43)*

In fact, in section 3.4 we will show that the physical meaning of this relation is, from a topological point of view, that *the displacement field* $\vec{u}_E(\vec{\xi},t)$ *is continuous,* in other words that there are *no discontinuities of the displacement field* $\vec{u}_E(\vec{\xi},t)$, meaning that *there are no dislocations inside the medium.* We revisit the relations of spatial derivatives of the topological tensors introduced in section 1.4, and we will apply the same manipulations as in 1.4. This condition gives us a complete set of compatibility relations in eulerian coordinates.

$$\begin{cases} \overrightarrow{\mathrm{rot}}\,\vec{\beta}_i = \overrightarrow{\mathrm{rot}}\,\vec{\beta}_i^{(\delta)} = 0 \\ \mathrm{div}\,\vec{\omega} = \mathrm{div}\,\vec{\omega}^{(\delta)} = 0 \end{cases} \qquad (2.44)$$

$$\begin{cases} \overrightarrow{\mathrm{rot}}\left[\overrightarrow{\mathrm{rot}}\,\vec{\varepsilon}_i\right]^{\mathrm{T}} = 0 \\ \overrightarrow{\mathrm{rot}}\left[\overrightarrow{\mathrm{rot}}\,\vec{\alpha}_i\right]^{\mathrm{T}} - \dfrac{1}{3}\overrightarrow{\mathrm{rot}}\,\overrightarrow{\mathrm{rot}}\left(\vec{e}_i\tau\right) = 0 \end{cases} \qquad (2.45)$$

We will show now the consequences of the condition of geometric compatibility in eulerian coordinates:

Consequence 1: existence of potential field of displacement

The condition of geometro-compatibility *(2.43)* implies that $\vec{\beta}_i$ and $\vec{\beta}_i^{(\delta)}$ derive from gradients of scalar fields $\vec{u}_E(\vec{\xi},t)$ and $\vec{u}_E^{(\delta)}(\vec{r},t)$ defined respectively in frame $Q\xi_1\xi_2\xi_3$ and $Ox_1x_2x_3$, which allows us to write without loss of generality

$$\overrightarrow{\mathrm{rot}}\,\vec{\beta}_i = \overrightarrow{\mathrm{rot}}\,\vec{\beta}_i^{(\delta)} = 0 \;\Rightarrow\; \begin{cases} \vec{\beta}_i = -\overrightarrow{\mathrm{grad}}\,u_{Ei}(\vec{\xi},t) = -\sum\limits_k \dfrac{\partial u_{Ei}}{\partial \xi_k}\,\vec{\mathbf{e}}_k \\ \vec{\beta}_i^{(\delta)} = -\overrightarrow{\mathrm{grad}}\,u_{Ei}^{(\delta)}(\vec{r},t) = -\sum\limits_k \dfrac{\partial u_{Ei}^{(\delta)}}{\partial x_k}\,\vec{e}_k \end{cases} \qquad (2.46)$$

With the scalar fields $u_{Ei}(\vec{\xi},t)$ and $u_{Ei}^{(\delta)}(\vec{r},t)$ necessary to the description of the tensors $\vec{\beta}_i$ and $\vec{\beta}_i^{(\delta)}$, 2 vectorial fields can be defined $\vec{u}_E(\vec{\xi},t)$ and $\vec{u}_E(\vec{\xi},t)$ such that

$$\vec{u}_E(\vec{\xi},t) = \sum_k u_{Ei}\,(\vec{\xi},t)\,\vec{\mathbf{e}}_k \quad \text{and} \quad \vec{u}_E^{(\delta)}(\vec{r},t) = \sum_k u_{Ei}^{(\delta)}(\vec{r},t)\,\vec{e}_k \qquad (2.47)$$

As a consequence, the distortion tensors $\vec{\beta}_i$ and $\vec{\beta}_i^{(\delta)}$ in a geometro-compatible solid lattice, meaning which satisfies the relation of geometric compatibility *(2.44)*, are indeed the *tensor gradients* of *continuous vectorial fields* $\vec{u}_E(\vec{\xi},t)$ and $\vec{u}_E^{(\delta)}(\vec{r},t)$. Dimensional analysis shows us that vectors $\vec{u}_E$ and $\vec{u}_E^{(\delta)}$ have dimensions of displacements. As a consequence, $\vec{u}_E(\vec{\xi},t)$ will be called the *potential field of global displacement of the solid* and $\vec{u}_E^{(\delta)}(\vec{r},t)$ the *potential field of local displacement of the solid.* These fields of potential of displacement correspond, up to a continuous vectorial field $\vec{u}_0(t)$, to the fields of displacement we have already discussed in the eulerian coordinates case (figure 2.5).

It is clear that the field of global displacement $\vec{u}_E(\vec{\xi},t)$ can be expressed in the local frame

$Ox_1x_2x_3$ as a function of $\vec{r}$ and t, simply by applying the relation $\vec{\beta}_i\,(\vec{r},t)=-\overrightarrow{\mathrm{grad}}\,u_{E\,i}(\vec{r},t)$ in the frame $Ox_1x_2x_3$. Using the decompositions of $\vec{\beta}_i$ and $\vec{\beta}_i^{(\delta)}$ described in tables 2.3 et 2.4, all the tensors of distortion can be directly derived from the vectorial fields of displacement $\vec{u}_E(\vec{\xi},t)$ and $\vec{u}_E^{(\delta)}(\vec{r},t)$ in the frame $Ox_1x_2x_3$

$$\begin{cases} \vec{\beta}_i = -\overrightarrow{\mathrm{grad}}\,u_{Ei} \quad and \quad \vec{\beta}_i^{(\delta)} = -\overrightarrow{\mathrm{grad}}\,u_{Ei}^{(\delta)} \\ \vec{\omega} = -\dfrac{1}{2}\overrightarrow{\mathrm{rot}}\,\vec{u}_E \quad and \quad \vec{\omega}^{(\delta)} = -\dfrac{1}{2}\overrightarrow{\mathrm{rot}}\,\vec{u}_E^{(\delta)} \\ \tau = -\mathrm{div}\,\vec{u}_E = -\mathrm{div}\,\vec{u}_E^{(\delta)} \end{cases} \tag{2.48a}$$

$$\begin{cases} \vec{\varepsilon}_i = -\overrightarrow{\mathrm{grad}}\,u_{Ei} + \dfrac{1}{2}\vec{e}_i \wedge \overrightarrow{\mathrm{rot}}\,\vec{u}_E = -\overrightarrow{\mathrm{grad}}\,u_{Ei}^{(\delta)} + \dfrac{1}{2}\vec{e}_i \wedge \overrightarrow{\mathrm{rot}}\,\vec{u}_E^{(\delta)} \\ \vec{\alpha}_i = -\overrightarrow{\mathrm{grad}}\,u_{Ei} + \dfrac{1}{2}\vec{e}_i \wedge \overrightarrow{\mathrm{rot}}\,\vec{u}_E + \dfrac{1}{3}\vec{e}_i\,\mathrm{div}\,\vec{u}_E = -\overrightarrow{\mathrm{grad}}\,u_{Ei}^{(\delta)} + \dfrac{1}{2}\vec{e}_i \wedge \overrightarrow{\mathrm{rot}}\,\vec{u}_E^{(\delta)} + \dfrac{1}{3}\vec{e}_i\,\mathrm{div}\,\vec{u}_E^{(\delta)} \end{cases} \tag{2.48b}$$

These relations make use of the fact that we can decompose the displacements $\vec{u}_E(\vec{\xi},t)$ and $\vec{u}_E^{(\delta)}(\vec{r},t)$ in, for example, a purely divergent part, which we will mentally associate with the expansion of the volume, and a purely rotational part, which will associate to global and local rotations, $\vec{\omega}$ and $\vec{\omega}^{(\delta)}$ respectively, by deformation of the solid. It should be noted that the shear tensor of the solid is included in the above.

One should also remark that we have used a negative sign in the defining relations $\vec{\beta}_i = -\overrightarrow{\mathrm{grad}}\,u_{Ei}$ and $\vec{\beta}_i^{(\delta)} = -\overrightarrow{\mathrm{grad}}\,u_{Ei}^{(\delta)}$. This motivation becomes clear when one compares figures 2.4 and 2.5. Indeed, we can see that the displacement field written in lagrangian coordinates $\vec{u}_L$ and eulerian coordinates $\vec{u}_E$ are identical but with opposite signs, so that it is necessary to define the relations *(2.46)* in eulerian coordinates. For example imagine a *divergent* field $\vec{u}_L$ in the lagrangian view and coordinates, the divergence is computed as $\tau = \mathrm{div}\,\vec{u}_L > 0$. The same $\vec{u}_E$ field from the eulerian point of view is *convergent* and $\tau = -\mathrm{div}\,\vec{u}_E > 0$.

Consequence 2: interpretation of the tensors of local distortion

From the fields of eulerian virtual displacement $\vec{u}^{(\delta)}(\vec{r},t)$, it is easy to give a physical interpretation of each tensor that arises from local distortion. Consider 3 points $A_0^{(i)}$ of the initial lattice undeformed, marked by 3 initial infinitesimal vectors $\vec{a}_0^{(i)} = d\vec{r}^{(i)}$, which are respectively parallel to the 3 axes of coordinates in frame $Ox_1x_2x_3$ (figure 2.6). In this local frame we map the 3 vectors $\vec{a}^{(i)}$ attached to each point $A^{(i)}$ of the real, deformed, lattice. On this visual frame we can see each of the vectors $\vec{a}^{(i)}$ of the deformed solid. It is possible to project on these 3 axes of coordinates the variation of the displacement field under deformation $d\vec{u}^{(\delta)(i)}$. The projections are given by the equations $\vec{a}_0^{(i)} = \vec{a}^{(i)} + d\vec{u}^{(\delta)(i)} = d\vec{r}^{(i)} + d\vec{u}^{(\delta)(i)}$. We can link these 3 projections to an *elongation* $\Delta a_i^{(i)}$ in the direction of the initial vector $\vec{a}_0^{(i)}$ and 2 angles of rotation ψ_{ji} and ψ_{ki} around the 2 remaining axes which are perpendicular to the initial vector $\vec{a}_0^{(i)}$. (In order to lighten figure 2.6, the notation $\vec{u}^{(\delta)(i)}$ is shortened to $\vec{u}^{d(i)}$).

The 3 relative elongations and the 6 relative angles generated by the 3 $d\vec{u}^{(\delta)(i)}$ are linked to the components of the local distortion tensor $\vec{\beta}_i^{(\delta)}$ by deformation

$$\beta_{ii}^{(\delta)} = -\frac{\partial u_i^{(\delta)}}{\partial x_i} = \frac{-du_i^{(\delta)(i)}}{dx_i^{(i)}} = \frac{\Delta a_i^{(i)}}{a_i^{(i)}} = \Delta\left(\ln a_i^{(i)}\right) = \ln a_i^{(i)} - \ln a_{0i}^{(i)} = \ln\left(\frac{a_i^{(i)}}{a_{0i}^{(i)}}\right) \tag{2.49}$$

$$\beta_{ji}^{(\delta)} = -\frac{\partial u_j^{(\delta)}}{\partial x_i} = \frac{-du_j^{(\delta)(i)}}{dx_i^{(i)}} = \tan\psi_{ji} \tag{2.50}$$

From this we deduct that the *diagonal* components $\beta_{ii}^{(\delta)}$ measure the elongation relative to the 3 directions of the coordinate system, while the *non-diagonal* components $\beta_{ji}^{(\delta)}$ measure the tangents of the angles of local rotation. It is essentially the same interpretation we had given for the $\beta_{ii}^{(\delta)}$ and $\beta_{ji}^{(\delta)}$ in section 1.5.

This visual is translated in crisp form if we decompose the local vector of rotation $\vec{\omega}^{(\delta)}$ and the non-diagonal components of the tensor of deformation $\vec{\varepsilon}_i$ and shear $\vec{\alpha}_i$ as functions of the angles of rotation ψ_{jk} et ψ_{kj}, as well as the scalar of expansion τ as a function of the relative elongations $a_i^{(i)} / a_{0i}^{(i)}$, like so

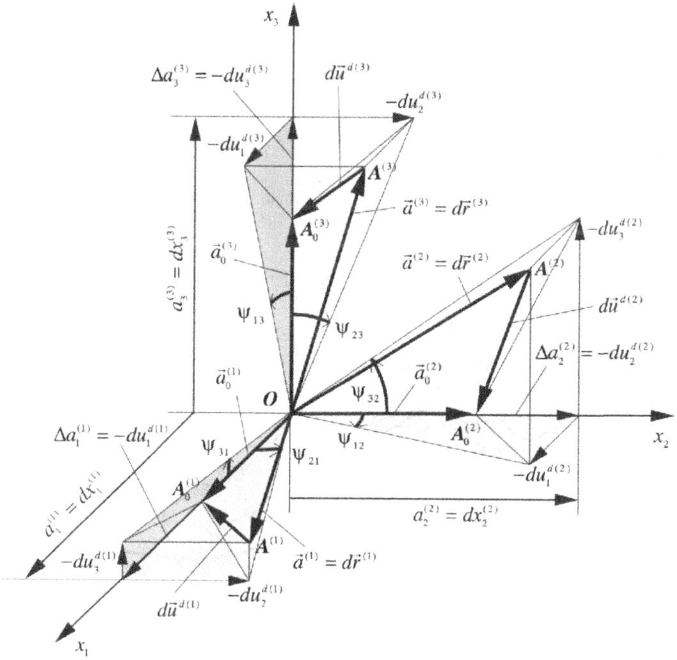

Figure 2.6 - *visual decomposition of the displacements in the local frame* $Ox_1x_2x_3$

$$\begin{cases} \omega_i^{(\delta)} = \frac{1}{2}\left(\tan\psi_{jk} - \tan\psi_{kj}\right) \\ \varepsilon_{jk} = \alpha_{jk} = \frac{1}{2}\left(\tan\psi_{jk} + \tan\psi_{kj}\right) \\ \tau = \sum_i \ln\left(\frac{a_i^{(i)}}{a_{0i}^{(i)}}\right) = \ln\left(\frac{\prod a_i^{(i)}}{\prod a_{0i}^{(i)}}\right) \end{cases} \tag{2.51}$$

Consequence 3: relation between the displacement field and the velocity field

In the absolute referential $Q\xi_1\xi_2\xi_3$, the global potential field of displacement $\vec{u}_E(\vec{\xi},t)$ must be linked to the field of absolute velocity $\vec{\phi}$. The link between the two can be seen on figure 2.7.
To establish the map between the two, we follow the movement of a point O, a node of a solid lattice, in the absolute frame. Along the path, we associate vector-points $\vec{\xi}_1$ and $\vec{\xi}_2$ corresponding respectively to the positions of point O at time t and at a slightly later moment $t+\delta t$.
The velocity is, by definition, the difference between the 2 vectors of global displacement $\vec{u}_E(\vec{\xi}_1,t)$ and $\vec{u}_E(\vec{\xi}_2,t)$ divided by the time δt, as δt becomes infinitesimally small. This reads

$$\vec{\phi}(\vec{\xi}_1,t)\delta t + \vec{u}_E(\vec{\xi}_2,t+\delta t) = \vec{u}_E(\vec{\xi}_1,t) \tag{2.52}$$

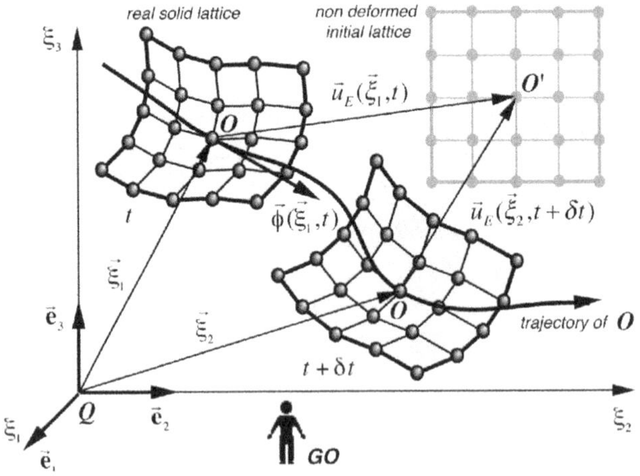

***Figure 2.7** - relation between the field of absolute velocity $\vec{\phi}$ and the field of global displacement $\vec{u}$*

Since $\vec{\xi}_2 = \vec{\xi}_1 + \vec{\phi}(\vec{\xi}_1,t)\delta t$, the former relation can be transformed in the following way

$$\vec{\phi}(\vec{r}_1,t)=\frac{\vec{u}_E(\vec{\xi}_1,t)-\vec{u}_E(\vec{\xi}_2,t+\delta t)}{\delta t}=\frac{\vec{u}_E(\vec{\xi}_1,t)-\vec{u}_E(\vec{\xi}_1+\vec{\phi}\delta t,t+\delta t)}{\delta t}$$
$$=\frac{\vec{u}_E(\vec{\xi}_1,t)-\vec{u}_E(\vec{\xi}_1,t+\delta t)-\dfrac{\partial\vec{u}_E}{\partial\xi_i}\phi_i\delta t}{\delta t}=\frac{\vec{u}_E(\vec{\xi}_1,t)-\vec{u}_E(\vec{\xi}_1,t+\delta t)}{\delta t}-\left(\vec{\phi}\vec{\nabla}\right)\vec{u}_E \tag{2.53}$$

By taking the limit $\delta t\rightarrow 0$, the field of absolute velocity $\vec{\phi}$ becomes equal to the material derivative of the field of global displacement $\vec{u}_E$, but with a change of sign

$$\vec{\phi}(\vec{\xi},t)=-\left[\frac{\partial\vec{u}_E}{\partial t}+\left(\vec{\phi}\vec{\nabla}\right)\vec{u}_E\right]=-\frac{d\vec{u}_E}{dt} \tag{2.54}$$

In the local frame $Ox_1x_2x_3$, the field of local displacement $\vec{u}_E^{(\delta)}(\vec{r},t)$ is directly linked to the field of local relative velocity $\vec{\varphi}$. This link can be seen on figure 2.8. Let's consider the movement in a local frame of a point B of a solid lattice. Along the path of this point we identify 2 points $\vec{r}_1$ and $\vec{r}_2$ taken at a time t and a later time $t+\delta t$.
As δt goes to 0 we have

$$\vec{\varphi}(\vec{r}_1,t)\delta t+\vec{u}_E^{(\delta)}(\vec{r}_2,t+\delta t)=\vec{u}_E^{(\delta)}(\vec{r}_1,t) \tag{2.55}$$

$$\vec{\varphi}(\vec{r}_1,t)=\frac{\vec{u}_E^{(\delta)}(\vec{r}_1,t)-\vec{u}_E^{(\delta)}(\vec{r}_2,t+\delta t)}{\delta t}=\frac{\vec{u}_E^{(\delta)}(\vec{r}_1,t)-\vec{u}_E^{(\delta)}(\vec{r}_1+\vec{\varphi}\delta t,t+\delta t)}{\delta t}$$
$$=\frac{\vec{u}_E^{(\delta)}(\vec{r}_1,t)-\vec{u}_E^{(\delta)}(\vec{r}_1,t+\delta t)-\dfrac{\partial\vec{u}_E^{(\delta)}}{\partial x_i}\varphi_i\delta t}{\delta t}=\frac{\vec{u}_E^{(\delta)}(\vec{r}_1,t)-\vec{u}_E^{(\delta)}(\vec{r}_1,t+\delta t)}{\delta t}-\left(\vec{\varphi}\vec{\nabla}\right)\vec{u}_E^{(\delta)} \tag{2.56}$$

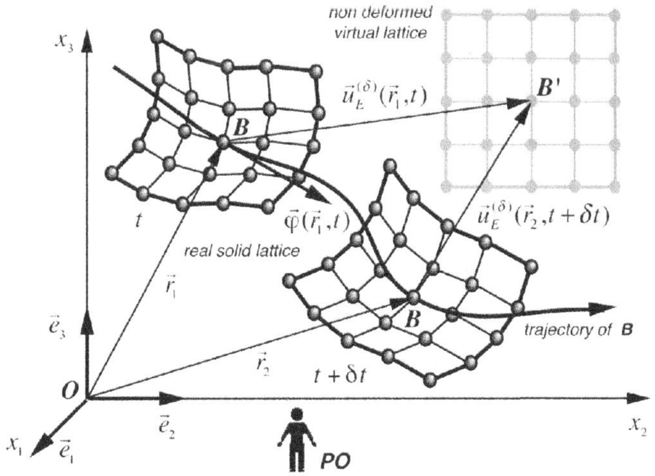

Figure 2.8 - *relation between the local field velocity $\vec{\varphi}$ and the field of local displacement $\vec{u}_E^{(\delta)}$*

By using the relation $\vec{r}_2 = \vec{r}_1 + \vec{\varphi}(\vec{r}_1, t)\delta t$, and by taking the limit $\delta t \to 0$, the field of local velocity $\vec{\varphi}$ becomes equal to the material derivative of the field of local displacement $\vec{u}_E^{(\delta)}$, but with a sign change

$$\vec{\varphi}(\vec{r}, t) = -\left[\frac{\partial \vec{u}_E^{(\delta)}}{\partial t} + \left(\vec{\varphi}\vec{\nabla}\right)\vec{u}_E^{(\delta)}\right] = -\frac{d\vec{u}_E^{(\delta)}}{dt} \tag{2.57}$$

In the local frame $Ox_1x_2x_3$, the partial derivative of $\vec{u}_E^{(\delta)}$ with respect to time also has an interesting physical interpretation which we can discuss with figure 2.9. We will focus this time on the nodes of the real lattice which pass through a fixed point A of the local frame, marked by a position vector $\vec{r}_A$. At a given instant t, it is node (particle) C of the lattice which coincides with fixed (abstract) point A, and then at a later instant $t + \delta t$, it is node B. We define the local vectors of displacement associated with point A at the 2 moments t and $t + \delta t$, denoted $\vec{u}_C^{(\delta)}(\vec{r}_A, t)$ and $\vec{u}_B^{(\delta)}(\vec{r}_A, t + \delta t)$, as pointing respectively to points C' and B' of the initial non-deformed lattice. The end of this vector traces a path in the virtual space of the initial lattice with a velocity we denote $\vec{\psi}(\vec{r}_A, t)$ thus defined by the relation

$$\vec{\psi}(\vec{r}_A, t) = \frac{\vec{u}_B^{(\delta)}(\vec{r}_A, t + \delta t) - \vec{u}_C^{(\delta)}(\vec{r}_A, t)}{\delta t} = \frac{\partial \vec{u}^{(\delta)}(\vec{r}_A, t)}{\partial t} \tag{2.58}$$

We can conclude that the partial derivative with respect to time of the local displacement vector $\vec{u}_E^{(\delta)}$ represents the velocity with which the extremity of this vector traces its path through the original (not deformed) virtual lattice when it is centered on a fixed point A in the local frame $Ox_1x_2x_3$.

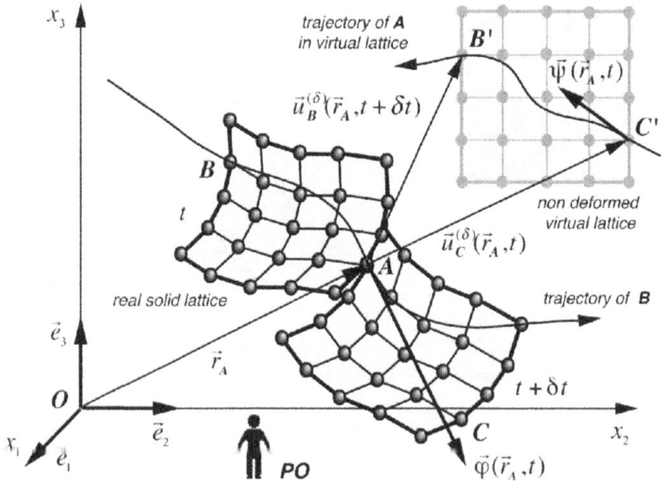

Figure 2.9 *- interpretation of the partial derivative of the local field $\vec{u}_E^{(\delta)}$ with respect to time*

Consequence 4: relation between the 2 potential fields of displacement

Relations *(2.54)* and *(2.57)* can be linked thanks to equation *(2.16)*, which gives us an interesting relation between $\vec{u}_E$ et $\vec{u}_E^{(\delta)}$

$$\frac{d\vec{u}_E}{dt} = -\vec{\phi}(\vec{\xi},t) = -\vec{\phi}(\vec{r},t) = -\vec{\varphi}(\vec{r},t) - \vec{\phi}_O(t) - \dot{\vec{\omega}}_O(t) \wedge \vec{r} = \frac{d\vec{u}_E^{(\delta)}}{dt} - \vec{\phi}_O(t) - \dot{\vec{\omega}}_O(t) \wedge \vec{r} \qquad (2.59)$$

This relation holds because displacements $\vec{u}_E$ et $\vec{u}_E^{(\delta)}$ written in the local frame $Ox_1x_2x_3$ have to be related to each other by

$$\vec{u}_E^{(\delta)}(\vec{r},t) = \vec{u}_E(\vec{r},t) + \vec{\omega}_O(t) \wedge \vec{r} + \vec{u}_{E0}(t) \qquad (2.60)$$

Chapter 3

Contortions of a lattice

In a solid lattice, the fields of distortion $\vec{\beta}_i$, $\vec{\varepsilon}_i$, $\vec{\alpha}_i$, $\vec{\omega}$, τ represent the distortions, deformations, shears, rotations and volume expansions that happen at every point and throughout the cells of the lattice. If the unit cell of a lattice experiences distortions, which can vary from one point of the lattice to another, effects will appear that are more macroscopic than the flexions and torsions of the solid and which are linked to the continuity of the lattice. These "*curvatures*" of the solid, resulting from a field of distortions are called the *contortions of the lattice*. We will describe this curvature in the first part of the chapter.
In the second part, we interpret the conditions of geometro-compatibility of a lattice by showing that they guarantee that the topology of the lattice is compact and connected.

3.1 - Rotational (curl) and divergence of a tensor

In order to derive the contortion tensors of a solid lattice, we will first find the relations between the first and second spatial derivatives of the various topological tensors. These relations will be *established independently of the fact that $\vec{\beta}_i$ can be defined either from the eulerian displacement field $\vec{u}(\vec{r},t)$* or $\vec{u}^{(\delta)}(\vec{r},t)$ in a geometro-compatible lattice, simply by applying the vectorial operators to the distortion tensors.

The «rotational of a tensor» ("curl of a tensor")

By using the decomposition of the $\vec{\beta}_i$ tensors as documented in figure 1.8, it is possible to calculate the relations that can exist between the spatial derivatives of first order of these tensors. We simply apply the operators of vectorial analysis and their laws of distributivity. Applying the rotational operator to expression *(1.36)* of the symmetric tensor $\vec{\varepsilon}_i$ implies the relation

$$\overrightarrow{\mathrm{rot}}\,\vec{\varepsilon}_i = \overrightarrow{\mathrm{rot}}\,\vec{\alpha}_i - \frac{1}{3}\vec{e}_i \wedge \overrightarrow{\mathrm{grad}}\,\tau = (\vec{e}_i\vec{\nabla})\vec{\omega} + \overrightarrow{\mathrm{rot}}\,\vec{\beta}_i - \vec{e}_i\,\mathrm{div}\,\vec{\omega} \qquad (3.1)$$

By using the following transposition rules

$$\left[(\vec{e}_i\vec{\nabla})\vec{\omega}\right]^{\mathrm{T}} = \overrightarrow{\mathrm{grad}}\,\omega_i \quad \text{and} \quad \left[-\vec{e}_i \wedge \overrightarrow{\mathrm{grad}}\,\tau\right]^{\mathrm{T}} = \vec{e}_i \wedge \overrightarrow{\mathrm{grad}}\,\tau \qquad (3.2)$$

the previous expression for $\overrightarrow{\mathrm{rot}}\,\vec{\varepsilon}_i$ can be transposed

$$\left[\overrightarrow{\mathrm{rot}}\,\vec{\varepsilon}_i\right]^{\mathrm{T}} = \left[\overrightarrow{\mathrm{rot}}\,\vec{\alpha}_i\right]^{\mathrm{T}} + \frac{1}{3}\vec{e}_i \wedge \overrightarrow{\mathrm{grad}}\,\tau = \overrightarrow{\mathrm{grad}}\,\omega_i + \left[\overrightarrow{\mathrm{rot}}\,\vec{\beta}_i\right]^{\mathrm{T}} - \vec{e}_i\,\mathrm{div}\,\vec{\omega} \qquad (3.3)$$

Relations *(3.1)* and *(3.3)* allow us to find the expression for the symmetrical part of the rotational of the symmetric tensor $\vec{\varepsilon}_i$, and as we can see it only depends on the transverse symmetric tensor $\vec{\alpha}_i$

$$\left[\overrightarrow{\text{rot}}\,\vec{\varepsilon}_i\right]^S = \left[\overrightarrow{\text{rot}}\,\vec{\alpha}_i\right]^S = \frac{1}{2}\left(\overrightarrow{\text{grad}}\,\omega_i + (\vec{e}_i\vec{\nabla})\vec{\omega}\right) + \left[\overrightarrow{\text{rot}}\,\vec{\beta}_i\right]^S - \vec{e}_i\,\text{div}\,\vec{\omega} \tag{3.4}$$

The anti-symmetric part of the rotational of a tensor

From the previous relations it is also possible to find direct relations between the anti-symmetric parts of the rotational of a tensor

$$\sum_k \vec{e}_k \wedge \overrightarrow{\text{rot}}\,\vec{\varepsilon}_k = \sum_k \vec{e}_k \wedge \overrightarrow{\text{rot}}\,\vec{\alpha}_k + \frac{2}{3}\overrightarrow{\text{grad}}\,\tau = \sum_k \vec{e}_k \wedge \overrightarrow{\text{rot}}\,\vec{\beta}_k + \overrightarrow{\text{rot}}\,\vec{\omega} \tag{3.5}$$

The trace of the rotational of a tensor and the divergence of its anti-symmetric part

The computation of the divergences of the following expressions, represent respectively the anti-symmetric parts of tensors β_{ij}, ε_{ij} et α_{ij}

$$\sum_k \vec{e}_k \wedge \vec{\beta}_k = -2\vec{\omega} \quad \text{and} \quad \sum_k \vec{e}_k \wedge \vec{\varepsilon}_k \equiv 0 \quad \text{and} \quad \sum_k \vec{e}_k \wedge \vec{\alpha}_k \equiv 0 \tag{3.6}$$

they allow us to find a set of relations regarding the traces of the rotational of these tensors.

$$\sum_k \vec{e}_k \overrightarrow{\text{rot}}\,\vec{\beta}_k = \text{div}\left(-\sum_k \vec{e}_k \wedge \vec{\beta}_k\right) = 2\,\text{div}\,\vec{\omega} \tag{3.7}$$

And we can observe, among other things, that the symmetric nature of tensors $\vec{\varepsilon}_i$ et $\vec{\alpha}_i$ implies that the trace of their rotational is null

$$\begin{cases} \sum_k \vec{e}_k \overrightarrow{\text{rot}}\,\vec{\varepsilon}_k = \text{div}\left(-\sum_k \vec{e}_k \wedge \vec{\varepsilon}_k\right) \equiv 0 \\ \sum_k \vec{e}_k \overrightarrow{\text{rot}}\,\vec{\alpha}_k = \text{div}\left(-\sum_k \vec{e}_k \wedge \vec{\alpha}_k\right) \equiv 0 \end{cases} \tag{3.8}$$

The «divergence of a tensor»

We can find relations for the "*divergence*" of these tensors

$$\sum_k \vec{e}_k\,\text{div}\,\vec{\varepsilon}_k = \sum_k \vec{e}_k\,\text{div}\,\vec{\alpha}_k + \frac{1}{3}\overrightarrow{\text{grad}}\,\tau = \sum_k \vec{e}_k\,\text{div}\,\vec{\beta}_k + \overrightarrow{\text{rot}}\,\vec{\omega} \tag{3.9}$$

and we can show that the divergence of a tensor is directly linked to the anti-symmetric part of its rotational

$$\begin{cases} \sum_k \vec{e}_k\,\text{div}\,\vec{\beta}_k = -\sum_k \vec{e}_k \wedge \overrightarrow{\text{rot}}\,\vec{\beta}_k + \sum_k \overrightarrow{\text{rot}}\left(\vec{e}_k \wedge \vec{\beta}_k\right) + \sum_k \overrightarrow{\text{grad}}\left(\vec{e}_k\vec{\beta}_k\right) \\ \qquad = -\sum_k \vec{e}_k \wedge \overrightarrow{\text{rot}}\,\vec{\beta}_k - 2\,\overrightarrow{\text{rot}}\,\vec{\omega} + \overrightarrow{\text{grad}}\,\tau \\ \sum_k \vec{e}_k\,\text{div}\,\vec{\varepsilon}_k = -\sum_k \vec{e}_k \wedge \overrightarrow{\text{rot}}\,\vec{\varepsilon}_k + \overrightarrow{\text{grad}}\,\tau \\ \sum_k \vec{e}_k\,\text{div}\,\vec{\alpha}_k = -\sum_k \vec{e}_k \wedge \overrightarrow{\text{rot}}\,\vec{\alpha}_k \end{cases} \tag{3.10}$$

The contraction operations on the indices

The spatial derivatives of a tensor of order 2, β_{ij}, which we can represent under a vectorial form, have been manipulated thanks to vectorial analysis. These vectorial manipulations of tensorial objects are in fact *mathematical operations of contraction of indices*. From the 27 component of a 3rd order tensor $A_{ijk} = \partial\beta_{ij} / \partial\xi_k$ we can form tensors of order 2, vectors and scalars, as can be seen from relations *(3.3)*, *(3.5)* and *(3.7)* respectively.

It is possible to contract indices on tensors of order larger than 3 to form tensors of order 2, 1 (vectors) and 0 (scalar). For example the rotational in relation *(3.3)* is a tensor of order 2 (9 components) which results from the contraction of a tensor of order 4 (81 components), and which can be calculated, given that $\overrightarrow{\mathrm{rot}}\,\overrightarrow{\mathrm{grad}}\,\omega_i \equiv 0$

$$\overrightarrow{\mathrm{rot}}\left[\overrightarrow{\mathrm{rot}}\,\vec{\varepsilon}_i\right]^{\mathrm{T}} = \overrightarrow{\mathrm{rot}}\left\{\left[\overrightarrow{\mathrm{rot}}\,\vec{\beta}_i\right]^{\mathrm{T}} - \vec{e}_i\,\mathrm{div}\,\vec{\omega}\right\} = \overrightarrow{\mathrm{rot}}\left[\sum_k \vec{e}_k\left(\vec{e}_i\,\overrightarrow{\mathrm{rot}}\,\vec{\beta}_k\right) - \vec{e}_i\,\mathrm{div}\,\vec{\omega}\right] \tag{3.11}$$

3.2 - Links between the spatial derivatives in the case of geometro-compatibility

The condition of geometro-compatibility $\overrightarrow{\mathrm{rot}}\,\vec{\beta}_i = \overrightarrow{\mathrm{rot}}\,\vec{\beta}_i^{(\delta)} = 0$ implies the following relation

$$\overrightarrow{\mathrm{rot}}\,\vec{\beta}_i^{(\delta)} = \overrightarrow{\mathrm{rot}}\left(\vec{\alpha}_i + \vec{e}_i \wedge \vec{\omega}^{(\delta)} + \frac{1}{3}\tau\vec{e}_i\right) = \overrightarrow{\mathrm{rot}}\,\vec{\alpha}_i + \vec{e}_i\,\mathrm{div}\,\vec{\omega}^{(\delta)} - \left(\vec{e}_i\vec{\nabla}\right)\vec{\omega}^{(\delta)} + \frac{1}{3}\overrightarrow{\mathrm{grad}}\,\tau \wedge \vec{e}_i = 0 \tag{3.12}$$

With $\mathrm{div}\,\vec{\omega}^{(\delta)} = 0$, $\left(\vec{e}_i\vec{\nabla}\right)\vec{\omega}^{(\delta)} = \overrightarrow{\mathrm{grad}}\,\omega_i^{(\delta)} - \vec{e}_i \wedge \overrightarrow{\mathrm{rot}}\,\vec{\omega}^{(\delta)}$ and $\vec{\omega}^{(\delta)} = \vec{\omega} - \vec{\omega}_0(t)$, this relation leads directly to the following relations between the spatial derivatives of the various topological fields

$$\begin{cases} \overrightarrow{\mathrm{rot}}\,\vec{\varepsilon}_i = \overrightarrow{\mathrm{rot}}\,\vec{\alpha}_i - \frac{1}{3}\vec{e}_i \wedge \overrightarrow{\mathrm{grad}}\,\tau = \overrightarrow{\mathrm{grad}}\,\omega_i^{(\delta)} - \vec{e}_i \wedge \overrightarrow{\mathrm{rot}}\,\vec{\omega}^{(\delta)} = \overrightarrow{\mathrm{grad}}\,\omega_i - \vec{e}_i \wedge \overrightarrow{\mathrm{rot}}\,\vec{\omega} \\ -\sum_k \vec{e}_k \wedge \overrightarrow{\mathrm{rot}}\,\vec{\varepsilon}_k = \sum_k \vec{e}_k\,\mathrm{div}\,\vec{\alpha}_k - \frac{2}{3}\overrightarrow{\mathrm{grad}}\,\tau = -\overrightarrow{\mathrm{rot}}\,\vec{\omega}^{(\delta)} = -\overrightarrow{\mathrm{rot}}\,\vec{\omega} \\ \mathrm{div}\,\vec{\omega}^{(\delta)} = \mathrm{div}\,\vec{\omega} = 0 \end{cases} \tag{3.13}$$

We will see that these relations are very useful for the description of the evolution of geometro-compatible solid lattices. If, additionally, such a solid has a volume expansion τ independent of the space coordinates, the *spatial derivatives of the rotation vector* $\vec{\omega}$ *and of the local rotation vector* $\vec{\omega}^{(\delta)}$ *are only dependent on the spatial derivatives of the shear tensor* $\vec{\alpha}_i$

$$\tau = \tau(t) \neq \tau(\vec{r},t) \Rightarrow \begin{cases} \overrightarrow{\mathrm{rot}}\,\vec{\alpha}_i = \overrightarrow{\mathrm{grad}}\,\omega_i^{(\delta)} - \vec{e}_i \wedge \overrightarrow{\mathrm{rot}}\,\vec{\omega}^{(\delta)} = \overrightarrow{\mathrm{grad}}\,\omega_i - \vec{e}_i \wedge \overrightarrow{\mathrm{rot}}\,\vec{\omega} \\ \sum_k \vec{e}_k\,\mathrm{div}\,\vec{\alpha}_k = -\overrightarrow{\mathrm{rot}}\,\vec{\omega}^{(\delta)} = -\overrightarrow{\mathrm{rot}}\,\vec{\omega} \\ \mathrm{div}\,\vec{\omega}^{(\delta)} = \mathrm{div}\,\vec{\omega} = 0 \end{cases} \tag{3.14}$$

3.3 - Contortion tensors of a solid

The contortion tensor of a solid

Relation *(3.3)* allows us, at least in a compatible lattice, to link directly $\left[\overrightarrow{\mathrm{rot}}\,\vec{\varepsilon}_i\right]^{\mathrm{T}}$ to the gradient of the component of the rotation vector $\vec{\omega}$

$$\overrightarrow{\mathrm{rot}}\,\vec{\beta}_i = 0 \quad \Rightarrow \quad \left[\overrightarrow{\mathrm{rot}}\,\vec{\varepsilon}_i\right]^{\mathrm{T}} = \overrightarrow{\mathrm{grad}}\,\omega_i \tag{3.15}$$

this means that it is possible, in a compatible solid lattice, to find the local variations of the rotational field directly from the field of deformations. To that end, we define a new tensor, *the tensor of contortions* $\vec{\chi}_i$ in the following way:

$$\vec{\chi}_i = \left[\overrightarrow{\mathrm{rot}}\,\vec{\varepsilon}_i\right]^{\mathrm{T}} = \overrightarrow{\mathrm{grad}}\,\omega_i + \left[\overrightarrow{\mathrm{rot}}\,\vec{\beta}_i\right]^{\mathrm{T}} - \vec{e}_i\,\mathrm{div}\,\vec{\omega} \tag{3.16}$$

In a compatible medium, the physical significance of this tensor can be deduced directly from $\overrightarrow{\mathrm{grad}}\,\omega_i$. It is indeed relevant to measure the spatial variations of the rotational vector $\vec{\omega}$, since $\vec{\omega}$ is an invariant of the solid medium, meaning independent of the choice of frame used to calculate it. In a compatible medium, relation *(3.3)* takes the following form

$$\overrightarrow{\mathrm{rot}}\,\vec{\beta}_i = 0 \quad \Rightarrow \quad \vec{\chi}_i = \left[\overrightarrow{\mathrm{rot}}\,\vec{\varepsilon}_i\right]^{\mathrm{T}} = \overrightarrow{\mathrm{grad}}\,\omega_i = \left(\vec{e}_i\vec{\nabla}\right)\vec{\omega} + \vec{e}_i \wedge \overrightarrow{\mathrm{rot}}\,\vec{\omega} \tag{3.17}$$

The components of this tensors are linked to the spatial derivatives of the rotational vector $\vec{\omega}$, as these measure how rotation changes in different points of the solid, such as due to *torsions* or *flexions*. We can define a little more precisely and visually the components of the tensors by using 2 typical examples of spatial rotations $\vec{\omega}$.

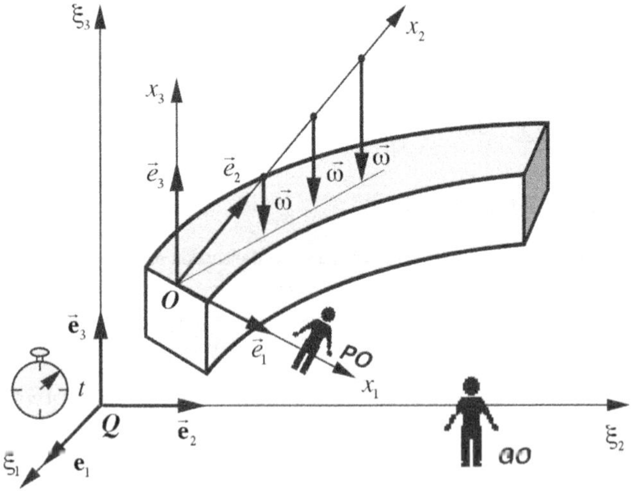

Figure 3.1 *- flexion of a solid medium*

In the first example, a flexed medium shows a rotational vector parallel to Ox_3 (pointing down in figure 3.1), and its amplitude increases along Ox_2, as shown in figure 3.1. In this case, there exists a *non-diagonal component* of the $\vec{\chi}_i$ tensor, which is not null

$$\chi_{23} = \frac{\partial \omega_3}{\partial x_2} \neq 0 \quad (3.18)$$

This component is associated with the *flexion* of the solid.

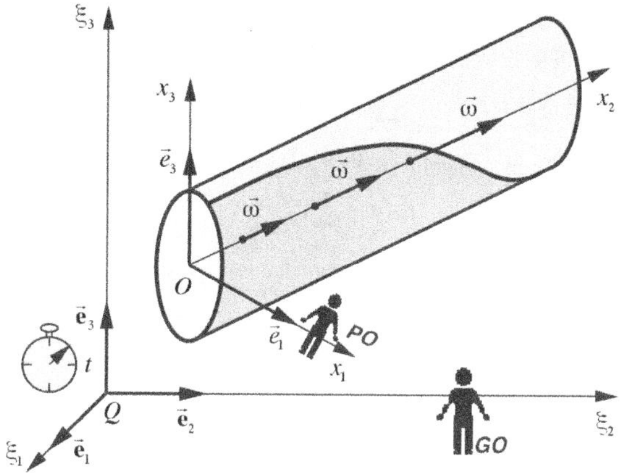

Figure 3.2 *- torsion of a solid medium*

In the second example of figure 3.2, we represent a *torsion* of the solid medium. The rotational vector is along the axis Ox_2 and increases in amplitude along Ox_2 as well. As a result, there is a *diagonal component* of tensor $\vec{\chi}_i$, which is non-null, a fact we translate as

$$\chi_{22} = \frac{\partial \omega_2}{\partial x_2} \neq 0 \quad (3.19)$$

From these two examples, it is clear that the components of the tensor $\vec{\chi}_i$ measure the *flexions* and *torsions* of the solid medium, via non-diagonal and diagonal components respectively. We will call those components the *"contortions" of the medium.* The rotational part of this new *contortion tensor* $\vec{\chi}_i$ satisfies an interesting relationship.

$$\overrightarrow{\mathrm{rot}}\, \vec{\chi}_i = \overrightarrow{\mathrm{rot}}\left[\overrightarrow{\mathrm{rot}}\, \vec{\varepsilon}_i\right]^{\mathrm{T}} = \overrightarrow{\mathrm{rot}}\left\{\left[\overrightarrow{\mathrm{rot}}\, \vec{\beta}_i\right]^{\mathrm{T}} - \vec{e}_i \,\mathrm{div}\, \vec{\omega}\right\} \quad (3.20)$$

Indeed, in a compatible medium, the contortion tensor $\vec{\chi}_i$ satisfies a compatibility relation that is equivalent to the compatibility relation of the deformation tensor $\vec{\varepsilon}_i$

$$\overrightarrow{\mathrm{rot}}\, \vec{\beta}_i = 0 \quad \Rightarrow \quad \overrightarrow{\mathrm{rot}}\, \vec{\chi}_i = \overrightarrow{\mathrm{rot}}\left[\overrightarrow{\mathrm{rot}}\, \vec{\varepsilon}_i\right]^{\mathrm{T}} = 0 \quad (3.21)$$

These relations are known as the *equations of De Saint-Venant.* Since the tensor $\vec{\varepsilon}_i$ is symmetrical, these encompass 6 non-trivial relations between the partial derivatives of order 2 of the ε_{ij}.

It should be noted that the trace of $\vec{\chi}_i$ is identically null, meaning that $\vec{\chi}_i$ is a *transverse tensor.* This is due to the simple fact that, as seen in relation *(1.70)*, the $\vec{\varepsilon}_i$ tensor is symmetrical

$$\chi = \sum_k \vec{e}_k \vec{\chi}_k = \sum_k \vec{e}_k \left[\overrightarrow{\mathrm{rot}}\, \vec{\varepsilon}_k\right]^{\mathrm{T}} = \sum_k \vec{e}_k \overrightarrow{\mathrm{rot}}\, \vec{\varepsilon}_k \equiv 0 \qquad (3.22)$$

The flexion vector of a solid

From the contortion tensor $\vec{\chi}_i$, it is possible to build an invariant vector $\vec{\chi}$ which depends on the anti-symmetric part of the $\vec{\chi}_i$ with the following construct

$$\vec{\chi} = \sum_k \vec{e}_k \wedge \vec{\chi}_k \qquad (3.23)$$

Since the anti-symmetric part of a tensor is, by definition, equal to the anti-symmetric part of its transpose, but with an opposite signe, we can deduct the following

$$\vec{\chi} = \sum_k \vec{e}_k \wedge \vec{\chi}_k = \sum_k \vec{e}_k \wedge \left[\overrightarrow{\mathrm{rot}}\, \vec{\varepsilon}_k\right]^{\mathrm{T}} = -\sum_k \vec{e}_k \wedge \overrightarrow{\mathrm{rot}}\, \vec{\varepsilon}_k \qquad (3.24)$$

We then use relation *(3.5)*, and we obtain the following relation on the $\vec{\chi}$ vector

$$\vec{\chi} = \sum_k \vec{e}_k \wedge \vec{\chi}_k = -\sum_k \vec{e}_k \wedge \overrightarrow{\mathrm{rot}}\, \vec{\varepsilon}_k = -\overrightarrow{\mathrm{rot}}\, \vec{\omega} - \sum_k \vec{e}_k \wedge \overrightarrow{\mathrm{rot}}\, \vec{\beta}_k \qquad (3.25)$$

Thus, in a compatible medium, $\vec{\chi}$ only depends on the rotation vector $\vec{\omega}$

$$\overrightarrow{\mathrm{rot}}\, \vec{\beta}_i = 0 \quad \Rightarrow \quad \vec{\chi} = -\overrightarrow{\mathrm{rot}}\, \vec{\omega} \qquad (3.26)$$

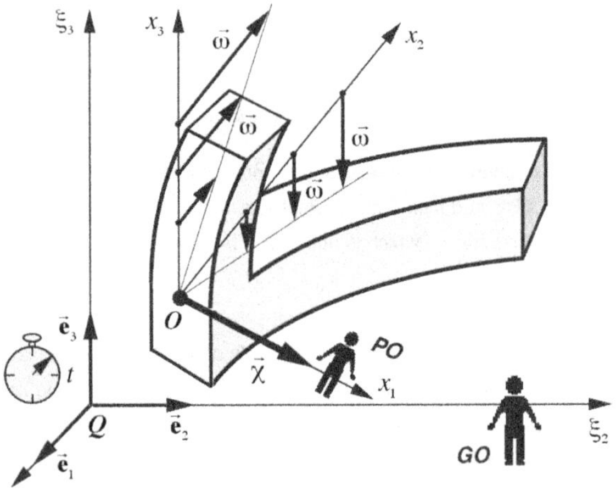

Figure 3.3 *- definition of the flexion vector* $\vec{\chi}$

Since $\vec{\chi}$ is defined as a rotational, it only contains components of the type $\partial\omega_i / \partial x_j$, with $i \neq j$, by definition of a rotational. Thus, it only depends on the non diagonal components of the $\vec{\chi}_i$ tensors, which as we have just seen, are the *flexion components* of the medium. We can verify this assertion on the example of the deformation shown on figure 3.3, in which $\vec{\omega}$ has components along the axis Ox_2 et Ox_3, that grow in the directions Ox_3, and

$$\vec{\omega} = \omega_3(x_2)\vec{e}_3 + \omega_2(x_3)\vec{e}_2 \qquad (3.27)$$

which gives us the following tensor $\vec{\chi}_i$

$$\vec{\chi}_1 = 0 \qquad \vec{\chi}_2 = \frac{\partial \omega_2}{\partial x_3}\vec{e}_3 \qquad \vec{\chi}_3 = \frac{\partial \omega_3}{\partial x_2}\vec{e}_2 \tag{3.28}$$

from which we deduct the vector $\vec{\chi}$, with the help of relation *(3.10)*

$$\vec{\chi} = \vec{e}_2 \wedge \frac{\partial \omega_2}{\partial x_3}\vec{e}_3 + \vec{e}_3 \wedge \frac{\partial \omega_3}{\partial x_2}\vec{e}_2 = \left(\frac{\partial \omega_2}{\partial x_3} - \frac{\partial \omega_3}{\partial x_2}\right)\vec{e}_1 \tag{3.29}$$

The $\vec{\chi}$ vector is perpendicular to the general curvature of the medium due to flexion. It is aligned with the direction of the 'radius of curvature'. It's magnitude is inversely proportional to the average radius of curvature in the directions Ox_3 et Ox_2. We will therefore call it the *flexion vector* $\vec{\chi}$ of our solid medium. The $\vec{\chi}$ vector satisfies an interesting relation concerning its divergence. Since the divergence of a rotational is null, we have, following *(3.25)*

$$\begin{aligned}
&\operatorname{div}\vec{\chi} = \operatorname{div}\left(\sum_k \vec{e}_k \wedge \vec{\chi}_k\right) = -\operatorname{div}\left(\sum_k \vec{e}_k \wedge \overrightarrow{\operatorname{rot}}\,\vec{\varepsilon}_k\right) = -\operatorname{div}\left(\sum_k \vec{e}_k \wedge \overrightarrow{\operatorname{rot}}\,\vec{\beta}_k\right) \\
&\Rightarrow \quad \operatorname{div}\vec{\chi} = -\sum_k \vec{e}_k \overrightarrow{\operatorname{rot}}\,\vec{\chi}_k = \sum_k \vec{e}_k \overrightarrow{\operatorname{rot}}\,\overrightarrow{\operatorname{rot}}\,\vec{\varepsilon}_k = \sum_k \vec{e}_k \overrightarrow{\operatorname{rot}}\,\overrightarrow{\operatorname{rot}}\,\vec{\beta}_k
\end{aligned} \tag{3.30}$$

This relation shows that $\operatorname{div}\vec{\chi}$ is in fact equal to the *trace* of $\overrightarrow{\operatorname{rot}}\,\vec{\chi}_k$ with a changed sign.
In a compatible medium, this relation shows that the flexion vector $\vec{\chi}$ also obeys a compatibility relation which assures that the field of flexion is non divergent:

$$\overrightarrow{\operatorname{rot}}\,\vec{\beta}_i = 0 \quad \Rightarrow \quad \operatorname{div}\vec{\chi} = 0 \tag{3.31}$$

The torsion tensor of a solid: the transverse symmetrical part of the contortion tensor

Since the contortion tensor $\vec{\chi}_i$ is a transverse tensor (of null trace), its symmetrical part $\left[\vec{\chi}_i\right]^S$ is a transverse symmetrical tensor, which cannot be decomposed further. We compute it from the following relations

$$\left[\vec{\chi}_i\right]^S = \vec{\chi}_i + \frac{1}{2}\vec{e}_i \wedge \left(\sum_k \vec{e}_k \wedge \vec{\chi}_k\right) = \vec{\chi}_i + \frac{1}{2}\vec{e}_i \wedge \vec{\chi} \tag{3.32}$$

This transverse symmetrical tensor satisfies the following relations, which are easy to derive using the transpose operation $2\left[\vec{\chi}_i\right]^S = \vec{\chi}_i + \left[\vec{\chi}_i\right]^T$

$$\left[\vec{\chi}_i\right]^S = \left[\overrightarrow{\operatorname{rot}}\,\vec{\varepsilon}_i\right]^S = \left[\overrightarrow{\operatorname{rot}}\,\vec{\alpha}_i\right]^S = \overrightarrow{\operatorname{grad}}\,\omega_i - \frac{1}{2}\vec{e}_i \wedge \overrightarrow{\operatorname{rot}}\,\vec{\omega} + \left[\overrightarrow{\operatorname{rot}}\,\vec{\beta}_i\right]^S - \vec{e}_i \operatorname{div}\vec{\omega} \tag{3.33}$$

The transverse symmetrical part $\left[\vec{\chi}_i\right]^S$ of the contortion tensor therefore represents the torsions of the solid, since it is obtained from the contortion tensor from which we have subtracted the anti-symmetrical part representing the flexions. We thus call $\left[\vec{\chi}_i\right]^S$ the *tensor of torsion* of the solid.

The complete description of the distortions and contortions of a geometro-compatible solid lattice exhibits here a symmetry perfectly described visually in figure 3.3. We also give table 3.1 with the relation which permits a complete topological description of the contortions and the compatibility of a solid lattice. In this table you will find the terms which correspond to the rotation of the distortion tensor and the divergence of the rotation vector, which are by definition null in a compatible medium.

It should be noted here that the transverse symmetrical part of $\vec{\chi}_i$ does not depend directly on the volume expansion τ, but exclusively on the shear tensor $\vec{\alpha}_i$. This means that the *torsions*

$\left[\vec{\chi}_i\right]^S$ *of a solid are intimately linked to the pure shear of the solid.* Another important remark can be made about the topological contortions of a medium: since the gradient of the scalar of volume expansion τ does not figure in the anti-symmetric part of the contortion tensor $\vec{\chi}_i$, and furthermore, since this anti-symmetric part is represented by the flexion vector $\vec{\chi}$, it is rather evident that *flexions $\vec{\chi}$ of a solid are linked to the volume expansion τ of said medium.*

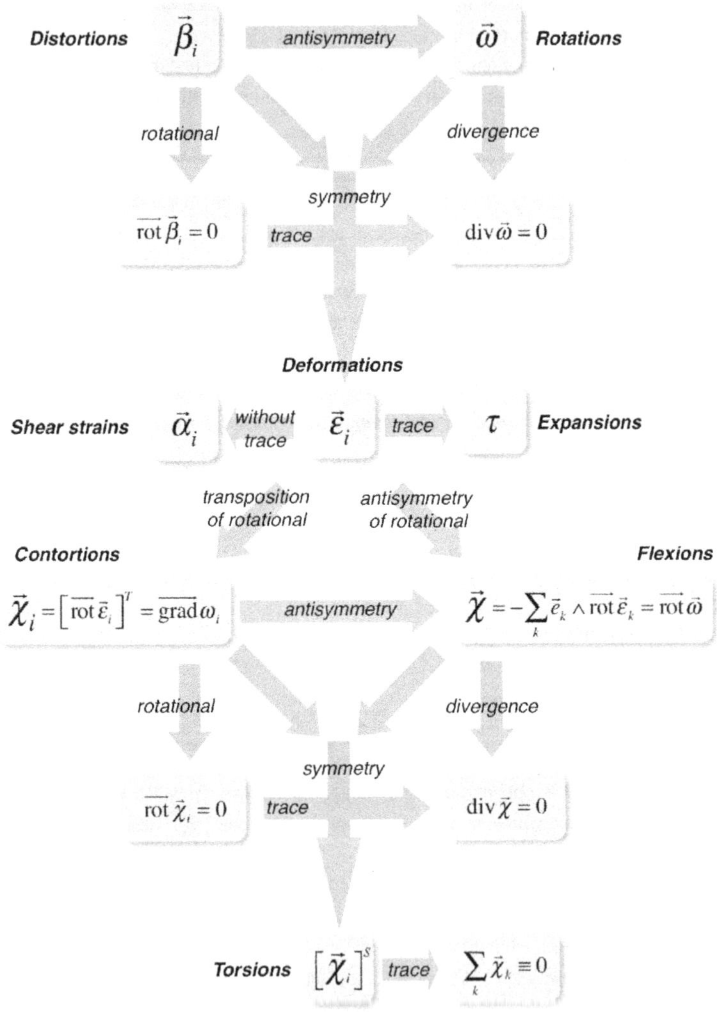

Figure 3.4 *- complete description of the distortions and contortions of a compatible solid lattice*

Table 3.1 - The contortions and compatibility of a solid lattice

Transverse tensor of contortion

$$\vec{\chi}_i = \left[\overrightarrow{\text{rot}}\,\vec{\varepsilon}_i\right]^{\text{T}} = \left[\overrightarrow{\text{rot}}\,\vec{\alpha}_i\right]^{\text{T}} + \frac{1}{3}\vec{e}_i \wedge \overrightarrow{\text{grad}}\,\tau = \overrightarrow{\text{grad}}\,\omega_i \underbrace{\boxed{+\left[\overrightarrow{\text{rot}}\,\vec{\beta}_i\right]^{\text{T}} - \vec{e}_i \,\text{div}\,\vec{\omega}}}_{=0} \quad (1)$$

Flexion vector

$$\vec{\chi} = \sum_k \vec{e}_k \wedge \vec{\chi}_k = -\sum_k \vec{e}_k \wedge \overrightarrow{\text{rot}}\,\vec{\varepsilon}_k = -\sum_k \vec{e}_k \wedge \overrightarrow{\text{rot}}\,\vec{\alpha}_k - \frac{2}{3}\overrightarrow{\text{grad}}\,\tau$$

$$= \sum_k \vec{e}_k \,\text{div}\,\vec{\alpha}_k - \frac{2}{3}\overrightarrow{\text{grad}}\,\tau = -\overrightarrow{\text{rot}}\,\vec{\omega} \underbrace{\boxed{-\sum_k \vec{e}_k \wedge \overrightarrow{\text{rot}}\,\vec{\beta}_k}}_{=0} \quad (2)$$

Null trace of the contortion tensor

$$\chi = \sum_k \vec{e}_k \vec{\chi}_k = \sum_k \vec{e}_k \left[\overrightarrow{\text{rot}}\,\vec{\varepsilon}_k\right]^{\text{T}} = \sum_k \vec{e}_k \overrightarrow{\text{rot}}\,\vec{\varepsilon}_k \equiv 0 \quad (3)$$

Transverse symmetrical part of torsion

$$\left[\vec{\chi}_i\right]^{\text{S}} = \vec{\chi}_i + \frac{1}{2}\vec{e}_i \wedge \vec{\chi} = \left[\overrightarrow{\text{rot}}\,\vec{\varepsilon}_i\right]^{\text{S}} = \left[\overrightarrow{\text{rot}}\,\vec{\alpha}_i\right]^{\text{S}}$$

$$= \overrightarrow{\text{grad}}\,\omega_i - \frac{1}{2}\vec{e}_i \wedge \overrightarrow{\text{rot}}\,\vec{\omega} + \underbrace{\boxed{+\left[\overrightarrow{\text{rot}}\,\vec{\beta}_i\right]^{\text{S}} - \vec{e}_i \,\text{div}\,\vec{\omega}}}_{=0} \quad (4)$$

Geometro-compatibility of distortions and rotations

$$\vec{\beta}_i = \overrightarrow{\text{grad}}\,u_i \quad (5) \quad \Rightarrow \quad \boxed{\overrightarrow{\text{rot}}\,\vec{\beta}_i} = 0 \quad (6)$$

$$\vec{\omega} = \frac{1}{2}\overrightarrow{\text{rot}}\,\vec{u} \quad (7) \quad \Rightarrow \quad \text{div}\,\vec{\omega} = \boxed{\frac{1}{2}\sum_k \vec{e}_k \overrightarrow{\text{rot}}\,\vec{\beta}_k} = 0 \quad (8)$$

Geometro-compatibility of contortions and flexions

$$\overrightarrow{\text{rot}}\,\vec{\chi}_i = \overrightarrow{\text{rot}}\left[\overrightarrow{\text{rot}}\,\vec{\varepsilon}_i\right]^{\text{T}} = \overrightarrow{\text{rot}}\left[\overrightarrow{\text{rot}}\,\vec{\alpha}_i\right]^{\text{T}} - \frac{1}{3}\overrightarrow{\text{rot}}\,\overrightarrow{\text{rot}}\left(\vec{e}_i\tau\right)$$

$$= \boxed{\overrightarrow{\text{rot}}\left\{\left[\overrightarrow{\text{rot}}\,\vec{\beta}_i\right]^{\text{T}} - \vec{e}_i \,\text{div}\,\vec{\omega}\right\}} = 0 \quad (9)$$

$$\text{div}\,\vec{\chi} = -\sum_k \vec{e}_k \overrightarrow{\text{rot}}\,\vec{\chi}_k = \sum_k \vec{e}_k \overrightarrow{\text{rot}}\,\overrightarrow{\text{rot}}\,\vec{\varepsilon}_k = \sum_k \vec{e}_k \overrightarrow{\text{rot}}\,\overrightarrow{\text{rot}}\,\vec{\alpha}_k - \frac{2}{3}\Delta\tau$$

$$= \boxed{\sum_k \vec{e}_k \overrightarrow{\text{rot}}\,\overrightarrow{\text{rot}}\,\vec{\beta}_k} = \boxed{-\text{div}\left(\sum_k \vec{e}_k \wedge \overrightarrow{\text{rot}}\,\vec{\beta}_k\right)} = 0 \quad (10)$$

3.4 - Physical interpretation of the equations of compatibility

The diverse equations of compatibility have a physical meaning we will now explore in this section.

The absence of dislocations and the continuity of the field of displacement.

The compatibility condition $\overrightarrow{\mathrm{rot}}\,\vec{\beta}_i = 0$ for the distortion tensor $\vec{\beta}_i$ implies that the field has nice smoothness properties, in the mathematical sense. To show this, let's calculate $d\vec{u}$ over a closed contour C anywhere inside the lattice

$$\oint_C d\vec{u} = \sum_k \vec{e}_k \oint_C du_k = \sum_k \vec{e}_k \oint_C \overrightarrow{\mathrm{grad}}\, u_k d\vec{r} = -\sum_k \vec{e}_k \oint_C \vec{\beta}_k d\vec{r} \tag{3.34}$$

By using the theorem of the rotational, an integral form of compatibility is obtained over any surface S whose boundary is C

$$\oint_C d\vec{u} = -\sum_k \vec{e}_k \iint_S \overrightarrow{\mathrm{rot}}\vec{\beta}_k d\vec{S} = 0 \quad ; \quad \forall C \tag{3.35}$$

Let's consider the displacement $\vec{u}$ associated with a closed contour C inside the lattice, such as that represented in figure 3.5. For this closed contour from $\boldsymbol{A}$ to $\boldsymbol{B}$, we have

$$\oint_C d\vec{u} = \vec{u}_B - \vec{u}_A = \vec{\boldsymbol{B}} = 0 \tag{3.36}$$

And thus, if the medium presents a distortion field which satisfies the compatibility condition $\overrightarrow{\mathrm{rot}}\,\vec{\beta}_i = 0$, the closing vector $\vec{\boldsymbol{B}}$, called a *Burgers vector* [1], of contour C' generated by the sum of the $\vec{u}$ vectors is null. This effectively signifies, from a topological point of view, that there are *no discontinuities of displacement, called dislocations* [2] *inside the medium.*

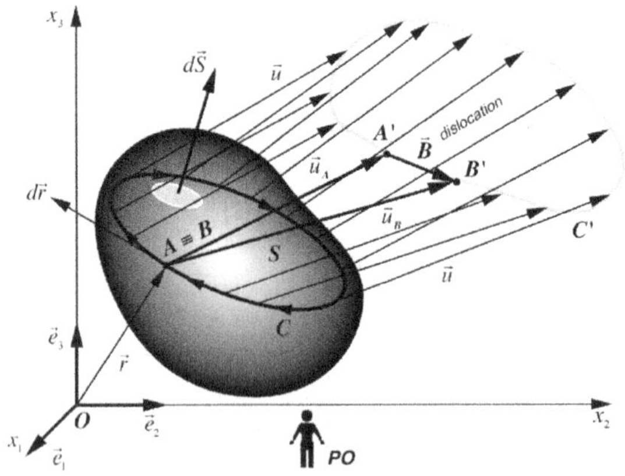

Figure 3.5 - *Interpretation of the condition of compatibility* $\overrightarrow{\mathrm{rot}}\,\vec{\beta}_i = 0$

---

1 *J. M. Burgers, Proc. Kon. Ned. Akad. Weten, vol.42, p. 293, 378, 1939*

2 *A. E. H. Love, «A Treatise on Mathematical Theory of Elasticity», 3rd edition, Cambridge U.P., 1920*

The existence of a displacement field $\vec{u}$ without any discontinuity allows us to say the following:
- The *topology is connected.* This means, from a physical standpoint, that there are no stacking of distortions resulting in a tear and slide of a part,
- The *topology is compact.* This means, from a physical standpoint, that there were no formation of holes inside the medium. In short, the condition $\overrightarrow{\text{rot}}\,\vec{\beta}_i = 0$ *assures the solidity and continuity of the medium.*

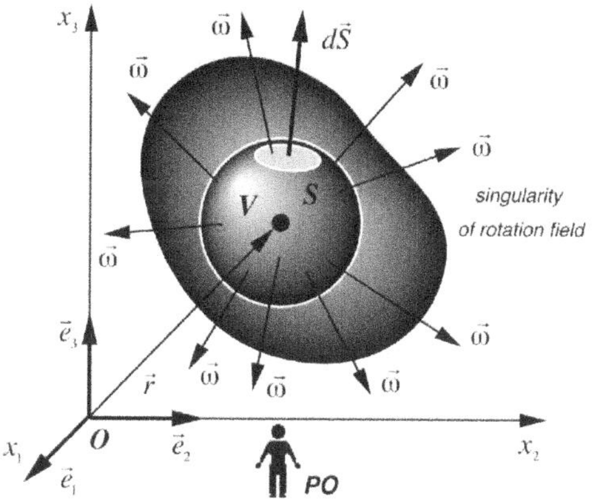

***Figure 3.6** - interpretation of the condition of compatibility* $\text{div}\,\vec{\omega} = 0$

The absence of divergence singularities of the rotation field

To find the significance of the condition of compatibility $\text{div}\,\vec{\omega} = 0$ for the vector of rotation $\vec{\omega}$, the flux of the vector of rotation $\vec{\omega}$, meaning the quantity $\vec{\omega} d\vec{S} = \vec{\omega}\vec{n} dS = \omega_\perp dS$ defined by the scalar product of $\vec{\omega}$ with the element of surface $d\vec{S}$, is integrated on a closed boundary surface S of a volume V of the solid (figure 3.6).
By applying the theorem of divergence we find the integral relations for a medium satisfying the compatibility condition associated with the rotation vector

$$\oiint_S \omega_\perp dS = \oiint_S \vec{\omega} d\vec{S} = \iiint_V \text{div}\,\vec{\omega} dV = 0 \quad ; \quad \forall S \tag{3.37}$$

Thus, the condition of compatibility for the vector of rotation $\vec{\omega}$, which stipulates that the field of rotations is non-divergent, implies that *there are no rotational singularities of the field of rotation* $\vec{\omega}$ in the solid, so no divergent torsions as seen in figure 3.6.

The absence of disclinations and the continuity of the rotation field by deformation

The condition of compatibility $\overrightarrow{\text{rot}}\,\vec{\chi}_i = 0$ of the tensor of contortion $\vec{\chi}_i$ also presents an interesting interpretation. If $d\vec{\omega}$ is integrated on a closed contour C in a compatible medium, it follows

$$\oint_C d\vec{\omega}^{(\varepsilon)} = \sum_k \vec{e}_k \oint_C d\omega_k^{(\varepsilon)} = \sum_k \vec{e}_k \oint_C \overrightarrow{\mathrm{grad}}\, \omega_k^{(\varepsilon)} d\vec{r} = \sum_k \vec{e}_k \oint_C \vec{\chi}_k d\vec{r} \tag{3.38}$$

In this relation we have introduced a rotational vector $\vec{\omega}^{(\varepsilon)}$ which represents the *rotations directly deduced from the tensor of deformations* $\vec{\varepsilon}_i$, which is identified with the local vector of rotation $\vec{\omega}$ in compatible medium. It follows that

$$\oint_C d\vec{\omega}^{(\varepsilon)} = \sum_k \vec{e}_k \iint_S \overrightarrow{\mathrm{rot}}\vec{\chi}_k\, d\vec{S} = \sum_k \vec{e}_k \iint_S \overrightarrow{\mathrm{rot}}\left[\overrightarrow{\mathrm{rot}}\,\vec{\varepsilon}_k\right]^{\mathrm{T}} d\vec{S} = 0 \quad ; \quad \forall C \tag{3.39}$$

By representing the field of rotation $\vec{\omega}^{(\varepsilon)}$ on a closed contour C in the medium (figure 3.7), we have that, from A to B

$$\oint_C d\vec{\omega}^{(\varepsilon)} = \vec{\omega}_B^{(\varepsilon)} - \vec{\omega}_A^{(\varepsilon)} = \vec{\Omega} = 0 \tag{3.40}$$

In this relation, the condition of compatibility $\overrightarrow{\mathrm{rot}}\,\vec{\chi}_i = 0$ implies that the closing vector $\vec{\Omega}$, called the *Frank vector*[3], is null, which signifies from a topological standpoint that there are no *rotational discontinuities by deformation,* these are called disclinations [4]. Since the compatibility condition for the contortion tensor $\vec{\chi}_i$ is also the compatibility condition for the deformation tensor $\vec{\varepsilon}_i$, we can conclude that the equations of *de Saint-Venant (3.21)* for $\vec{\varepsilon}_i$ are equivalent to the continuity of rotations by deformation.

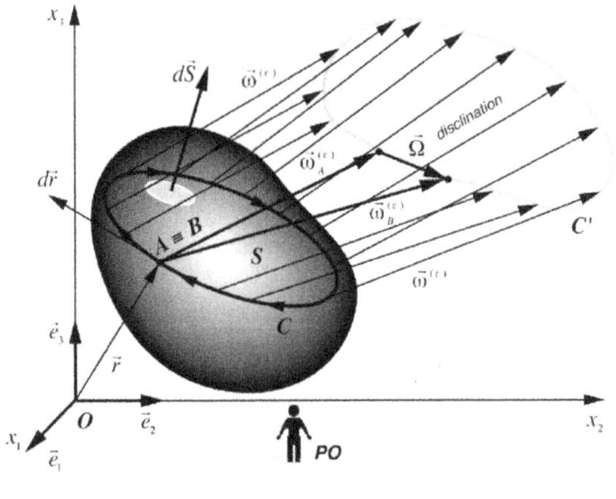

Figure 3.7 *- interpretation of the condition of compatibility* $\overrightarrow{\mathrm{rot}}\,\vec{\chi}_i = 0$

On the absence of divergent singularities within the flexion field

The condition of compatibility $\mathrm{div}\,\vec{\chi} = 0$ of the vector of flexion $\vec{\chi}$ can be interpreted in the following way. By integration of the flux of $\vec{\chi}$ over a closed boundary surface S around a given

[3] *R. DeWit, «Theory of disclinations II, III and IV», J. of the Nat. Bureau of Standards A, vol. 77A, p.49-100, p. 359-368, and p. 607-658, 1973*

[4] *F. C. Frank, Disc. Faraday Soc., vol. 25, p. 19, 1958*

volume V of the medium (figure 3.8), and by using the theorem of divergence and the definition of $\vec{\chi}$ from the deformation tensor $\vec{\varepsilon}_i$, it follows that

$$\oiint_S \chi_\perp dS = \oiint_S \vec{\chi} d\vec{S} = \iiint_V \operatorname{div}\vec{\chi} dV = \sum_k \vec{e}_k \iiint_V \overrightarrow{\operatorname{rot}}\,\overrightarrow{\operatorname{rot}}\,\vec{\varepsilon}_k dV = 0 \quad ; \quad \forall S \tag{3.41}$$

Therefore the compatibility condition on $\vec{\chi}$ implies that the field of flexion is non-divergent, and that there exists no singularity of the field of flexion $\vec{\chi}$ in the medium, such as the one represented on figure 3.8.

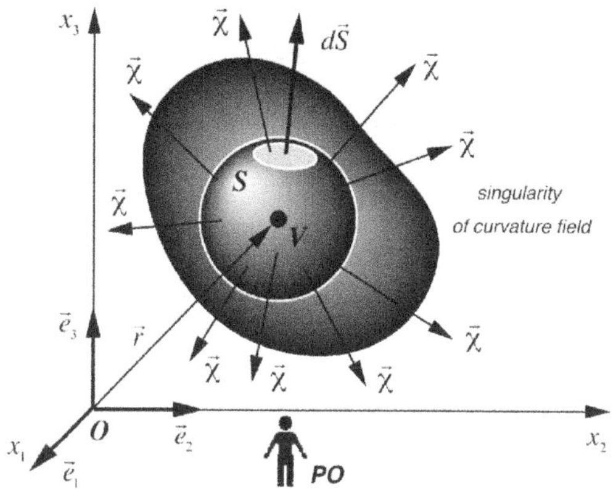

***Figure 3.8** - interpretation of compatibility condition* $\operatorname{div}\vec{\chi} = 0$

One should note that the condition of compatibility for the flexion tensor $\vec{\chi}$ corresponds indeed to the condition of *compatibility for the anti-symmetric part of tensor* $\overrightarrow{\operatorname{rot}}\,\vec{\varepsilon}_i$ since

$$\operatorname{div}\vec{\chi} = -\operatorname{div}\left(\sum_k \vec{e}_k \wedge \overrightarrow{\operatorname{rot}}\,\vec{\varepsilon}_k\right) = \sum_k \vec{e}_k \overrightarrow{\operatorname{rot}}\,\overrightarrow{\operatorname{rot}}\,\vec{\varepsilon}_k \tag{3.42}$$

3.5 - Pass-through conditions for a compatible interface

The *pass-through conditions* for the topological tensors across a compatible interface between 2 mediums, meaning an interface for which the fields of displacement on each side of the surface are the same, can be deducted from the conditions of compatibility. All one has to do is write down the integration of the condition of compatibility on a really slim volume, which moves with the interface, and where the 2 faces are on each side of the interface (figure 3.9).

The condition of compatibility for the distortion tensor $\vec{\beta}_i$ implies the following relations on a mobile volume V

$$\iiint_V \overrightarrow{\operatorname{rot}}\,\vec{\beta}_i\, dV = \oiint_S d\vec{S} \wedge \vec{\beta}_i = \iint_{S_1}\left(\vec{n} \wedge \vec{\beta}_i^{(1)}\right)dS - \iint_{S_2}\left(\vec{n} \wedge \vec{\beta}_i^{(2)}\right)dS = 0 \tag{3.43}$$

This means that the tangential components of the $\vec{\beta}_i$ are the same on each side of the interface.

$$\vec{n} \wedge \vec{\beta}_i^{(1)} = \vec{n} \wedge \vec{\beta}_i^{(2)} \tag{3.44}$$

The topological interpretation of this condition of passage is simple. Indeed, remember that, in a compatible medium, the tensor $\vec{\beta}_i$ derives from the gradient of a field of displacement $\vec{u}$, by the relation $\vec{\beta}_i = \overrightarrow{\text{grad}}\, u_i$, it is possible to re-write the integrals on surfaces S_1 and S_2 under the form

$$\iint_{S_1} \left(\vec{n} \wedge \overrightarrow{\text{grad}}\, u_i^{(1)}\right) dS - \iint_{S_2} \left(\vec{n} \wedge \overrightarrow{\text{grad}}\, u_i^{(2)}\right) dS = 0 \qquad (3.45)$$

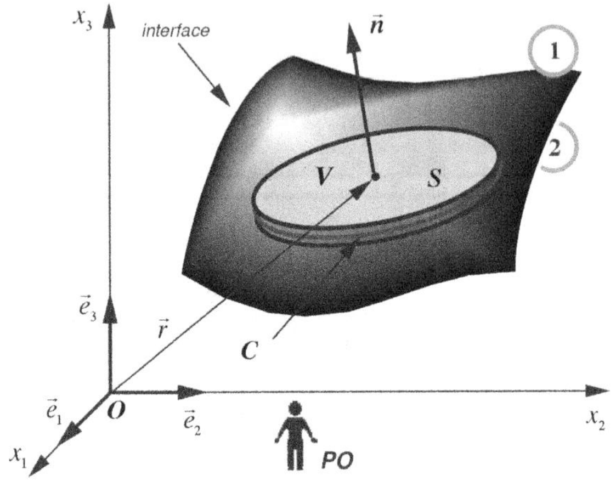

Figure 3.9 *- compatible pass-through conditions between 2 solid medium*

The theorem of the gradient applied to this last relation leads us to

$$\oint_C u_i^{(1)} d\vec{r} - \oint_C u_i^{(2)} d\vec{r} = 0 \qquad (3.46)$$

in which C is the mobile contour situated at the intersection off the mobile surface with the 2 mediums (figure 3.9). For this condition to be satisfied for any contour C we must have

$$u_i^{(1)} = u_i^{(2)} \quad \Rightarrow \quad \vec{u}^{(1)} = \vec{u}^{(2)} \qquad (3.47)$$

In other words the conditions of passage *(3.47)* implies that the displacement $\vec{u}$ on bots sides of the interface are equal, and thus that the 2 mediums in contact cannot 'unglue' nor slip one against the other, defining the "compatible interface".

The relation of compatibility for the vector of rotation $\vec{\omega}$ implies that the following relations

$$\iiint_V \text{div}\,\vec{\omega}\, dV = \oiint_S \vec{\omega} d\vec{S} = \iint_{S_1} \vec{\omega}^{(1)} \vec{n} dS - \iint_{S_2} \vec{\omega}^{(2)} \vec{n} dS = 0 \qquad (3.48)$$

can only be satisfied if

$$\vec{\omega}^{(1)} \vec{n} = \vec{\omega}^{(2)} \vec{n} \qquad (3.49)$$

The interpretation of this pass-through condition is self evident since it assures that the normal component of the field of local rotation is conserved on each side of the interface between 2 mediums; it prevents a sliding at the interface by rotation of one medium with respect to another.

The following integral relations at the interface can be deducted from the condition of compatibility for the tensor of contortion $\vec{\chi}_i$

$$\iiint_V \overrightarrow{\operatorname{rot}}\,\vec{\chi}_i dV = \oiint_S d\vec{S} \wedge \vec{\chi}_i = \iint_{S_1} \left(\vec{n} \wedge \vec{\chi}_i^{(1)}\right) dS - \iint_{S_2} \left(\vec{n} \wedge \vec{\chi}_i^{(2)}\right) dS = 0 \tag{3.50}$$

So that they are satisfied, it must be so that the tangential components of the $\vec{\chi}_i$ be conserved on each side of the interface. This can be written as

$$\vec{n} \wedge \vec{\chi}_i^{(1)} = \vec{n} \wedge \vec{\chi}_i^{(2)} \tag{3.51}$$

To find a topological interpretation for this condition of passage, one has to decompose the surface integral in 2 parts and remember that, in a compatible medium, the tensor $\vec{\chi}_i$ is directly linked to the gradients of the components of the field of rotation $\vec{\omega}^{\varepsilon}$ associated with the deformations $\vec{\varepsilon}_i$. So we have

$$\iint_{S_1} \left(\vec{n} \wedge \overrightarrow{\operatorname{grad}}\,\omega_i^{\varepsilon(1)}\right) dS - \iint_{S_2} \left(\vec{n} \wedge \overrightarrow{\operatorname{grad}}\,\omega_i^{\varepsilon(2)}\right) dS = 0 \tag{3.52}$$

which with the gradient theorem becomes

$$\oint_C \omega_i^{\varepsilon(1)} d\vec{r} - \oint_C \omega_i^{\varepsilon(2)} d\vec{r} = 0 \quad \Rightarrow \quad \omega_i^{\varepsilon(1)} = \omega_i^{\varepsilon(2)} \quad \Rightarrow \quad \vec{\omega}^{\varepsilon(1)} = \vec{\omega}^{\varepsilon(2)} \tag{3.53}$$

In other words the pass-through condition *(3.53)* assures us that the rotations $\vec{\omega}^{\varepsilon}$ implicitly contained in the tensor of deformations $\vec{\varepsilon}_i$ (from which we deduct the tensor of contortions $\vec{\chi}_i$) are equal on each side of the interface between the 2 mediums.

The equation of compatibility for the flexion vector $\vec{\chi}$ implies the following integral relation on the interface

$$\iiint_V \operatorname{div}\vec{\chi}\, dV = \oiint_S \vec{\chi} d\vec{S} = \iint_{S_1} \vec{\chi}^{(1)} \vec{n} dS - \iint_{S_2} \vec{\chi}^{(2)} \vec{n} dS = 0 \tag{3.54}$$

from which we deduct that the normal component of the flexion vector must be conserved on each part of the interface

$$\vec{\chi}^{(1)} \vec{n} = \vec{\chi}^{(2)} \vec{n} \tag{3.55}$$

3.6 - Examples of fields of flexion and torsion

From simple examples of field displacement $\vec{u}^{(\delta)}(\vec{r},t)$ that belong to known deformations of the medium, this paragraph will show that the topological tensors $\vec{\chi}_i$ et $\vec{\chi}$ perfectly describe the contortions, meaning the flexions and torsions that can appear inside the lattice.

The pure flexions of a solid lattice

Imagine the following static field displacement $\vec{u}^{(\delta)}(\vec{r})$ in the neighborhood of the origin of local referential $Ox_1x_2x_3$, which only present one component in the direction of the Ox_3

$$\vec{u}^{(\delta)}(\vec{r}) = \left(R - \sqrt{R^2 + \alpha x_1^2 + \beta x_2^2}\right) \vec{e}_3 \quad \text{for} \quad |x_1| < R \quad \text{and} \quad |x_2| < R \tag{3.56}$$

where the constants α and β can take the values -1, 0 or 3.

If the deformation of the horizontal planes of a solid submitted to such a field of displacement is represented graphically, we observe a *curvature of the planes of flexion* that depends on the value of the parameters α and β, as shown on the 4 examples of figure 3.10.

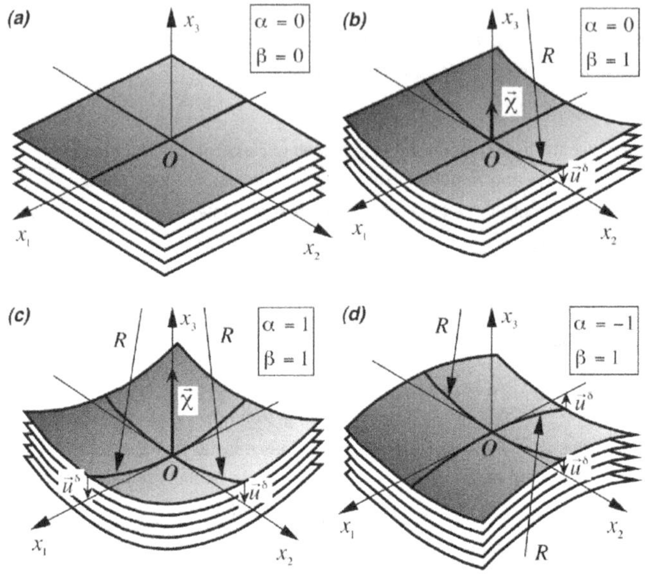

Figure 3.10 *- curvature by pure flexion in a solid lattice*

We can verify that, for the mathematical expression chosen here for the displacement field, there corresponds a radius of curvature R in the planes Ox_1x_3 and Ox_2x_3, in which the curvature is oriented towards the top or the bottom depending on whether the parameters α and β are positive or negative, respectively. If α and β are both null, there is no deformation (figure 3.10a). If $\alpha = 0$ and $\beta = 1$, the horizontal planes are portions of a cylinder of radius R (figure 3.10b). If α and β both are equal to 1, it is a portion of sphere or radius R (figure 3.10c), and if $\alpha = -1$ et $\beta = 1$, the horizontal planes represent a saddle point which is at the origin of coordinates (figure 3.10d).

With relations *(2.48)*, the tensors of distortion of the solid can be deduced from the displacement vector $\vec{u}^{(\delta)}(\vec{r})$. The volume expansion τ of the solid is null, because $\tau = -\operatorname{div}\vec{u}^{(\delta)} = 0$, and the local field of rotation $\vec{\omega}^{(\delta)}$ does not present a component along the axis Ox_3

$$\vec{\omega}^{(\delta)} = -\frac{1}{2}\overrightarrow{\operatorname{rot}}\,\vec{u}^{(\delta)} = \frac{\beta x_2}{2\sqrt{R^2 + \alpha x_1^2 + \beta x_2^2}}\vec{e}_1 - \frac{\alpha x_1}{2\sqrt{R^2 + \alpha x_1^2 + \beta x_2^2}}\vec{e}_2 \qquad (3.57)$$

One can verify that the expression for field $\vec{\omega}^{(\delta)}$ satisfies the condition of compatibility $\operatorname{div}\vec{\omega}^{(\delta)} = 0$. The tensor of deformation $\vec{\varepsilon}_i$ is equal to the shear tensor $\vec{\alpha}_i$, due to the fact that $\operatorname{div}\vec{u}^{(\delta)} = 0$. As a consequence

$$\begin{cases} \vec{\varepsilon}_1 = \vec{\alpha}_1 = \dfrac{\alpha x_1}{2\sqrt{R^2 + \alpha {x_1}^2 + \beta {x_2}^2}}\vec{e}_3 \\ \vec{\varepsilon}_2 = \vec{\alpha}_2 = \dfrac{\beta x_2}{2\sqrt{R^2 + \alpha {x_1}^2 + \beta {x_2}^2}}\vec{e}_3 \\ \vec{\varepsilon}_3 = \vec{\alpha}_3 = \dfrac{\alpha x_1}{2\sqrt{R^2 + \alpha {x_1}^2 + \beta {x_2}^2}}\vec{e}_1 + \dfrac{\beta x_2}{2\sqrt{R^2 + \alpha {x_1}^2 + \beta {x_2}^2}}\vec{e}_2 \end{cases} \tag{3.58}$$

The tensor of contortions $\vec{\chi}_i$ can also be deducted, and we note that it is of null trace, as we had deduced in *(3.22)*

$$\begin{cases} \vec{\chi}_1 = \overrightarrow{\text{grad}}\, {\omega_1}^{(\delta)} = \dfrac{\alpha\beta x_1\ x_2}{2\left(R^2 + \alpha {x_1}^2 + \beta {x_2}^2\right)^{3/2}}\vec{e}_1 + \dfrac{\beta\left(R^2 + \alpha {x_1}^{22}\right)}{2\left(R^2 + \alpha {x_1}^2 + \beta {x_2}^2\right)^{3/2}}\vec{e}_2 \\ \vec{\chi}_2 = \overrightarrow{\text{grad}}\, {\omega_2}^{(\delta)} = \dfrac{\alpha\left(R^2 + \beta {x_2}^2\right)}{2\left(R^2 + \alpha {x_1}^2 + \beta {x_2}^2\right)^{3/2}}\vec{e}_1 + \dfrac{\alpha\beta x_1\ x_2}{2\left(R^2 + \alpha {x_1}^2 + \beta {x_2}^2\right)^{3/2}}\vec{e}_2 \\ \vec{\chi}_3 = \overrightarrow{\text{grad}}\, {\omega_3}^{(\delta)} = 0 \end{cases} \tag{3.59}$$

And for the flexion vector $\vec{\chi}$, it follows

$$\vec{\chi} = -\overrightarrow{\text{rot}}\ \vec{\omega}^{(\delta)} = \sum_k \vec{e}_k \wedge \vec{\chi}_k = \frac{(\alpha+\beta)R^2 + \alpha\beta\left(x_1^2 + x_2^2\right)}{2\left(R^2 + \alpha x_1^2 + \beta x_2^2\right)^{3/2}}\ \vec{e}_3 \tag{3.60}$$

The norm of the flexion vector $\vec{\chi}$ measured at point O of our system of coordinates is directly linked to the *inverse of the radius of curvature* R, in the following way

$$\left|\vec{\chi}\right|_{x_1 = x_2 = 0} = \begin{cases} 0 & \text{for} \quad \alpha = 0 \quad \text{and} \quad \beta = 0 \\ 1/2R & \text{for} \quad \alpha = 0 \quad \text{and} \quad \beta = 1 \\ 1/R & \text{for} \quad \alpha = 1 \quad \text{and} \quad \beta = 1 \\ 0 & \text{for} \quad \alpha = -1 \quad \text{and} \quad \beta = 1 \end{cases} \tag{3.61}$$

and in such a way that the norm of the flexion vector in a given point of space is equal to the inverse of radius R of a cylinder (or a sphere) which would be centered at a distance R of this point in the direction of the flexion vector, and which would be tangential at this point to the deformed surface.

The pure torsion of a solid lattice

Consider now the field of local static displacement $\vec{u}^{(\delta)}(\vec{r})$ in the local frame $Ox_1x_2x_3$, where α, β and γ are constants

$$\vec{u}^{(\delta)}(\vec{r}) = \alpha x_2 x_3 \vec{e}_1 + \beta x_1 x_3 \vec{e}_2 + \gamma x_1 x_2 \vec{e}_3 \tag{3.62}$$

It is possible to obtain the vector of torsion (local rotation) $\vec{\omega}^{(\delta)}$ and the scalar of volume expansion τ associated with this field of displacement

$$\begin{cases} \vec{\omega}^{(\delta)} = -\frac{1}{2}\overrightarrow{\mathrm{rot}}\,\vec{u}^{(\delta)} = \frac{1}{2}\left[(\beta-\gamma)x_1\vec{e}_1 + (\gamma-\alpha)x_2\vec{e}_2 + (\alpha-\beta)x_3\vec{e}_3\right] \\ \tau = -\mathrm{div}\,\vec{u}^{(\delta)} = 0 \end{cases} \tag{3.63}$$

This field of displacement is thus non divergent and it leads to a field of rotation whose components increase or diminish along the coordinate axis, which is characteristic of a *pure torsion of the solid* (which corresponds in fact exactly to the case represented in figure 3.2 if we assume $\beta = 0$ et $\alpha = -\gamma$). Furthermore, we can verify that the vector of local rotation $\vec{\omega}^{(\delta)}$ satisfies the geometro-compatibility condition, which is normal since the divergence of a rotational is null

$$\mathrm{div}\,\vec{\omega}^{(\delta)} = \frac{1}{2}\left[(\beta-\gamma)+(\gamma-\alpha)+(\alpha-\beta)\right] = 0 \tag{3.64}$$

For a tensor of local distortions $\vec{\beta}_i^{(\delta)}$, it follows

$$\begin{cases} \vec{\beta}_1^{(\delta)} = -\overrightarrow{\mathrm{grad}}\,u_1^{(\delta)} = -\alpha x_3\vec{e}_2 - \alpha x_2\vec{e}_3 \\ \vec{\beta}_2^{(\delta)} = -\overrightarrow{\mathrm{grad}}\,u_2^{(\delta)} = -\beta x_3\vec{e}_1 - \beta x_1\vec{e}_3 \\ \vec{\beta}_3^{(\delta)} = -\overrightarrow{\mathrm{grad}}\,u_3^{(\delta)} = -\gamma x_2\vec{e}_1 - \gamma x_1\vec{e}_2 \end{cases} \tag{3.65}$$

and for the tensors of deformation $\vec{\varepsilon}_i$ and shear $\vec{\alpha}_i$

$$\begin{cases} \vec{\varepsilon}_1 = \vec{\alpha}_1 = \vec{\beta}_1^{(\delta)} - \vec{e}_1 \wedge \vec{\omega}^{(\delta)} = -(\alpha+\beta)x_3\vec{e}_2/2 - (\alpha+\gamma)x_2\vec{e}_3/2 \\ \vec{\varepsilon}_2 = \vec{\alpha}_2 = \vec{\beta}_2^{(\delta)} - \vec{e}_2 \wedge \vec{\omega}^{(\delta)} = -(\alpha+\beta)x_3\vec{e}_1/2 - (\beta+\gamma)x_1\vec{e}_3/2 \\ \vec{\varepsilon}_3 = \vec{\alpha}_3 = \vec{\beta}_3^{(\delta)} - \vec{e}_3 \wedge \vec{\omega}^{(\delta)} = -(\alpha+\gamma)x_2\vec{e}_1/2 - (\beta+\gamma)x_1\vec{e}_2/2 \end{cases} \tag{3.66}$$

The tensor of contortions is then written

$$\begin{cases} \vec{\chi}_1 = \overrightarrow{\mathrm{grad}}\,\omega_1^{(\delta)} = (\beta-\gamma)\vec{e}_1/2 \\ \vec{\chi}_2 = \overrightarrow{\mathrm{grad}}\,\omega_2^{(\delta)} = (\gamma-\alpha)\vec{e}_2/2 \\ \vec{\chi}_3 = \overrightarrow{\mathrm{grad}}\,\omega_3^{(\delta)} = (\alpha-\beta)\vec{e}_3/2 \end{cases} \tag{3.67}$$

This tensor is purely diagonal and of null trace. It is therefore equal, in this particular case, to the *tensor of torsion* $\left[\vec{\chi}_i\right]^S$. Regarding the vector of flexion, it is in this case null

$$\vec{\chi} = -\overrightarrow{\mathrm{rot}}\;\vec{\omega}^{(\delta)} = \sum_k \vec{e}_k \wedge \vec{\chi}_k = 0 \tag{3.68}$$

Chapter 4

Newtonian dynamics and eulerian thermo-kinetics

If we suppose that the lattice under consideration behaves in a newtonian fashion in the absolute frame of ***GO***, we can introduce in a rigorous manner the dynamics and thermo-kinetics of the lattice in the eulerian coordinates. We will do so on the frame of 3 axioms, which are classical and well-known: the definition of the kinetic energy of a particle and the 2 first principles of thermodynamics. With this rigorous if axiomatic approach, one is led to define average quantities on each site of the lattice, and corresponding sources and flux, which must satisfy 3 *continuity principles that are mandatory in solid newtonian lattices.*

4.1 - The principle of newtonian dynamics

The axiom of newtonian dynamics

It was shown in chapter 1 that a lattice of solid particles in space exhibits a collective movement which corresponds to the global movements of translation, rotation and deformation of the medium in the space of the observer, and which can be described in eulerian coordinates, by a local average velocity $\vec{\phi}(\vec{r},t)$.

We postulate here, axiomatically, that this collective lattice obeys the laws of *newtonian dynamics*. This means that, to the collective movement of particles at velocity $\vec{\phi}$, there corresponds a kinetic energy. For each particle, this kinetic energy can be written, according to newtonian mechanics as:

$$e_{cin} = \frac{1}{2} m \vec{\phi}^2 = \text{kinetic energy per particle} \qquad (4.1)$$

This expression of the kinetic energy of a particle in the medium involves a physical invariant scalar that is proper to the particle: its *inertial mass* also called *rest mass,* m.

The equation of local continuity of the inertial mass in the absolute frame of GO

In eulerian coordinates, one has to find an equation that translates the continuity of the mass in the medium in the frame of the ***GO***. To write this equation, we introduce the *density of mass* (mass per unit volume) $\rho(\vec{r},t)$ of the medium, so that the total mass $\boldsymbol{M}$ contained in any given volume V_m , which can be mobile in the medium, is simply obtained by integration of the density of mass over said volume, by definition

$$\boldsymbol{M} = \iiint_{V_m} \rho \, dV \qquad (4.2)$$

On this volume V_m , the equation of continuity of mass is written on the frame of the temporal derivative of $\boldsymbol{M}$, under the simple form

$$\frac{d}{dt}\iiint_{V_m}\rho\, dV = \iiint_{V_m} S_m\, dV - \oint\!\!\oint_{S_m} \vec{J}_m d\vec{S} \tag{4.3}$$

What this equation translates is the fact that, within a given volume V_m, the change in mass must correspond to either a *source density of mass* S_m, representing the creation or annihilation of mass at each point within said volume (the first term), or to the fact that mass leaves the volume. Mass leaving a volume is equivalent to mass passing-through the boundary surface of said volume. This is then captured as a surface exchange term between the inside and outside of the volume and is represented by the *surface flux of mass* $\vec{J}_m$, with respect to the surface S_m (figure 4.1). Equation *(4.3)* can also be transformed by using the derivative of an integral on a mobile volume and the divergence theorem to give us

$$\iiint_{V_m}\left[\frac{\partial\rho}{\partial t} + \operatorname{div}\left(\rho\vec{\phi}\right)\right]dV = \iiint_{V_m} S_m\, dV - \iiint_{V_m} \operatorname{div}\vec{J}_m\, dV \tag{4.4}$$

We can deduce the local expression for the continuity condition in the $Q\xi_1\xi_2\xi_3$ frame

$$\frac{\partial\rho}{\partial t} = S_m - \operatorname{div}\left(\rho\vec{\phi} + \vec{J}_m\right) \tag{4.5}$$

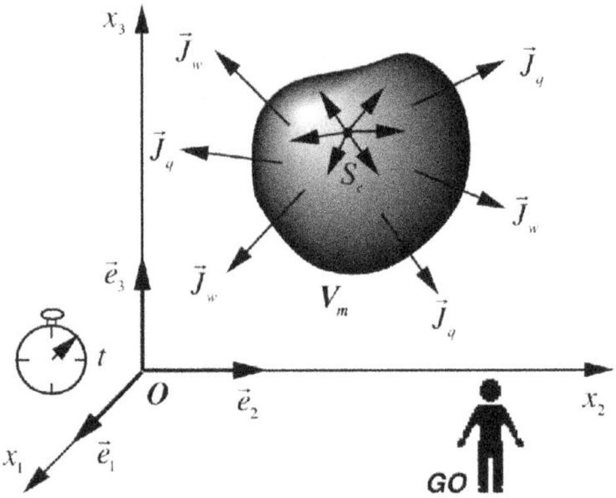

Figure 4.1 *- mass flux $\vec{J}_m$, work flux $\vec{J}_w$, heat flux $\vec{J}_q$, source of mass S_m and source of entropy density S_e on a given mobile volume V_m of medium*

The equation of local continuity of inertial mass in a local frame of PO

We can develop the same calculations we just did but in the local frame $Ox_1x_2x_3$ of ***PO***. With regards to the temporal derivative of ρ, meaning in a fixed point $\vec{r}$ of the local frame, the equation of continuity must involve the local velocity $\vec{\varphi}$ instead of the absolute velocity $\vec{\phi}$, and we obtain the following equation in $Ox_1x_2x_3$, which can also be expressed with the absolute velocity $\vec{\phi}$ by using the relation *(2.16)*

$$\frac{\partial\rho}{\partial t} = S_m - \operatorname{div}\left(\rho\vec{\varphi} + \vec{J}_m\right) = S_m - \operatorname{div}\left[\rho\left(\vec{\phi} - \vec{\phi}_O(t) - \vec{\omega}_O(t)\wedge\vec{r}\right) + \vec{J}_m\right] \tag{4.6}$$

The eulerian equation of continuity for inertial mass

We can revisit the 2 equations of local continuity of inertial mass in $Q\xi_1\xi_2\xi_3$ and $Ox_1x_2x_3$, and try to derive a continuity equation of the eulerian type, which would only depend on material derivatives, and as a consequence be true both in the absolute frame of ***GO*** and the mobile frame of ***PO***. We rewrite relation (4.5) in the following way

$$\frac{\partial \rho}{\partial t} = S_m - \rho \operatorname{div}\vec{\phi} - \vec{\phi}\,\overrightarrow{\operatorname{grad}}\,\rho - \operatorname{div}\vec{J}_m \tag{4.7}$$

By using the equation of compatibility *(1.42)* for τ, we obtain

$$\frac{\partial \rho}{\partial t} + \vec{\phi}\,\overrightarrow{\operatorname{grad}}\,\rho = S_m - \rho\left(\frac{d\tau}{dt} + \frac{S_n}{n}\right) - \operatorname{div}\vec{J}_m \tag{4.8}$$

The first member of this equation is the material derivative of ρ, so the equation of eulerian continuity for the inertial mass reads

$$\frac{d\rho}{dt} = S_m - \rho\left(\frac{d\tau}{dt} + \frac{S_n}{n}\right) - \operatorname{div}\vec{J}_m \tag{4.9}$$

One would obtain exactly the same results by starting from relation *(4.6)* in the frame $Ox_1x_2x_3$.

The newtonian concept of linear momentum ("quantité de mouvement" in french)

The equation *(4.3)* can be written in a fixed volume V_f of the observer. The temporal derivative of the total mass M contained in a fixed volume V_f is then written

$$\frac{d}{dt}\iiint_{V_f} \rho\, dV = \iiint_{V_f} S_m\, dV - \oiint_{S_f} \vec{P}_m d\vec{S} \tag{4.10}$$

in which the *absolute surface flux of mass* $\vec{P}_m$ with respect to the observer, actually represents an important characteristic measure of newtonian dynamics, the *volume density of linear momentum*. By introducing the *average linear momentum* $\vec{p}$ *of each elementary site of the lattice*, defined as

$$\vec{p} = \vec{P}_m / n \tag{4.11}$$

equation *(4.10)*, which corresponds to the temporal derivative of an integral on a fixed volume in space, can be simply written as

$$\iiint_{V_f} \frac{\partial \rho}{\partial t} dV = \iiint_{V_f} S_m\, dV - \iiint_{V_f} \operatorname{div}(n\vec{p})\, dV \tag{4.12}$$

which leads us to a second expression for the continuity of mass

$$\frac{\partial \rho}{\partial t} = S_m - \operatorname{div}(n\vec{p}) \tag{4.13}$$

Following *(4.5)* et *(4.10)*, the flux of mass can then be represented in 2 equivalent ways

$$\vec{P}_m = n\vec{p} = \rho\vec{\phi} + \vec{J}_m \tag{4.14}$$

which signifies that the absolute transport of mass $\vec{P}_m = n\vec{p}$ is equal to a sum of mass transport $\rho\vec{\phi}$ *by the mobile medium* with velocity $\vec{\phi}$ and an extra term of transport of mass $\vec{J}_m$ via another physical process, such as the self-diffusion in solid lattices which we will introduce later.

4.2 - The principle of energy continuity

To introduce the concept of energy, we must find the total energy contained in a volume V_m of the medium, mobile within it (figure 4.1), meaning having a displacement velocity $\vec{\phi}(\vec{r},t)$. It was shown in section 1.2 that the velocity of each particle i of the medium can simply be written

$$\vec{v}_i = \vec{\phi} + \Delta\vec{v}_i \tag{4.15}$$

where the $\Delta\vec{v}_i$ are the random fluctuations of the velocity of the particle, which we will associate later to thermal agitation.

To the collective movement of particles with velocity $\vec{\phi}$ corresponds, if we admit the postulate of newtonian dynamics, a kinetic energy that can be expressed in the form of a *volume density of kinetic energy* $T(\vec{r},t)$.

The axiom of the first principle of thermodynamics

To the random fluctuations of $\Delta\vec{v}_i$, as well as the interactions that can exist between the particles of the lattice must correspond a kinetic and an internal potential energy. This is precisely the axiom of *first principle of phenomenological thermodynamics*, which postulates, for a given physical system, a state function U, called *internal energy of the system*, which is such that, for all infinitesimal transformation of the system, the following relation holds

$$dU = \delta W + \delta Q \tag{4.16}$$

where δQ represents all exchanges of *heat* between the system and the external world, and δW all exchanges of *work* between the system and the external world.

Along its spatio-temporal evolution, it is certain that the medium will, in general, find itself out of thermodynamical equilibrium, and it is thus necessary to generalize the concept of internal energy by introducing a quantity of local internal energy which will depend on time, under the form of a *volume density of internal energy* $U(\vec{r},t)$.

The eulerian equation of continuity of energy

The total energy E contained in a volume V_m is then equal to the integral on the volume V_m of the sum of kinetic energy density T and internal energy density U, which is to say

$$E = T + U = \iiint_{V_m} (T + U)\, dV \tag{4.17}$$

According to the first principle of thermodynamics, this energy is a conserved quantity inside an isolated V_m. So any variation in said energy can only be due to external influences coming from outside the volume V_m, translating the fact that this volume is in fact *non-isolated*. These variations are due either to work exchange or heat exchange. By definition those exchanges either result from a pass through the boundary surface S_m around the volume V_m, or result from work due to an external work performed, on the internal medium, by an outside field (such as gravity, electro-magnetism, etc). The exchange of work and heat across the surface S_m can be represented as a form of *surface flux of work* $\vec{J}_w$ and of *surface flux of heat* $\vec{J}_q$. As for the volume work due to outside forces, it is introduced in the form of a volume source from external fields, and it is introduced in the form of a *volume source of work of external forces* S_w^{ext}.

The principle of conservation of energy of a given volume V_m can be written as the equality of the temporal derivation of total energy contained in said volume V_m and the different contributions to the volume, which comprises the work volume sources we just discussed and the pass-through contributions which necessarily cross the boundary surface , by definition of a boundary surface S_m

$$\frac{d}{dt}\iiint_{V_m}(T+U)dV = \iiint_{V_m} S_w^{ext}\, dV - \oiint_{S_m}\left(\vec{J}_w + \vec{J}_q\right)d\vec{S} \tag{4.18}$$

Using the theorem of divergence *(A.43)* and the derivation of an integral on a mobile volume *(A. 58)*, this equation can be transformed into the following

$$\iiint_{V_m}\left[\frac{\partial T}{\partial t}+\frac{\partial U}{\partial t}+\operatorname{div}\left(T\vec{\phi}+U\vec{\phi}\right)\right]dV = \iiint_{V_m}\left[S_w^{ext}-\operatorname{div}\left(\vec{J}_w+\vec{J}_q\right)\right]dV \tag{4.19}$$

which under local form, gives us

$$\frac{\partial T}{\partial t}+\frac{\partial U}{\partial t}+\operatorname{div}\left(T\vec{\phi}\right)+\operatorname{div}\left(U\vec{\phi}\right) = S_w^{ext}-\operatorname{div}\vec{J}_w-\operatorname{div}\vec{J}_q \tag{4.20}$$

It will be shown, later in this book, that for convenience reasons, it is judicious to transform this equation, which contains volume densities T and U, by introducing the measures e_{cin} et u, defined respectively as the *medium kinetic energy* and the *medium internal energy,* defined per elementary site of the lattice

$$e_{cin} = \frac{T}{n} \quad \text{and} \quad u = \frac{U}{n} \tag{4.21}$$

Thanks to these definitions, the first part of equation *(4.20)* can be transformed. Indeed, by using the equation of continuity *(1.12)* for n and making explicit the material derivative of u, as well as a term associated to the source S_n of the sites of the lattice, we deduce

$$\frac{\partial U}{\partial t}+\operatorname{div}\left(U\vec{\phi}\right) = n\left[\frac{\partial u}{\partial t}+(\vec{\phi}\vec{\nabla})u\right]+uS_n = n\frac{du}{dt}+uS_n \tag{4.22}$$

Applying transformation *(4.20)*, we obtain the final formulation of the *first principle*, or *principle of continuity of energy* in eulerian coordinates and in its local form

$$n\frac{du}{dt}+n\frac{de_{cin}}{dt} = S_w^{ext}-\operatorname{div}\vec{J}_w-\operatorname{div}\vec{J}_q-uS_n-e_{cin}S_n \tag{4.23}$$

Whether it be written in the absolute referential $Q\xi_1\xi_2\xi_3$ of the ***GO*** or in the local mobile frame $Ox_1x_2x_3$ of the ***PO***, the form of this equation of continuity remains identical to itself. Only the expression of the material derivative (absolute derivative as opposed to partial) changes since it depends on the absolute velocity $\vec{\phi}$ of the lattice in the frame $Q\xi_1\xi_2\xi_3$ and the local relative velocity $\vec{\varphi}$ of the lattice in the mobile frame $Ox_1x_2x_3$.

4.3 - The principle of entropy continuity

The axiom of second principle of thermodynamics

The *second principle of phenomenological thermodynamics* postulates the existence, for a given physical system, of a state function S, called the *entropy* of the system. This state function characterizes the disorder within the system, and is such that all infinitesimal transformation of the system satisfy the following inequality relation

$$dS \geq \frac{\delta Q}{T} \tag{4.24}$$

where δQ represents the sum of heat exchanges between the system and the outside world and T the temperature at which these exchanges happen. During its evolution in space and time, the system is, in general, out of thermodynamical equilibrium, such that it is necessary to generalize the concept of entropy, by introducing the notion of a *local* entropy variable which depends on time, the *volume density of entropy* $S(\vec{r},t)$.

The eulerian equation of continuity of entropy

By using the same mobile volume $\boldsymbol{V_m}$ defined in the previous paragraph, the volume density of entropy $S(\vec{r},t)$ allows us to calculate the total entropy $\boldsymbol{S}$ of a system by relation

$$\boldsymbol{S} = \iiint_{V_m} S\,dV \tag{4.25}$$

The second principle can then be generalized as a *principle of entropy in eulerian coordinates* by writing an integral relation for the temporal derivative of the total entropy $\boldsymbol{S}$ on a mobile volume

$$\frac{d}{dt}\iiint_{V_m} S\,dV = \iiint_{V_m} S_e\,dV - \oiint_{S_m} \frac{\vec{J}_q}{T}\,d\vec{S} \tag{4.26}$$

in which the exchanges of heat with the exterior are introduced via a *surface flux of heat* $\vec{J}_q$ on the boundary of said volume.

The irreversibility of thermodynamic transformations of the medium which is translated through the inequality $dS \geq \delta Q / T$ in phenomenological thermodynamics, was introduced here under the form of a source of volume entropy S_e within the volume $\boldsymbol{V_m}$. The notion of reversibility or irreversibility of processes of the medium is then described by a macroscopic source term σ of entropy

$$\sigma = \iiint_{V_m} S_e\,dV \tag{4.27}$$

which is such that the evolution of a system will be reversible for $\sigma = 0$ and irreversible for $\sigma > 0$.

The integral relation *(4.26)* can also be written under another form, by using the temporal derivative of an integral on a mobile volume and by making use of the theorem of divergence

$$\iiint_{V_m}\left[\frac{\partial S}{\partial t} + \operatorname{div}\left(S\vec{\phi}\right)\right]dV = \iiint_{V_m} S_e\,dV - \iiint_{V_m} \operatorname{div}\left(\frac{\vec{J}_q}{T}\right)dV \tag{4.28}$$

We can extract a local equation from the second principle of volume density of entropy S

$$\frac{\partial S}{\partial t} + \operatorname{div}\left(S\vec{\phi}\right) = S_e - \operatorname{div}\left(\frac{\vec{J}_q}{T}\right) \tag{4.29}$$

By introducing the value s representing the *average entropy per elementary site of the lattice*, defined by the relation

$$s = \frac{S}{n} \tag{4.30}$$

and by using the transformations described in *(4.22)*, the final formulation of the *second principle*, or *principle of eulerian continuity of entropy* in eulerian coordinates, is written

$$n\frac{ds}{dt} = S_e - \operatorname{div}\left(\frac{\vec{J}_q}{T}\right) - s\,S_n \tag{4.31}$$

Whether it be in the absolute frame $Q\xi_1\xi_2\xi_3$ of the ***GO*** or in the local mobile sphere $Ox_1x_2x_3$ of the ***PO***, this equation of continuity remains identical. Only the expression of the material derivative changes since it depends on the absolute velocity $\vec{\phi}$ of the lattice in the frame $Q\xi_1\xi_2\xi_3$ and on the local relative velocity $\vec{\varphi}$ of the lattice in the mobile frame $Ox_1x_2x_3$.

The 3 fundamental physics principles of continuity of newtonian dynamics and eulerian thermo-kinetics

Therefore, by starting from the axioms of newtonian dynamics and the classic phenomenological thermodynamic, 3 physical principles of continuity were postulated to describe the *"thermo-geometro-dynamic"* in eulerian coordinates, as following
- an eulerian principle of continuity of the density of inertial mass,
- an eulerian principle of continuity of the average energy per site of the lattice,
- an eulerian principle of continuity of the average entropy per site of the lattice.

As we will see during the next chapters, these 3 principles, which are summed up in table 4.1, are *the only physical principles needed for a complete galilean description of the newtonian geometro-dynamic and the phenomenological description of deformable medium in Euler coordinates, whether those are solids or fluids.*

Table 4.1 – Newtonian dynamics and eulerian thermo-kinetics

Absolute flux of mass $\vec{P}_m$ and average linear momentum $\vec{p}$ per site of the lattice

$$\vec{P}_m = n\vec{p} = \rho\vec{\phi} + \vec{J}_m \quad (1)$$

Principle of eulerian continuity for the density of inertial mass

$$\frac{d\rho}{dt} = S_m - \rho\left(\frac{d\tau}{dt} + \frac{S_n}{n}\right) - \operatorname{div}\vec{J}_m \quad (2)$$

Principle of eulerian continuity of the average energy per site of the lattice

$$n\frac{du}{dt} + n\frac{de_{cin}}{dt} = S_w^{ext} - \operatorname{div}\vec{J}_w - \operatorname{div}\vec{J}_q - uS_n - e_{cin}S_n \quad (3)$$

Principle of eulerian continuity of the average entropy per site of the lattice

$$n\frac{ds}{dt} = S_e - \operatorname{div}\left(\frac{\vec{J}_q}{T}\right) - s\,S_n \quad (4)$$

Chapter 5

Physical properties of a newtonian lattice

In this chapter, the fundamental physical properties of a solid lattice are shown. We will look at the *mechanical properties* and the *properties of mass transport.* Starting from a thermo-kinetic equation of the local solid lattice, we will show how the *elasticity* of the lattice gives rise to thermodynamical potentials called *stress tensors.* We then introduce the notion of *dissipative mechanical phenomenas* in lattices, by decomposing the tensor of distortions in dissipative and non-dissipative components. This leads to 2 phenomenological descriptions of possible dissipations: *anelasticity* and *plasticity.* We describe *self-diffusion,* the transport of mass through the lattice either in the form of *vacancies or self-interstitials.* The newtonian principle of mass conservation implies new fundamental equations for the lattice called *diffusion equations,* as well as new thermodynamic potentials called *chemical potentials.* Finally we derive the quantities associated with the newtonian behavior of the lattice, meaning its *average linear momentum per site,* its *average kinetic energy per site* and *the source of external work,* as in the case where the lattice is submitted to a constant gravity field.

5.1 - Elasticity of a lattice

In the local frame $Ox_1x_2x_3$, we can characterize the state of a solid in the most possible general way, first by giving its global tensor of distortion $\vec{\beta}_i$, and second, with regards to its thermal energy, by the local value of its entropy s. The average internal energy in each site of the lattice is, as defined here, a function of the local distortion and local entropy

$$u = u(\beta_{ij}, s) \tag{5.1}$$

The actual shape of this state function depends on the type of solid lattice we are considering. What we are doing with it is to describe the energy of a given solid phenomenologically, and therefore, we will call it the *phenomenological state function* of the solid.

Expression *(5.1)* of the internal energy is not the usual form found in textbooks to describe common solid latices. Usually only the components of the tensor of deformation are used to compute internal energy, not the β_{ij} components of the tensor of distortion. We will discuss this point further in the book. For now, let's just say that, *à priori,* the description of internal energy using the tensor of distortion $\vec{\beta}_i$ is more general than the one usually obtained from the components of the tensor of deformation.

The distortion stress tensor and the thermo-kinetic equation of elasticity

Since the internal energy u is a function of the scalar of entropy s and the components β_{ij} of the tensor of distortion, its differential form is

$$du = \frac{\partial u}{\partial \beta_{ij}} d\beta_{ij} + \frac{\partial u}{\partial s} ds \tag{5.2}$$

We introduce the *mechanical potential*, the conjugate of the tensor of distortion β_{ij}, and we will call it *distortion stress tensor* Σ_{ij}. We also introduce the *thermal potential*, the conjugate of entropy s, which is simply the *temperature* T. The differential can be written

$$ndu = \Sigma_{ij} d\beta_{ij} + nTds \tag{5.3}$$

The comparison between the 2 previous expressions of the differential of u shows the *phenomenological state equations of the solid*, which is to say the relations that gives us the stress tensor Σ_{ij} and temperature T as functions of the tensor of distortion β_{ij} and the entropy s

$$\begin{cases} \Sigma_{ij} = n\dfrac{\partial u}{\partial \beta_{ij}} = \Sigma_{ij}(\beta_{kl}, s) \\ T = \dfrac{\partial u}{\partial s} = T(\beta_{kl}, s) \end{cases} \tag{5.4}$$

As with the topological tensors, a vectorial notation of the stress tensor Σ_{ij} can be introduced here

$$\vec{\Sigma}_i = \sum_j \Sigma_{ij} \vec{e}_j \tag{5.5}$$

The differential of the internal energy u can then be written with a scalar product in the following form

$$ndu = \vec{\Sigma}_k d\vec{\beta}_k + nTds \tag{5.6}$$

From which we deduce the *thermo-kinetic equation of elasticity* of the lattice

$$n\frac{du}{dt} = \vec{\Sigma}_k \frac{d\vec{\beta}_k}{dt} + nT\frac{ds}{dt} \tag{5.7}$$

The deformation stress tensor and the torque torsor

The equations of thermodynamic obtained in the previous paragraph were written as functions of the components β_{ij} of the tensor of distortion. It could be interesting to rewrite them in terms of the tensor of deformation $\vec{\varepsilon}_i$ and the torsor of rotation $\vec{\omega}$. We recall,

$$\vec{\beta}_i = \vec{\varepsilon}_i + \vec{e}_i \wedge \vec{\omega} \tag{5.8}$$

which we introduce in the differential form of the internal energy u, and we have

$$\begin{aligned} ndu &= \vec{\Sigma}_k d\vec{\beta}_k + nTds = \vec{\Sigma}_k d\vec{\varepsilon}_k + \vec{\Sigma}_k \left(\vec{e}_k \wedge d\vec{\omega}\right) + nTds \\ &= \vec{\Sigma}_k d\vec{\varepsilon}_k + \left(\vec{\Sigma}_k \wedge \vec{e}_k\right) d\vec{\omega} + nTds \end{aligned} \tag{5.9}$$

The mechanical potential which is conjugated to the vector of rotation $\vec{\omega}$ will be called the *torsor of moments* or *torque torsor* $\vec{m}$, which we will then define as

$$\vec{m} = -\sum_k \vec{e}_k \wedge \vec{\Sigma}_k \tag{5.10}$$

It is linked to the anti-symmetric part of the stress tensor $\vec{\Sigma}_i$. In terms of this torque torsor, the differential u takes the form

$$ndu = \vec{\Sigma}_k d\vec{\varepsilon}_k + \vec{m}d\vec{\omega} + nTds = \Sigma_{ij}d\varepsilon_{ij} + m_k d\omega_k + nTds \tag{5.11}$$

Thus the internal energy u is also a function of the tensor of deformation $\vec{\varepsilon}_i$, of the vector of rotation $\vec{\omega}$ and of entropy s

$$u = u(\varepsilon_{ij}, \omega_k, s) \tag{5.12}$$

Its differential is

$$ndu = n\frac{\partial u}{\partial \varepsilon_{ij}}d\varepsilon_{ij} + n\frac{\partial u}{\partial \omega_k}d\omega_k + n\frac{\partial u}{\partial s}ds \tag{5.13}$$

The tensor of deformations $\vec{\varepsilon}_i$ is, by definition, a symmetrical tensor. We will then operate a few transformations on the differential of the internal energy u in order to explicitly show this symmetry which will be very useful later on

$$\begin{aligned} ndu &= \frac{1}{2}\Sigma_{ij}d\varepsilon_{ij} + \frac{1}{2}\Sigma_{ji}d\varepsilon_{ji} + m_k d\omega_k + nTds = \frac{1}{2}\left(\Sigma_{ij} + \Sigma_{ji}\right)d\varepsilon_{ij} + m_k d\omega_k + nTds \\ &= \frac{n}{2}\left(\frac{\partial u}{\partial \varepsilon_{ij}} + \frac{\partial u}{\partial \varepsilon_{ji}}\right)d\varepsilon_{ij} + n\frac{\partial u}{\partial \omega_k}d\omega_k + n\frac{\partial u}{\partial s}ds \end{aligned} \tag{5.14}$$

We have a symmetric mechanical potential, which is conjugated to the tensor of deformations ε_{ij}, we will call it the *deformation stress tensor* σ_{ij}, and it is defined by

$$\sigma_{ij} = \frac{1}{2}\left(\Sigma_{ij} + \Sigma_{ji}\right) \tag{5.15}$$

This new mechanical potential allows us to write the differential of u simply as

$$ndu = \sigma_{ij}d\varepsilon_{ij} + m_k d\omega_k + nTds \tag{5.16}$$

We define the state equations of the media in the representation $\left\{\vec{\varepsilon}_i, \vec{\omega}\right\}$

$$\begin{cases} \sigma_{ij} = \dfrac{n}{2}\left(\dfrac{\partial u}{\partial \varepsilon_{ij}} + \dfrac{\partial u}{\partial \varepsilon_{ji}}\right) = \sigma_{ij}(\varepsilon_{lm}, \omega_n, s) \\ m_k = n\dfrac{\partial u}{\partial \omega_k} = m_k(\varepsilon_{lm}, \omega_n, s) \\ T = \dfrac{\partial u}{\partial s} = T(\varepsilon_{lm}, \omega_n, s) \end{cases} \tag{5.17}$$

The deformation stress tensor σ_{ij} can be written from the tensor Σ_{ij}

$$\vec{\sigma}_i = \sum_j \sigma_{ij}\vec{e}_j = \vec{\Sigma}_i - \frac{1}{2}\left(\vec{e}_i \wedge \vec{m}\right) \tag{5.18}$$

The relation $\vec{\varepsilon}_i = \vec{\beta}_i - \vec{e}_i \wedge \vec{\omega}$ allows us to transform the differential of the internal energy u

$$\begin{aligned} ndu &= \vec{\Sigma}_k d\vec{\varepsilon}_k + \vec{m}d\vec{\omega} + nTds = \vec{\sigma}_k d\vec{\varepsilon}_k + \vec{m}d\vec{\omega} + nTds \\ &= \vec{\sigma}_k d\vec{\beta}_k + \left(\vec{e}_k \wedge \vec{\sigma}_k\right)d\vec{\omega} + \vec{m}d\vec{\omega} + nTds \end{aligned} \tag{5.19}$$

Since $\vec{\sigma}_i$ is symmetric we have, by definition

$$\sum_k \vec{e}_k \wedge \vec{\sigma}_k = 0 \tag{5.20}$$

In such a way that the differential of the internal energy u can be written as

$$n du = \vec{\sigma}_k d\vec{\beta}_k + \vec{m} d\vec{\omega} + nTds \tag{5.21}$$

This new form of the differential of u depends on the differential $d\vec{\beta}_i$ of the tensor of distortion. This new form was made possible by the introduction of the symmetric stress tensor $\vec{\sigma}_i$. The advantage of this expression is to be able to rewrite the *equation of thermo-kinetic* only using the material derivatives of $\vec{\beta}_i$ and of $\vec{\omega}$, which are linked by the simple geometro-kinetic relations

$$n\frac{du}{dt} = \vec{\sigma}_k \frac{d\vec{\beta}_k}{dt} + \vec{m}\frac{d\vec{\omega}}{dt} + nT\frac{ds}{dt} \tag{5.22}$$

The equation *(5.22)* shows that the temporal variations of the internal energy u along the path is associated with variations of the elastic energy of deformation, the potential energy of rotation and the thermal energy. Furthermore the variations of the elastic energy of deformation can be written in many equivalent ways

$$\vec{\Sigma}_k \frac{d\vec{\varepsilon}_k}{dt} = \vec{\sigma}_k \frac{d\vec{\varepsilon}_k}{dt} = \vec{\sigma}_k \frac{d\vec{\beta}_k}{dt} \tag{5.23}$$

The shear stress tensor, the torque torsor and the scalar of pressure

We use the decomposition of the tensor of deformation $\vec{\varepsilon}_i$ into the shear tensor $\vec{\alpha}_i$ and the scalar of volume expansion τ

$$\vec{\varepsilon}_i = \vec{\alpha}_i + \vec{e}_i \tau / 3 \tag{5.24}$$

to rewrite the differential of u in the following way

$$n du = \vec{\Sigma}_k d\vec{\varepsilon}_k + \vec{m} d\vec{\omega} + nTds = \vec{\Sigma}_k d\vec{\alpha}_k + \frac{1}{3}\vec{e}_k \vec{\Sigma}_k d\tau + \vec{m} d\vec{\omega} + nTds \tag{5.25}$$

This allows us to introduce the conjugate to the scalar of volume expansion τ, which is called the *pressure* p

$$p = -\frac{1}{3}\sum_k \vec{e}_k \vec{\Sigma}_k = -\frac{1}{3}\sum_k \vec{e}_k \vec{\sigma}_k = -\frac{1}{3}\sum_k \Sigma_{kk} = -\frac{1}{3}\sum_k \sigma_{kk} \tag{5.26}$$

So the differential of u is written

$$n du = \vec{\Sigma}_k d\vec{\alpha}_k + \vec{m} d\vec{\omega} - p d\tau + nTds = \Sigma_{ij} d\alpha_{ij} + m_k d\omega_k - p d\tau + nTds \tag{5.27}$$

Internal energy u can be expressed as a function of α_{ij}, ω_k and τ

$$u = u(\alpha_{ij}, \omega_k, \tau, s) \tag{5.28}$$

with the differential

$$n du = n\frac{\partial u}{\partial \alpha_{ij}} d\alpha_{ij} + n\frac{\partial u}{\partial \omega_k} d\omega_k + n\frac{\partial u}{\partial \tau} d\tau + n\frac{\partial u}{\partial s} ds \tag{5.29}$$

The shear tensor $\vec{\alpha}_i$ is symmetric, so

$$n du = \frac{1}{2}\Sigma_{ij} d\alpha_{ij} + \frac{1}{2}\Sigma_{ji} d\alpha_{ji} + m_k d\omega_k - p d\tau + nTds = \frac{1}{2}\left(\Sigma_{ij} + \Sigma_{ji}\right) d\alpha_{ji} + m_k d\omega_k - p d\tau + nTds \tag{5.30}$$

We introduce the *Kronecker symbol* δ_{ij}, such that $\delta_{ij} = 0$ if $i \neq j$ and $\delta_{ij} = 1$ if $i = j$, the sum $\delta_{ij} d\alpha_{ji} = d\alpha_{ii}$ is null, because the shear tensor $\vec{\alpha}_i$ is a null trace tensor, by construction. This allows us to add a term that contains the trace of the stress tensor Σ_{ij} in the differential of u

$$\begin{aligned} ndu &= \frac{1}{2}\left(\Sigma_{ij} + \Sigma_{ji}\right)d\alpha_{ji} - \frac{1}{3}\delta_{ij}d\alpha_{ji}\sum_k \Sigma_{kk} + m_k d\omega_k - pd\tau + nTds \\ &= \left[\frac{1}{2}\left(\Sigma_{ij} + \Sigma_{ji}\right) - \frac{1}{3}\delta_{ij}\sum_k \Sigma_{kk}\right]d\alpha_{ij} + m_k d\omega_k - pd\tau + nTds \end{aligned} \tag{5.31}$$

which can be written as

$$ndu = \left[\frac{n}{2}\left(\frac{\partial u}{\partial \alpha_{ji}} + \frac{\partial u}{\partial \alpha_{ji}}\right) - \frac{n}{3}\delta_{ij}\sum_k \frac{\partial u}{\partial \alpha_{kk}}\right]d\alpha_{ij} + n\frac{\partial u}{\partial \omega_k}d\omega_k + n\frac{\partial u}{\partial \tau}d\tau + n\frac{\partial u}{\partial s}ds \tag{5.32}$$

There appears a new mechanic potential, symmetrical and trace-less, which is conjugated to the shear tensor α_{ij}, which we will call the shear stress tensor s_{ij}, defined by the following relations

$$s_{ij} = \frac{1}{2}\left(\Sigma_{ij} + \Sigma_{ji}\right) - \frac{1}{3}\delta_{ij}\sum_k \Sigma_{kk} = \sigma_{ij} + \delta_{ij}p \tag{5.33}$$

By comparing the differential of u obtained thanks to the potential s_{ij}

$$ndu = s_{ij}d\alpha_{ij} + m_k d\omega_k - pd\tau + nTds \tag{5.34}$$

we can define the following state equations in the representation $\{\vec{\alpha}_i, \vec{\omega}, \tau\}$

$$\begin{cases} s_{ij} = \dfrac{n}{2}\left(\dfrac{\partial u}{\partial \alpha_{ij}} + \dfrac{\partial u}{\partial \alpha_{ji}}\right) - \dfrac{n}{3}\delta_{ij}\displaystyle\sum_k \dfrac{\partial u}{\partial \alpha_{kk}} = s_{ij}(\alpha_{lm}, \omega_n, \tau, s) \\ m_k = n\dfrac{\partial u}{\partial \omega_k} = m_k(\alpha_{lm}, \omega_n, \tau, s) \\ p = -n\dfrac{\partial u}{\partial \tau} = p(\alpha_{lm}, \omega_n, \tau, s) \\ T = \dfrac{\partial u}{\partial s} = T(\alpha_{lm}, \omega_n, \tau, s) \end{cases} \tag{5.35}$$

The shear stress tensor we just defined can be written in a simple manner in the vectorial form

$$\vec{s}_i = \sum_j s_{ij}\vec{e}_j = \vec{\sigma}_i + p\vec{e}_i \tag{5.36}$$

This tensor is symmetric and with null trace, so the following equations hold

$$\sum_k \vec{e}_k \wedge \vec{s}_k \equiv 0 \quad \text{and} \quad \sum_k \vec{e}_k \vec{s}_k \equiv 0 \tag{5.37}$$

The differential of u is expressed in terms of the shear tensor $\vec{\alpha}_i$

$$ndu = \vec{\Sigma}_k d\vec{\alpha}_k + \vec{m}d\vec{\omega} - pd\tau + nTds = \vec{s}_k d\vec{\alpha}_k + \vec{m}d\vec{\omega} - pd\tau + nTds \tag{5.38}$$

We decompose $\vec{\alpha}_i$

$$\vec{\alpha}_i = \vec{\varepsilon}_i - \frac{1}{3}\tau\vec{e}_i = \vec{\beta}_i - \vec{e}_i \wedge \vec{\omega} - \frac{1}{3}\tau\vec{e}_i \tag{5.39}$$

and exhibit, thanks to *(5.38)*, a direct dependence to $d\vec{\beta}_i$

$$
\begin{aligned}
ndu &= \vec{s}_k d\vec{\beta}_k - \vec{s}_k(\vec{e}_k \wedge d\vec{\omega}) - \vec{s}_k\left(\frac{1}{3}\vec{e}_k d\tau\right) + \vec{m}d\vec{\omega} - pd\tau + nTds \\
&= \vec{s}_k d\vec{\beta}_k + \left(\sum_k \vec{e}_k \wedge \vec{s}_k\right)d\vec{\omega} - \frac{1}{3}\left(\sum_k \vec{e}_k\vec{s}_k\right)d\tau + \vec{m}d\vec{\omega} - pd\tau + nTds \\
&= \vec{s}_k d\vec{\beta}_k + \vec{m}d\vec{\omega} - pd\tau + nTds
\end{aligned}
\tag{5.40}
$$

From this we deduce the thermo-kinetic equation in the representation $\{\vec{\alpha}_i, \vec{\omega}, \tau\}$

$$
n\frac{du}{dt} = \vec{s}_k \frac{d\vec{\beta}_k}{dt} + \vec{m}\frac{d\vec{\omega}}{dt} - p\frac{d\tau}{dt} + nT\frac{ds}{dt} \tag{5.41}
$$

We should note here that each term of the thermo-kinetic equation corresponds respectively to the temporal variations of the energy of elastic shear, global rotation, volume expansion and thermal energy. Furthermore this equation of thermo-kinetic depends on the temporal derivatives of the 3 topological tensors, which we can derive from the velocity field thanks to the geometro-kinetic equations given in table 1.1. Finally the following identities hold

$$
\vec{\Sigma}_k \frac{d\vec{\alpha}_k}{dt} = \vec{\sigma}_k \frac{d\vec{\alpha}_k}{dt} = \vec{s}_k \frac{d\vec{\alpha}_k}{dt} = \vec{s}_k \frac{d\vec{\varepsilon}_k}{dt} = \vec{s}_k \frac{d\vec{\beta}_k}{dt} \tag{5.42}
$$

The decompositions of the stress tensor $\vec{\Sigma}_i$, represented schematically in figure 5.1, follow a procedure completely equivalent, bar a few coefficients, to that used for the decomposition of the tensor of distortion.

Choosing a topological representation

As we have seen, in all generality, the thermo-kinetics of a deformable medium can be described in the three, equivalent, topological representations $\{\vec{\beta}_i\}$, $\{\vec{\varepsilon}_i, \vec{\omega}\}$ or $\{\vec{\alpha}_i, \vec{\omega}, \tau\}$, which each have their conjugate stress potentials, namely $\{\vec{\Sigma}_i\}$, $\{\vec{\sigma}_i, \vec{m}\}$ et $\{\vec{s}_i, \vec{m}, p\}$ respectively. The choice of one representation over another to describe a deformable medium depends on its nature and is a question of convenience as regards to the writing of the internal energy u. Let's give some explicit examples:

- in the case of a fluid, the internal energy u has to be expressed as an explicit function of the volume expansion τ (or of average volume v) and the entropy s, in such a way that the only non-null conservative mechanical potential in a fluid is the pressure

$$
u = u(\tau, s) \quad \text{or} \quad u = u(v, s) \quad \Rightarrow \quad p = -n\frac{\partial u}{\partial \tau} = -\frac{\partial u}{\partial v} \tag{5.43}
$$

which leads us to the following geometro-kinetic equation

$$
u = u(\tau, s) \quad \text{or} \quad u = u(v, s) \quad \Rightarrow \quad n\frac{du}{dt} = -p\frac{d\tau}{dt} + nT\frac{ds}{dt} = -np\frac{dv}{dt} + nT\frac{ds}{dt} \tag{5.44}
$$

- in the case of a usual solid lattice, the potential part of the internal energy depends on the elastic deformations $\vec{\varepsilon}_i$ of the medium, and only on those. It follows that the only non-null mechanical potential for a lattice is generally the symmetrical stress tensor $\vec{\sigma}_i$

$$
u = u(\varepsilon_{ij}, s) \quad \Rightarrow \quad \sigma_{ij} = \frac{n}{2}\left(\frac{\partial u}{\partial \varepsilon_{ji}} + \frac{\partial u}{\partial \varepsilon_{ij}}\right) \quad \Rightarrow \quad n\frac{du}{dt} = \vec{\sigma}_k \frac{d\vec{\varepsilon}_k}{dt} + nT\frac{ds}{dt} \tag{5.45}
$$

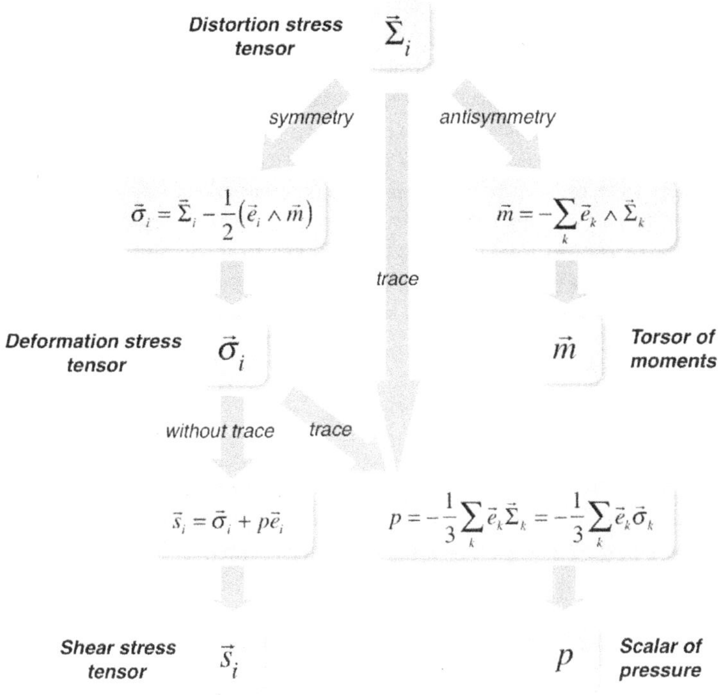

Figure 5.1 *- schematics of the decomposition of the stress tensors*

- in the case of an isotropic lattice, it is preferable to separate the deformations due to shear and the deformations due to volume expansion and to write the internal energy u as a function of the tensor of shear $\vec{\alpha}_i$, the scalar of volume expansion τ and the entropy s. This leads us to the fact that the non-null mechanical potentials are the shear stress tensor $\vec{s}_i$ and the scalar of pressure p

$$u = u(\alpha_{ij}, \tau, s) \quad \Rightarrow \quad s_{ij} = \frac{n}{2}\left(\frac{\partial u}{\partial \alpha_{ji}} + \frac{\partial u}{\partial \alpha_{ij}}\right) - \frac{n}{3}\delta_{ij}\sum_k \frac{\partial u}{\partial \alpha_{kk}} \quad \text{et} \quad p = -n\frac{\partial u}{\partial \tau} \tag{5.46}$$

the thermo-kinetic equation reads

$$u = u(\alpha_{ij}, \tau, s) \quad \Rightarrow \quad n\frac{du}{dt} = \vec{s}_k \frac{d\vec{\alpha}_k}{dt} - p\frac{d\tau}{dt} + nT\frac{ds}{dt} \tag{5.47}$$

Generally there is no direct dependence of the internal energy u on the vector of *global* rotation $\vec{\omega}$, as a global rotation does not deform the medium (merely rotates it). Consequently, the torque torsor $\vec{m}$ is null everywhere. However here we have voluntarily introduced the torque torsor $\vec{m}$ in order to stay general, and to account for the fact that we can easily imagine u as depending on the global vector of rotation $\vec{\omega}$ of the lattice, for example in the case where there

is a coupling of vectorial nature in the lattice having certain directional properties (such as an axial magnetic moment or an electrical axial polarization) and a certain external field defined in an absolute frame (such as a magnetic field or an electric field for example) where a preferred axis would appear and the symmetry of rotation would be broken by the external potential.
While this is true of a *global* rotation, we will show later, that under certain particular conditions, the stress tensor of elastic shear $\vec{s}_i$ can be replaced by a torsor of moments or torque torsor $\vec{m}$ conjugated to the vector of *local* rotation $\vec{\omega}^{(\delta)}$.
Another important remark can be made here. In our description in eulerian coordinates, the mechanic potentials are given by state equations that use the density n of sites of the lattice as a multiplication factor. This density n is directly related to the volume expansion τ of the solid or to the average volume v of the sites of the lattice via relations $n = 1/v = n_0 e^{-\tau}$, which implies that the mechanical potentials will depend in this case non-linearly on the expansion τ of the solid.

5.2 - Anelasticity and plasticity of a lattice

The only way to introduce dissipative processes in a solid lattice of particles is to postulate that the distortions by elastic deformation of the lattice are accompanied by dissipative distortions. It is these dissipative distortions that allow, at the macroscopic scale, relative movement with large distances between particles, by sliding planes of particles one on top of the other. We write in a generic way the total distortions $\vec{\beta}_i$ in the local frame

$$\vec{\beta}_i = \vec{\beta}_i^{(\delta)} + \vec{e}_i \wedge \vec{\omega}_O(t) = \vec{\beta}_i^{el} + \vec{\beta}_i^{dis} + \vec{e}_i \wedge \vec{\omega}_O(t) \tag{5.48}$$

It is clear that the variations of elastic distortions $\vec{\beta}_i^{el}$ contribute to a work of elastic deformation that we will write δW_{el},

$$\delta W_{el} = \vec{s}_k d\vec{\beta}_k^{el} + \vec{m} d\vec{\omega}^{el} - p d\tau^{el} \tag{5.49}$$

At the same time the dissipative distortions $\vec{\beta}_i^{dis}$ will be responsible for a variation of work δW_{dis} and a variation of heat δQ_{dis}, which we will thus write

$$\delta W_{dis} + \delta Q_{dis} = \vec{s}_k d\vec{\beta}_k^{dis} + \vec{m} d\vec{\omega}^{dis} - p d\tau^{dis} \tag{5.50}$$

Two possible phenomenologies for dissipation

Relation *(5.50)* can lead to 2 different dissipative phenomenologies, depending on whether the variation of work is null or not:
- if variation of work δW_{dis} is null, then the dissipative distortions do not store potential energy in the network, so *they cannot be recovered.* This thermodynamically irreversible phenomena will be called the *plasticity of the lattice.* The *plastic distortions* satisfy the following relation

$$\delta Q_{pl} = \vec{s}_k d\vec{\beta}_k^{pl} + \vec{m} d\vec{\omega}^{pl} - p d\tau^{pl} \tag{5.51}$$

- if variation of work δW_{dis} is not null, the dissipative distortions do store potential energy in the lattice, so *they are this time recoverable.* However this phenomena is *thermodynamically irre-*

versible since it dissipates heat. It will be called the *anelasticity of the lattice*, and the *anelastic distortions* that are associated with it satisfy the following equation

$$\delta W_{an} + \delta Q_{an} = \vec{s}_k d\vec{\beta}_k^{an} + \vec{m} d\vec{\omega}^{an} - p d\tau^{an} \tag{5.52}$$

Since the anelastic distortions contribute both to variations of work and heat, it is necessary to introduce a way to decompose the previous expression to separate δW_{an} and δQ_{an}. The only logical way to proceed here is to write

$$\begin{cases} \vec{s}_k = \vec{s}_k^{\,cons} + \vec{s}_k^{\,dis} \\ \vec{m} = \vec{m}^{cons} + \vec{m}^{dis} \\ p = p^{cons} + p^{dis} \end{cases} \tag{5.53}$$

and so to decompose the stress tensors in a *conservative part*, which is responsible for the source of work, and a *dissipative part*, which is responsible for the source of heat. This decomposition of stress tensors allows us to write the individual variations of work and heat linked to anelasticity in the following form

$$\begin{cases} \delta W_{an} = \vec{s}_k^{\,cons} d\vec{\beta}_k^{an} + \vec{m}^{cons} d\vec{\omega}^{an} - p^{cons} d\tau^{an} \\ \delta Q_{an} = \vec{s}_k^{\,dis} d\vec{\beta}_k^{an} + \vec{m}^{dis} d\vec{\omega}^{an} - p^{dis} d\tau^{an} \end{cases} \tag{5.54}$$

We can here make a hypothesis of anelasticity, which will allow to simplify the description of dissipative solid medium going forward. Indeed if we suppose that the trace of the anelastic distortion tensor is null, we can immediately conclude that there are no anelastic component of volume expansion, by definition, and therefore that the anelastic deformations are only made of shear

Hypothesis: $\tau^{an} = \sum_k \vec{\beta}_k^{an} \vec{e}_k \equiv 0$ (5.55)

The thermo-kinetic equation of an anelastic and plastic lattice

By using relations *(5.49)* and *(5.54)*, and by using the hypothesis *(5.55)*, the variation of work $\delta W_{déf}$ associated with the deformations of a lattice that has elasticity, anelasticity and plasticity is written

$$\delta W_{déf} = \delta W_{él} + \delta W_{an} = \vec{s}_k d\vec{\beta}_k^{el} + \vec{m} d\vec{\omega}^{el} - p d\tau^{el} + \vec{s}_k^{\,cons} d\vec{\beta}_k^{an} + \vec{m}^{cons} d\vec{\omega}^{an} \tag{5.56}$$

The total variation of the work of deformation $\delta W_{déf}$ will be found in the expression for the thermo-kinetic equation for the internal energy u

$$n\frac{du}{dt} = \vec{s}_k \frac{d\vec{\beta}_k^{el}}{dt} + \vec{m}\frac{d\vec{\omega}^{el}}{dt} - p\frac{d\tau^{el}}{dt} + \vec{s}_k^{\,cons}\frac{d\vec{\beta}_k^{an}}{dt} + \vec{m}^{cons}\frac{d\vec{\omega}^{an}}{dt} + nT\frac{ds}{dt} \tag{5.57}$$

This thermo-kinetic equation could be written by replacing the $\vec{\beta}_i$ with $\vec{\alpha}_i$, thanks to property *(5.42)*, so the internal energy function u of a self-diffusive lattice with anelasticity and plasticity is a function of the following state variables

$$u = u\left(\alpha_{ij}^{el}, \alpha_{ij}^{an}, \omega_k^{el}, \omega_k^{an}, \tau^{el}, s\right) \tag{5.58}$$

A first statement can be made about this state function: it does not depend on plastic distortions of the lattice. As a matter of fact, the plastic distortions of the network only appear in the total variation of heat $\delta Q_{déf}$ associated with the dissipative processes of anelastic and plastic deformation

$$\delta Q_{déf} = \delta Q_{an} + \delta Q_{pl} = \vec{s}_k^{dis} d\vec{\beta}_k^{an} + \vec{m}^{dis} d\vec{\omega}^{an} + \vec{s}_k d\vec{\beta}_k^{pl} + \vec{m} d\vec{\omega}^{pl} - p d\tau^{pl} \tag{5.59}$$

It is clear that in the description of the anelastic and plastic lattice, this variation of heat will have to appear in the heat equation of the lattice. It is responsible for a source of entropy reflecting the thermodynamical irreversibility of the dissipative processes of anelasticity and plasticity.

The constitutive equations of an anelastic and plastic lattice

Besides the simplifying hypothesis *(5.55)*, there are no restrictions to decomposing the dissipative distortions in relation *(5.48)* in an anelastic part and a plastic part, so that we can write in all generality, the following constitutive equations of a solid lattice that is both anelastic and plastic

$$\begin{cases} \vec{\beta}_i = \vec{\beta}_i^{(\delta)} + \vec{e}_i \wedge \vec{\omega}_0(t) = \vec{\beta}_i^{el} + \vec{\beta}_i^{an} + \vec{\beta}_i^{pl} + \vec{e}_i \wedge \vec{\omega}_0(t) \\ \vec{\omega} = \vec{\omega}^{(\delta)} + \vec{\omega}_0(t) = \vec{\omega}^{el} + \vec{\omega}^{an} + \vec{\omega}^{pl} + \vec{\omega}_0(t) = -\dfrac{1}{2}\sum_k \vec{e}_k \wedge \vec{\beta}_k \\ \tau = \tau^{el} + \tau^{pl} = \sum_k \vec{\beta}_k \vec{e}_k \qquad (\tau^{an} \equiv 0 \textit{ by hypothesis}) \\ \vec{\varepsilon}_i = \vec{\varepsilon}_i^{el} + \vec{\varepsilon}_i^{an} + \vec{\varepsilon}_i^{pl} = \vec{\beta}_i - \vec{e}_i \wedge \vec{\omega} \\ \vec{\alpha}_i = \vec{\alpha}_i^{el} + \vec{\alpha}_i^{an} + \vec{\alpha}_i^{pl} = \vec{\varepsilon}_i - \dfrac{1}{3}\tau \vec{e}_i \end{cases} \tag{5.60}$$

It is clear that the distortions $\vec{\beta}_i$, $\vec{\omega}$ and τ always satisfy, in the local frame, both the equations of geometro-kinetic of table 1.1 that give them in terms of the velocity field $\vec{\phi}$ inside the solid, as well as the equations of geometro-compatibility $\overrightarrow{\text{rot}}\,\vec{\beta}_i = 0$ and $\text{div}\,\vec{\omega} = 0$.

Existence of sources and sinks of sites of lattice in the presence of plasticity

We saw in relation *(1.18)*, that the equation of continuity for n can be represented in the form of a derivative along the trajectory

$$\frac{1}{n}\frac{dn}{dt} = \frac{d}{dt}\left(\ln n + cste\right) = \frac{S_n}{n} - \text{div}\,\vec{\phi} \tag{5.61}$$

By definition there is a link between the volume expansion and the density n of sites of the lattice. But, according to *(5.60)*, the volume expansion contains 2 components since $\tau = \tau^{el} + \tau^{pl}$. One can easily convince oneself that the density n of the lattice sites is not linked directly to the total volume expansion τ, *but only to its elastic part* τ^{el}. This assertion is equivalent to saying

$$\tau^{el} = -\ln\frac{n}{n_0} \quad \Rightarrow \quad n = n_0 \exp(-\tau^{el}) \quad \Rightarrow \quad \frac{1}{n}\frac{dn}{dt} = -\frac{d\tau^{el}}{dt} \tag{5.62}$$

From this assertion we deduce that the equation of continuity *(1.20)* must be written, in presence of elasticity and plasticity as

$$\frac{d\tau^{el}}{dt}=-\frac{S_n}{n}+\operatorname{div}\vec{\phi} \tag{5.63}$$

This relation is very important, as it shows that, in the *presence of plasticity*, the density n of lattice sites cannot obey the same principle of conservation. As a matter of fact, the phenomenas of plastic deformation behave as sources and sinks of lattice sites, which we will justify fully in the rest of this book when we introduce the notion of plastic charges. There appears a *source of lattice sites* S_n different than zero directly linked to the material derivative of the plastic volume expansion

$$S_n=n\frac{d\tau^{pl}}{dt} \quad\Rightarrow\quad \frac{d\tau^{el}}{dt}=-\frac{d\tau^{pl}}{dt}+\operatorname{div}\vec{\phi} \tag{5.64}$$

Thus, in the presence of plasticity the equations of geometro-kinetics will contain, implicitly, a source of lattice sites S_n which is non null and associated to $d\tau^{pl}/dt$. We will write them as

$$\frac{d\vec{\beta}_i}{dt}=\overrightarrow{\operatorname{grad}}\,\phi_i \quad;\quad \frac{d\vec{\omega}}{dt}=\frac{1}{2}\overrightarrow{\operatorname{rot}}\,\vec{\phi} \quad;\quad \frac{d\tau}{dt}=\operatorname{div}\vec{\phi} \tag{5.65}$$

5.3 - Self-diffusion in a lattice

There exists an important consequence to the equation of continuity *(5.61)* for the density n of sites of lattice: the *principle of newtonian conservation of mass* would be violated in the presence of non null site sources! To see this, one only has to multiply the equation of continuity *(1.7)* for n by mass m of the particles of the lattice. Since mass density is equal to $\rho=mn$ and the linear momentum is equal to $\vec{p}=m\vec{\phi}$ in a lattice without point defects, we have the following equation

$$\frac{\partial(mn)}{\partial t}=\frac{\partial\rho}{\partial t}=mS_n-\operatorname{div}\left(mn\vec{\phi}\right)=\rho\frac{d\tau^{pl}}{dt}-\operatorname{div}\left(n\vec{p}\right) \tag{5.66}$$

which has an effective mass source linked to the plasticity of the lattice. It is not possible to admit the non-conservation of mass in a newtonian network, so in a lattice which does not contain point defects, plastic distortions which give us a source of lattice sites S_n different from zero, with $d\tau^{pl}/dt$ non null, are impossible.

On the other hand we can show now that in the case of a network containing self-diffusion due to point defects, plastic distortions lead to a source of lattice sites S_n which is different from zero do not violate the principle of mass conservation and become consequently, possible.

Self-diffusion of particles in a lattice

An intrinsic point defect of *vacancy* type consists of a site of the lattice which does not contain a particle (figure 5.2). It is a *"hole"* in the lattice. An intrinsic point defect of *interstitial* type is a particle which is found in the solid lattice, but which does not occupy a regular site of the lattice (figure 5.2). It is an *"extra"* particle in the lattice.

It is easy to see that the presence of such point defects leads to a mass transport phenomena via the self-diffusion of said defects. We show these 2 mechanisms in a lattice moving with absolute velocity $\vec{\phi}$ (figure 5.2). The movement of a vacancy L (L for *lacuna*, meaning a missing particle in the lattice) with relative velocity $\Delta\vec{\varphi}_L$ with respect to the lattice in a given direction leads to a relative flux of mass in the opposite direction with velocity $-\Delta\vec{\varphi}_L$, while the movement of an interstitial I with relative velocity $\Delta\vec{\varphi}_I$ with respect to the lattice in a given direction, gives rise to a flux of mass in the same direction with velocity $\Delta\vec{\varphi}_I$.

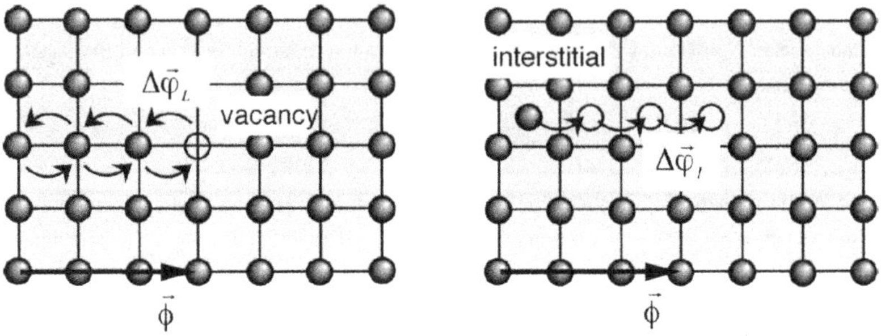

Figure 5.2 *- movement of a hole or interstitial in a solid lattice*

To mathematically transcribe the existence of these punctual defects, one must introduce a notion of volume density of vacancies, n_L, and interstitials, n_I

$$n_L = n_L(\vec{\xi}, t) \quad \text{and} \quad n_I = n_I(\vec{\xi}, t) \quad \text{in} \quad Q\xi_1\xi_2\xi_3$$
$$n_L = n_L(\vec{r}, t) \quad \text{and} \quad n_I = n_I(\vec{r}, t) \quad \text{in} \quad Ox_1x_2x_3 \tag{5.67}$$

From these densities, it is possible to define the notion of *atomic concentrations* of vacancies and self-interstitials with respect to density n of lattice sites with the following

$$C_L = n_L / n \quad \text{and} \quad C_I = n_I / n \tag{5.68}$$

Contrary to appearances, there is a certain asymmetry between vacancies and self-interstitials, which is seen in the fact that the maximum atomic concentration of vacancies is always limited to 1, when all sites of the lattice are empty, while the atomic concentration of self-interstitials depends on the number of interstitial sites and the number of defects one can put per site.

At point $\vec{\xi}$ of absolute space $Q\xi_1\xi_2\xi_3$ and at instant t, the absolute velocity $\vec{\phi}_L$ of the vacancies and the absolute velocity $\vec{\phi}_I$ of interstitials can be expressed in terms of the relative velocities $\Delta\vec{\varphi}_L$ et $\Delta\vec{\varphi}_I$ of these defects with respect to the lattice as such

$$\vec{\phi}_L = \vec{\phi} + \Delta\vec{\varphi}_L \quad \text{and} \quad \vec{\phi}_I = \vec{\phi} + \Delta\vec{\varphi}_I \tag{5.69}$$

In the local frame $Ox_1x_2x_3$, velocity $\vec{\varphi}_L$ of the vacancies and velocity $\vec{\varphi}_I$ of the interstitials can be written

$$\vec{\varphi}_L = \vec{\varphi} + \Delta\vec{\varphi}_L \quad \text{and} \quad \vec{\varphi}_I = \vec{\varphi} + \Delta\vec{\varphi}_I \tag{5.70}$$

If we follow a piece of the lattice during its evolution in space (figure 5.3), the equations of continuity which characterize n, n_L et n_I on this lattice can be found, provided we write them as integrals on this mobile lattice. Indeed on this mobile volume, the number of sites does not change, a fact we translate in a relation valid both in $Q\xi_1\xi_2\xi_3$ as well as $Ox_1x_2x_3$

$$\frac{d}{dt}\iiint_{V_m} n\,dV = 0 \tag{5.71}$$

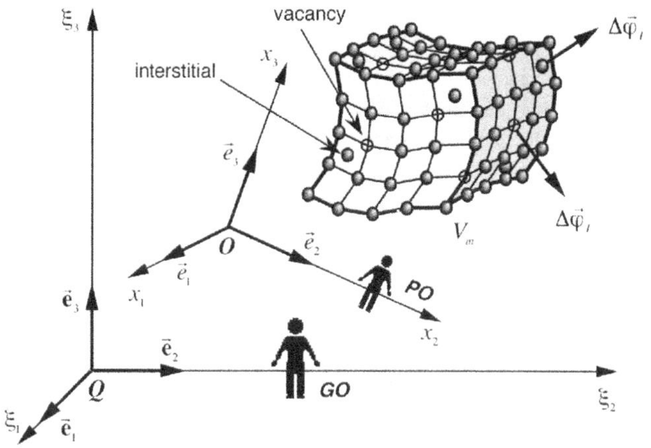

Figure 5.3 - *flux of vacancies and interstitials across the surface of a chunk of lattice*

Regarding the integral equation of evolution describing the number of vacancies or interstitials in a volume V_m, they can be written by taking into account, first the volume sources S_L and S_I associated with the creation and annihilation of vacancies and interstitials in the volume V_m, and second the flux of vacancies and interstitials entering or leaving said volume V_m via its boundary surface with relative velocity $\Delta\vec{\varphi}_L$ or $\Delta\vec{\varphi}_I$

$$\left\{\begin{array}{l} \dfrac{d}{dt}\iiint_{V_m} n_L\,dV = -\oiint_{S_m} n_L\Delta\vec{\varphi}_L\,d\vec{S} + \iiint_{V_m} S_L\,dV \\ \dfrac{d}{dt}\iiint_{V_m} n_I\,dV = -\oiint_{S_m} n_I\Delta\vec{\varphi}_I\,d\vec{S} + \iiint_{V_m} S_I\,dV \end{array}\right. \tag{5.72}$$

In their local form, in the absolute frame $Q\xi_1\xi_2\xi_3$, and by using the formulas of derivation of an integral on a mobile volume, the equations of continuity for the density of sites of lattice and for the density of vacancies and interstitials can be written as

$$\left\{\begin{array}{l} \dfrac{\partial n}{\partial t} + \operatorname{div}\left(n\vec{\phi}\right) = 0 \\ \dfrac{\partial n_L}{\partial t} + \operatorname{div}\left(n_L\vec{\phi}\right) = S_L - \operatorname{div}\left(n_L\Delta\vec{\varphi}_L\right) \\ \dfrac{\partial n_I}{\partial t} + \operatorname{div}\left(n_I\vec{\phi}\right) = S_I - \operatorname{div}\left(n_I\Delta\vec{\varphi}_I\right) \end{array}\right. \tag{5.73}$$

The equations of continuity can be written in their classic form, by making explicit the absolute velocity of vacancies and interstitials thanks to relations *(5.70)*

$$\begin{cases} \dfrac{\partial n}{\partial t} = -\operatorname{div}\left(n\vec{\phi}\right) \\ \dfrac{\partial n_L}{\partial t} = S_L - \operatorname{div}\left(n_L\vec{\phi}_L\right) \\ \dfrac{\partial n_I}{\partial t} = S_I - \operatorname{div}\left(n_I\vec{\phi}_I\right) \end{cases} \tag{5.74}$$

The equation of continuity for the density of lattice sites n does not contain a source term for lattice sites ($S_n = 0$), so that the topological equations of an elastic lattice with self-diffusion remain exactly the same to that established in the case of an elastic lattice. The first equations of continuity *(5.73)* obtained for n_L and n_I can be further transformed by replacing n_L and n_I by nC_L et nC_I and using the material derivative

$$\begin{cases} \dfrac{d(nC_L)}{dt} = S_L - nC_L \operatorname{div}\vec{\phi} - \operatorname{div}\left(n_L\Delta\vec{\varphi}_L\right) \\ \dfrac{d(nC_I)}{dt} = S_I - nC_I \operatorname{div}\vec{\phi} - \operatorname{div}\left(n_I\Delta\vec{\varphi}_I\right) \end{cases} \tag{5.75}$$

Thanks to the following relation, it is easy to verify that for $S_n = 0$

$$\frac{d(nC_L)}{dt} = n\frac{dC_L}{dt} + C_L\frac{dn}{dt} = n\frac{dC_L}{dt} - nC_L \operatorname{div}\vec{\phi} \tag{5.76}$$

Equations that directly describe the variations of the atomic concentrations C_L and C_I along the path, are obtained

$$\begin{cases} n\dfrac{dC_L}{dt} = S_L - \operatorname{div}\left(n_L\Delta\vec{\varphi}_L\right) \\ n\dfrac{dC_I}{dt} = S_I - \operatorname{div}\left(n_I\Delta\vec{\varphi}_I\right) \end{cases} \tag{5.77}$$

We define the *fluxes of diffusion* $\vec{J}_L$ and $\vec{J}_I$ of vacancies and interstitials with respect to the lattice with the following relations

$$\begin{cases} \vec{J}_L = nC_L\Delta\vec{\varphi}_L = nC_L\left(\vec{\phi}_L - \vec{\phi}\right) = nC_L\left(\vec{\varphi}_L - \vec{\varphi}\right) \\ \vec{J}_I = nC_I\Delta\vec{\varphi}_I = nC_I\left(\vec{\phi}_I - \vec{\phi}\right) = nC_I\left(\vec{\varphi}_I - \vec{\varphi}\right) \end{cases} \tag{5.78}$$

We will now rewrite the *equations of diffusion of vacancies and interstitials* in their final form

$$\begin{cases} n\dfrac{dC_L}{dt} = S_L - \operatorname{div}\vec{J}_L \\ n\dfrac{dC_I}{dt} = S_I - \operatorname{div}\vec{J}_I \end{cases} \tag{5.79}$$

These equations, which use the material derivative are valid both in the absolute frame $Q\xi_1\xi_2\xi_3$ as well as the local frame $Ox_1x_2x_3$. Indeed we can verify that the equations *(5.79)* can be deduced from the equations of continuity of integral relations *(5.72)*, but written in the local frame $Ox_1x_2x_3$, namely

$$\begin{cases} \dfrac{\partial n}{\partial t} = -\operatorname{div}(n\vec{\varphi}) \\ \dfrac{\partial n_L}{\partial t} = S_L - \operatorname{div}(n_L\vec{\varphi}_L) \\ \dfrac{\partial n_I}{\partial t} = S_I - \operatorname{div}(n_I\vec{\varphi}_I) \end{cases} \quad (5.80)$$

Chemical potential and thermo-kinetic equation in the presence of self-diffusion

In an elastic, anelastic and plastic lattice, with self-diffusion, it is clear that the atomic concentrations C_L and C_I of vacancies and self-interstitials must also influence the energetic state of the lattice, so that we must complete the state function of internal energy *(5.58)*, and write it now as a function of thermodynamical variables as follows

$$u = u\left(\alpha_{ij}^{el}, \omega_k^{el}, \tau^{el}, \alpha_{ij}^{an}, \omega_k^{an}, C_L, C_I, s\right) \quad (5.81)$$

The differential of the state function u along the path of the sites of lattice can be written

$$du = \frac{\partial u}{\partial \alpha_{ij}^{el}} d\alpha_{ij}^{el} + \frac{\partial u}{\partial \omega_k^{el}} d\omega_k^{el} + \frac{\partial u}{\partial \tau^{el}} d\tau^{el} + \frac{\partial u}{\partial \alpha_{ij}^{an}} d\alpha_{ij}^{an} + \frac{\partial u}{\partial \omega_k^{an}} d\omega_k^{an} + \frac{\partial u}{\partial s} ds + \frac{\partial u}{\partial C_L} dC_L + \frac{\partial u}{\partial C_I} dC_I \quad (5.82)$$

This differential allows us to introduce the mechanical potentials of stress, as well as the thermal potential of temperature $T = \partial u / \partial s$. By analogy, it is now possible to introduce new potentials associated to the partial derivatives $\partial u / \partial C_L$ and $\partial u / \partial C_I$. These potentials will be called the *chemical potentials* μ_L and μ_I of vacancies and interstitials, respectively. Bearing in mind relation *(5.42)*, we can now write the differential

$$n\,du = \vec{s}_k d\vec{\beta}_k^{el} + m_k \omega_k^{el} - p\,d\tau^{el} + \vec{s}_k^{cons} d\vec{\beta}_k^{an} + m_k^{cons} \omega_k^{an} + nT\,ds + n\mu_L dC_L + n\mu_I dC_I \quad (5.83)$$

The thermo-kinetic equation of such a solid lattice can be immediately deduced

$$n\frac{du}{dt} = \vec{s}_k \frac{d\vec{\beta}_k^{el}}{dt} + \vec{m}\frac{d\vec{\omega}^{el}}{dt} - p\frac{d\tau^{el}}{dt} + \vec{s}_k^{cons} \frac{d\vec{\beta}_k^{an}}{dt} + \vec{m}^{cons}\frac{d\vec{\omega}^{an}}{dt} + n\mu_L \frac{dC_L}{dt} + n\mu_I \frac{dC_I}{dt} + nT\frac{ds}{dt} \quad (5.84)$$

Application of the newtonian principle of mass conservation

Indeed, the equations of continuity *(5.74)* and *(5.80)* which were obtained in the case of a self-diffusive lattice can be written taking into account the source S_n of sites of the lattice, due to the plastic component of volume expansion of the solid lattice. The first equation describing n is written in $Q\xi_1\xi_2\xi_3$ and $Ox_1x_2x_3$ respectively

$$\frac{\partial n}{\partial t} = n\frac{d\tau^{pl}}{dt} - \operatorname{div}(n\vec{\phi}) = n\frac{d\tau^{pl}}{dt} - \operatorname{div}(n\vec{\varphi}) \quad (5.85)$$

while relations *(5.74)* and *(5.80)* for the density of vacancies n_L and interstitials n_I remain unchanged

$$\begin{cases} \dfrac{\partial n_L}{\partial t} = S_L - \operatorname{div}\left(n_L \vec{\phi}_L\right) = S_L - \operatorname{div}\left(n_L \vec{\varphi}_L\right) \\ \dfrac{\partial n_I}{\partial t} = S_I - \operatorname{div}\left(n_I \vec{\phi}_I\right) = S_I - \operatorname{div}\left(n_I \vec{\varphi}_I\right) \end{cases} \tag{5.86}$$

The density of mass ρ of the solid lattice containing n_L vacancies and n_I interstitials per unit volume can be simply expressed

$$\rho = m\left(n + n_I - n_L\right) = mn\left(1 + C_I - C_L\right) \tag{5.87}$$

We can establish an equation of continuity for the mass density ρ in the absolute frame $Q\xi_1\xi_2\xi_3$ from the equations *(5.85) and (5.86)*

$$\frac{\partial \rho}{\partial t} = \frac{\partial}{\partial t}\left[m\left(n + n_I - n_L\right)\right] = -\operatorname{div}\left[m\left(n\vec{\phi} + n_I\vec{\phi}_I - n_L\vec{\phi}_L\right)\right] + m\left(S_I - S_L\right) + mn\frac{d\tau^{pl}}{dt} \tag{5.88}$$

We can compare this equation with the fundamental equation of continuity of mass obtained from relations *(4.5)* and *(4.14)*

$$\frac{\partial \rho}{\partial t} = S_m - \operatorname{div}\left(n\vec{p}\right) = S_m - \operatorname{div}\left(\rho\vec{\phi} + \vec{J}_m\right) \tag{5.89}$$

From this comparison we have a first relation giving the source of mass S_m per unit volume

$$S_m = m\left(S_I - S_L\right) + mn\frac{d\tau^{pl}}{dt} = m\left(S_I - S_L\right) + mS_n \tag{5.90}$$

If we admit that *principle of conservation of mass* cannot be violated, which we capture with the following hypothesis

Hypothesis: $S_m \equiv 0$ *(5.91)*

then we have the following relation between S_n, S_L et S_I

$$S_n = n\frac{d\tau^{pl}}{dt} = S_L - S_I \tag{5.92}$$

In this expression, the terms of source of vacancies S_L and of interstitial source S_I contain in fact 2 contributions: *the spontaneous creation and annihilation of pairs of 'vacancy-interstitial'*, which we will write as S_{I-L}, and *the creation and annihilation of vacancies and/or interstitials by the process of plastic deformation*, which will be written S_L^{pl} and S_I^{pl}. The total sources S_L and S_I can be decomposed as such

$$S_L = S_{I-L} + S_L^{pl} \quad ; \quad S_I = S_{I-L} + S_I^{pl} \tag{5.93}$$

These values of S_L and S_I can be put in equation *(5.92)*

$$S_n = n\frac{d\tau^{pl}}{dt} = S_L^{pl} - S_I^{pl} \tag{5.94}$$

which translates the fact that plastic distortions with non-null trace, can either be sources or sinks of vacancies and/or interstitials.

Self-diffusion in the presence of sources of lattice sites by plasticity

From the equations of continuity *(5.86)* obtained for n_L and n_I, by replacing n_L and n_I by nC_L and nC_I, we have

$$\begin{cases} d(nC_L)/dt = S_L - nC_L \operatorname{div}\vec{\phi} - \operatorname{div}(n_L \Delta\vec{\varphi}_L) \\ d(nC_I)/dt = S_I - nC_I \operatorname{div}\vec{\phi} - \operatorname{div}(n_I \Delta\vec{\varphi}_I) \end{cases} \tag{5.95}$$

With relation *(5.66)* which contains a non null source S_n of lattice sites, we do the following operations.

$$\begin{cases} \dfrac{d(nC_L)}{dt} = n\dfrac{dC_L}{dt} + C_L\dfrac{dn}{dt} = n\dfrac{dC_L}{dt} + C_L S_n - C_L n \operatorname{div}\vec{\phi} \\ \dfrac{d(nC_I)}{dt} = n\dfrac{dC_I}{dt} + C_I\dfrac{dn}{dt} = n\dfrac{dC_I}{dt} + C_I S_n - C_I n \operatorname{div}\vec{\phi} \end{cases} \tag{5.96}$$

we introduce, again, the fluxes of diffusion $\vec{J}_L$ and $\vec{J}_I$ of vacancies and interstitials, with respect to the lattice, defined in relations *(5.59)*, and the equations of self-diffusion of vacancies and interstitials in the lattice take the following form which is valid both in the global frame $Q\xi_1\xi_2\xi_3$ or the local frame $Ox_1x_2x_3$

$$\begin{cases} n\dfrac{dC_L}{dt} = S_L - C_L S_n - \operatorname{div}\vec{J}_L \\ n\dfrac{dC_I}{dt} = S_I - C_I S_n - \operatorname{div}\vec{J}_I \end{cases} \tag{5.97}$$

Comparatively to the equations of self-diffusion *(5.60)* obtained previously, we have here the extra terms $-C_L S_n$ and $-C_I S_n$ associated with source S_n of lattice sites. By using relations *(5.94) and (5.95)*, we explicitly exhibit the sources and sinks of point defects

$$\begin{cases} n\dfrac{dC_L}{dt} = \left(S_{I\text{-}L} + S_L^{pl}\right) - C_L\left(S_L^{pl} - S_I^{pl}\right) - \operatorname{div}\vec{J}_L \\ n\dfrac{dC_I}{dt} = \left(S_{I\text{-}L} + S_I^{pl}\right) - C_I\left(S_L^{pl} - S_I^{pl}\right) - \operatorname{div}\vec{J}_I \end{cases} \tag{5.98}$$

5.4 - Newtonian dynamic of a lattice

If the lattice considered behaves in a newtonian manner in the absolute frame of the ***GO***, we can express its local linear momentum, its local kinetic energy and the source of work of external forces.

The linear momentum

From comparing *(5.88)* and *(5.89)*, there emerges another interesting relation giving the average linear momentum $\vec{p}$ per lattice site

$$\vec{p} = m\left(\vec{\phi} + C_I\vec{\phi}_I - C_L\vec{\phi}_L\right) = \frac{1}{n}\left(\rho\vec{\phi} + \vec{J}_m\right) \tag{5.99}$$

The linear momentum $\vec{p}$ can be written in a slightly different manner by decomposing the velocities $\vec{\phi}_L$ and $\vec{\phi}_I$ thanks to relations *(5.69)*

$$\vec{p} = m\left(\vec{\phi} + C_I - C_L\right)\vec{\phi} + \frac{m}{n}\left(nC_I\Delta\vec{\varphi}_I - nC_L\Delta\vec{\varphi}_L\right) \tag{5.100}$$

We introduce the diffusion fluxes $\vec{J}_L$ and $\vec{J}_I$ of vacancies and interstitials defined in *(5.78), and* we deduce that

$$\vec{p} = m\vec{\phi} + m\left(C_I - C_L\right)\vec{\phi} + \frac{m}{n}\left(\vec{J}_I - \vec{J}_L\right) \tag{5.101}$$

and thus we see that the average linear momentum transported per site of the lattice is the sum of the linear momentum $m\vec{\phi}$ associated with the absolute velocity of the lattice, the linear momentum $m\left(C_I - C_L\right)\vec{\phi}$ due to the transport of vacancies and interstitials by the lattice and finally the linear momentum $m\left(\vec{J}_I - \vec{J}_L\right)/n$ associated to the self-diffusion of vacancies and interstitials by the lattice

We use the mass density ρ *(5.87)*, to write $\vec{p}$ as

$$\vec{p} = \frac{1}{n}\left[\rho\vec{\phi} + m\left(\vec{J}_I - \vec{J}_L\right)\right] \tag{5.102}$$

By comparing this relation with relation *(5.90)*, we obtain

$$\vec{J}_m = m\left(\vec{J}_I - \vec{J}_L\right) \tag{5.103}$$

which also means that the average linear momentum $\vec{p}$ associated to each site of the lattice is the sum of a mass transport $\rho\vec{\phi}/n$ deriving from the local entrained movement of the lattice charged with vacancies and interstitials, and a mass transport $m\left(\vec{J}_I - \vec{J}_L\right)/n = \vec{J}_m/n$ due to the self-diffusion of intrinsic point defects of vacancy and interstitial types.

It is possible to derive the value of $\vec{p}$ by another mean. In the lattice there are n_I interstitial particles with relative velocity $\Delta\vec{\varphi}_I$, that transport the mass in the direction $\Delta\vec{\varphi}_I$. Furthermore, the movement of the n_L vacancies with velocity $\Delta\vec{\varphi}_L$ is associated with a mass flux of n_L particles of the lattice with velocity $-\Delta\vec{\varphi}_L$, that transport mass in the direction opposite to $\Delta\vec{\varphi}_L$. The rest of the particles of the lattice, meaning the $n - n_L$ particles situated on nodes, from which we have subtracted, again, the n_L particles situated on nodes associated with vacancy jumps, move with a null velocity with respect to the lattice, hence the following summary of the dynamic situation of the particles by unit volume of the media:

- There exists $n - 2n_L$ particles with null velocity with respect to the lattice and thus with absolute velocity $\vec{\phi}$,

- There exists n_I particles with relative velocity $\Delta\vec{\varphi}_I$ with respect to the lattice and thus with absolute velocity $\vec{\phi} + \Delta\vec{\varphi}_I = \vec{\phi}_I$,

- There exists n_L particles with relative velocity $-\Delta\vec{\varphi}_L$ with respect to the lattice, and thus with absolute velocity $\vec{\phi} - \Delta\vec{\varphi}_L = 2\vec{\phi} - \vec{\phi}_L$

The flux of mass, or density of linear momentum $\vec{P}_m$ at point $\vec{\xi}$ and at instant t in the absolute frame $Q\xi_1\xi_2\xi_3$, can be deduced from this dynamical situation of the particles, and we deduce the expressions *(5.99) to (5.102)*

$$\vec{P}_m = m(n-2n_L)\vec{\phi} + mn_L(2\vec{\phi}-\vec{\phi}_L) + mn_I\vec{\phi}_I$$
$$= m(n+n_I-n_L)\vec{\phi} + mn_I\Delta\vec{\varphi}_I - mn_L\Delta\vec{\varphi}_L = \rho\vec{\phi} + m(\vec{J}_I-\vec{J}_L) = n\vec{p} \quad (5.104)$$

Kinetic energy

We can also deduce the average kinetic energy e_{cin} per site of the lattice. Indeed, the volume density of kinetic energy E_{cin} is simply written from the absolute velocities and the densities of the various flavors of defects

$$E_{cin} = (n-2n_L)\frac{1}{2}m\vec{\phi}^2 + n_I\frac{1}{2}m\vec{\phi}_I^2 + n_L\frac{1}{2}m(2\vec{\phi}-\vec{\phi}_L)^2 \quad (5.105)$$

We deduce the average kinetic energy e_{cin} per site by division per n

$$e_{cin} = (1-2C_L)\frac{1}{2}m\vec{\phi}^2 + C_I\frac{1}{2}m\vec{\phi}_I^2 + C_L\frac{1}{2}m(2\vec{\phi}-\vec{\phi}_L)^2 \quad (5.106)$$

The average kinetic energy per site of the lattice is equal to the kinetic energy of a site of the lattice minus a fraction corresponding to the kinetic energy of a couple of particles subtracted from the lattice to have vacancy diffusion, plus a fraction corresponding to the kinetic energy of interstitial and a fraction corresponding to the kinetic energy of vacancy jumps.
The expression e_{cin} can be written by replacing $\vec{\phi}$ by $\vec{\phi}_L - \Delta\vec{\varphi}_L$ in the following term

$$\frac{1}{2}m(2\vec{\phi}-\vec{\phi}_L)^2 - m\vec{\phi}^2 = -\frac{1}{2}m(\vec{\phi}_L^2 - 2\Delta\vec{\varphi}_L^2) \quad (5.107)$$

which allows us to write

$$e_{cin} = \frac{1}{2}m\left[\vec{\phi}^2 + C_I\vec{\phi}_I^2 - C_L(\vec{\phi}_L^2 - 2\Delta\vec{\varphi}_L^2)\right] \quad (5.108)$$

The source of work of external forces

With respect to the source of work S_w^{ext} due to a field of external forces, it can be calculated if we know the nature of the field force. For example, if we suppose that the lattice is in a *constant gravity field* $\vec{g}$, we will have

$$S_w^{ext} = m\left[(n-2n_L)\vec{\phi}\vec{g} + n_I\vec{\phi}_I\vec{g} + n_L(2\vec{\phi}-\vec{\phi}_L)\vec{g}\right] = \vec{P}_m\vec{g} = n\vec{p}\vec{g} \quad (5.109)$$

By introducing the expression *(5.102)* for $\vec{p}$, we have the following expression for the source of work for a constant gravity field

$$S_w^{ext} = \rho\vec{\phi}\vec{g} + (\vec{J}_I - \vec{J}_L)m\vec{g} \quad (5.110)$$

Chapter 6

Evolution equations for a newtonian lattice

The introduction of the elasticity, of the dissipative mechanical processes of anelasticity and plasticity, of the processes of mass transport and of the newtonian dynamics of the lattice in the fundamental equations of continuity for energy and entropy, allows us to re-derive the entire set of equations that describe the spatiotemporal evolution of said lattice. These behavioral equations are of two kinds: the *fundamental equations* which stay identical to themselves regardless of the lattice under consideration and the *phenomenological equations* which do depend (and describe) the particular nature of the lattice under consideration.

6.1 - Equations of evolution of a lattice

The first part of the equation describing the principle of energy continuity in an elastic, anelastic and plastic solid lattice, in the presence of vacancies and self-interstitials, is written using the relations *(5.108)* for e_{cin} and *(5.84)* for du/dt, all the while assuming here, in all generality, that internal energy u can depend on $\vec{\omega}^{él}$ and $\vec{\omega}^{an}$

$$\begin{aligned} n\frac{de_{cin}}{dt}+n\frac{du}{dt} &= nm\vec{\phi}\frac{d\vec{\phi}}{dt}+n\left[\mu_I+\frac{1}{2}m\vec{\phi}_I^2\right]\frac{dC_I}{dt}+nmC_I\boxed{\vec{\phi}_I}\frac{d\vec{\phi}_I}{dt} \\ &+n\left[\mu_L-\frac{1}{2}m\left(\vec{\phi}_L^2-2\Delta\vec{\varphi}_L^2\right)\right]\frac{dC_L}{dt}-nmC_L\boxed{\vec{\phi}_L}\frac{d\vec{\phi}_L}{dt}+2nmC_L\Delta\vec{\varphi}_L\frac{d\Delta\vec{\varphi}_L}{dt} \\ &+\vec{s}_k\frac{d\vec{\beta}_k^{él}}{dt}+\vec{m}\frac{d\vec{\omega}^{él}}{dt}-p\frac{d\tau^{él}}{dt}+\vec{s}_k^{\,cons}\frac{d\vec{\beta}_k^{an}}{dt}+\vec{m}^{cons}\frac{d\vec{\omega}^{an}}{dt}+nT\frac{ds}{dt} \end{aligned} \qquad (6.1)$$

The terms in boxes can be decomposed with relations *(5.69)*, and it is possible to introduce the *generalized chemical energies* $\overset{*}{\mu}_L$ and $\overset{*}{\mu}_I$ of vacancies and interstitials, defined as the sum of chemical potentials and kinetic energies associated with each of the diffusing species

$$\overset{*}{\mu}_L=\mu_L-\frac{1}{2}m\left(\vec{\phi}_L^2-2\Delta\vec{\varphi}_L^2\right)=\mu_L+e_{cin}^L \quad \text{and} \quad \overset{*}{\mu}_I=\mu_I+\frac{1}{2}m\vec{\phi}_I^2=\mu_I+e_{cin}^I \qquad (6.2)$$

Thanks to relation *(5.60)*, the expression

$$\vec{s}_k\frac{d\vec{\beta}_k^{él}}{dt}+\vec{m}\frac{d\vec{\omega}^{él}}{dt}-p\frac{d\tau^{él}}{dt}+\vec{s}_k^{\,cons}\frac{d\vec{\beta}_k^{an}}{dt}+\vec{m}^{cons}\frac{d\vec{\omega}^{an}}{dt} \qquad (6.3)$$

can be replaced by the following, which has the advantage of explicitly showing $\vec{\beta}_i$, $\vec{\omega}$ and τ

$$\vec{s}_k\frac{d\vec{\beta}_k}{dt}+\vec{m}\frac{d\vec{\omega}}{dt}-p\frac{d\tau}{dt}+\left(-\vec{s}_k^{\,dis}\frac{d\vec{\beta}_k^{an}}{dt}-\vec{m}^{dis}\frac{d\vec{\omega}^{an}}{dt}-\vec{s}_k\frac{d\vec{\beta}_k^{pl}}{dt}-\vec{m}\frac{d\vec{\omega}^{pl}}{dt}+p\frac{d\tau^{pl}}{dt}\right) \qquad (6.4)$$

We then have

$$n\frac{de_{cin}}{dt}+n\frac{du}{dt}=nm\vec{\phi}\left(\frac{d\vec{\phi}}{dt}+C_I\frac{d\vec{\phi}_I}{dt}-C_L\frac{d\vec{\phi}_L}{dt}\right)+m\vec{J}_I\frac{d\vec{\phi}_I}{dt}-m\vec{J}_L\frac{d}{dt}\left(\vec{\phi}_L-2\Delta\vec{\varphi}_L\right)$$
$$+nT\frac{ds}{dt}+n\overset{*}{\mu}_I\frac{dC_I}{dt}+n\overset{*}{\mu}_L\frac{dC_L}{dt}+\vec{s}_k\frac{d\vec{\beta}_k}{dt}+\vec{m}\frac{d\vec{\omega}}{dt}-p\frac{d\tau}{dt}$$
$$+\left(-\vec{s}_k^{\,dis}\frac{d\vec{\beta}_k^{an}}{dt}-\vec{m}^{dis}\frac{d\vec{\omega}^{an}}{dt}-\vec{s}_k\frac{d\vec{\beta}_k^{pl}}{dt}-\vec{m}\frac{d\vec{\omega}^{pl}}{dt}+p\frac{d\tau^{pl}}{dt}\right) \tag{6.5}$$

In the second line of this expression, the material derivatives of the topological tensors and the atomic concentrations are replaceable using the equations of geometro-kinetic *(5.65)* and the equations of diffusion *(5.98)*. We then have

$$n\frac{de_{cin}}{dt}+n\frac{du}{dt}=\left[nm\frac{d\vec{\phi}}{dt}+nmC_I\frac{d\vec{\phi}_I}{dt}-nmC_L\frac{d\vec{\phi}_L}{dt}-\sum_k\vec{e}_k\operatorname{div}\vec{s}_k+\frac{1}{2}\overrightarrow{\operatorname{rot}}\,\vec{m}+\overrightarrow{\operatorname{grad}}p\right]\vec{\phi}$$
$$-\operatorname{div}\left[\overset{*}{\mu}_L\vec{J}_L+\overset{*}{\mu}_I\vec{J}_I-\phi_k\vec{s}_k-\frac{1}{2}\left(\vec{\phi}\wedge\vec{m}\right)+p\vec{\phi}\right]$$
$$+\left[\begin{array}{l}\overset{*}{\mu}_L\left(S_L-C_LS_n\right)+\overset{*}{\mu}_I\left(S_I-C_IS_n\right)\\[4pt] +\vec{J}_I\left(\overrightarrow{\operatorname{grad}}\overset{*}{\mu}_I+m\dfrac{d\vec{\phi}_I}{dt}\right)+\vec{J}_L\left(\overrightarrow{\operatorname{grad}}\overset{*}{\mu}_L-m\dfrac{d\left(\vec{\phi}_L-2\Delta\vec{\varphi}_L\right)}{dt}\right)\\[4pt] -\vec{s}_k^{\,dis}\dfrac{d\vec{\beta}_k^{an}}{dt}-\vec{m}^{dis}\dfrac{d\vec{\omega}^{an}}{dt}-\vec{s}_k\dfrac{d\vec{\beta}_k^{pl}}{dt}-\vec{m}\dfrac{d\vec{\omega}^{pl}}{dt}+p\dfrac{d\tau^{pl}}{dt}+nT\dfrac{ds}{dt}\end{array}\right] \tag{6.6}$$

This expression must be compared to the second part of the first principle *(4.23)*, in which the sources S_n is not null

$$n\frac{de_{cin}}{dt}+n\frac{du}{dt}=S_w^{ext}-\operatorname{div}\vec{J}_w-\operatorname{div}\vec{J}_q-uS_n-e_{cin}S_n \tag{6.7}$$

which can be written, by introducing a source of work *(5.110)* due to a constant gravity field

$$n\frac{de_{cin}}{dt}+n\frac{du}{dt}=\left[\rho\vec{g}\right]\vec{\phi}-\operatorname{div}\left[\vec{J}_w\right]+\left[\left(\vec{J}_I-\vec{J}_L\right)m\vec{g}-\operatorname{div}\vec{J}_q-uS_n-e_{cin}S_n\right] \tag{6.8}$$

From this comparison, we have 3 important equations that must be satisfied at all times in an elastic, anelastic, plastic and self-diffusing lattice.

The lattice's Newton equation

The comparison of the vectorial expressions in brackets, in the scalar product with $\vec{\phi}$, leads to the following equation

$$nm\frac{d\vec{\phi}}{dt}+nmC_I\frac{d\vec{\phi}_I}{dt}-nmC_L\frac{d\vec{\phi}_L}{dt}-\sum_k\vec{e}_k\operatorname{div}\vec{s}_k+\frac{1}{2}\overrightarrow{\operatorname{rot}}\,\vec{m}+\overrightarrow{\operatorname{grad}}\,p=\rho\vec{g} \tag{6.9}$$

By using the material derivative of the average linear momentum $\vec{p}$ per site given by expression *(5.99)*

$$\frac{d\vec{p}}{dt}=m\left(\frac{d\vec{\phi}}{dt}+C_I\frac{d\vec{\phi}_I}{dt}+\vec{\phi}_I\frac{dC_I}{dt}-C_L\frac{d\vec{\phi}_L}{dt}-\vec{\phi}_L\frac{dC_L}{dt}\right) \tag{6.10}$$

equation *(6.9)* becomes a dynamic equation for the linear momentum $\vec{p}$

$$n\frac{d\vec{p}}{dt}=\rho\vec{g}+\sum_k \vec{e}_k \operatorname{div}\vec{s}_k-\frac{1}{2}\overrightarrow{\operatorname{rot}}\ \vec{m}-\overrightarrow{\operatorname{grad}}p+nm\vec{\phi}_I\frac{dC_I}{dt}-nm\vec{\phi}_L\frac{dC_L}{dt} \qquad (6.11)$$

This equation is nothing more than the *Newton equation* applied to the lattice in eulerian coordinates. This equation describes the dynamics of the solid, which depends on gravity $\rho\vec{g}$, volume forces with stress potentials $\vec{s}_i$, $\vec{m}$ and p, and additional linear momentum brought about by increase of the number of point defects in the lattice. Expression *(6.11)* represents in fact the Newton equation we can obtain by using the thermo-kinetic equation *(5.84)*. However, we can verify while working on thermo-kinetic equations *(5.13)* and (5.7), that the same Newton equation can be written under two different forms, which use mechanical stress potentials $\vec{\sigma}_i$ and $\vec{\Sigma}_i$

$$n\frac{d\vec{p}}{dt}=\rho\vec{g}+\sum_k \vec{e}_k \operatorname{div}\vec{\sigma}_k-\frac{1}{2}\overrightarrow{\operatorname{rot}}\,\vec{m}+nm\vec{\phi}_I\frac{dC_I}{dt}-nm\vec{\phi}_L\frac{dC_L}{dt} \qquad (6.12)$$

$$n\frac{d\vec{p}}{dt}=\rho\vec{g}+\sum_k \vec{e}_k \operatorname{div}\vec{\Sigma}_k+nm\vec{\phi}_I\frac{dC_I}{dt}-nm\vec{\phi}_L\frac{dC_L}{dt} \qquad (6.13)$$

The flux of work and the surface force

The comparison of the vectorial terms of the divergence operator, leads to an expression for the *flux of work* within the lattice

$$\vec{J}_w=\mu_L^*\vec{J}_L+\mu_I^*\vec{J}_I-\phi_k\vec{s}_k-\frac{1}{2}\left(\vec{\phi}\wedge\vec{m}\right)+p\vec{\phi} \qquad (6.14)$$

This flux corresponds to an energy propagation within the solid. We will remark that the contribution of $\vec{m}$ is a vector totally analogous to the *Poynting vector* of electromagnetism. The expression *(6.14)* allows us to write the surface integral that appears in expression *(4.18)*

$$\begin{aligned}-\oiint_{S_m}\vec{J}_w\,d\vec{S}&=-\oiint_{S_m}\left(\mu_L^*\vec{J}_L+\mu_I^*\vec{J}_I-\phi_k\vec{s}_k-\left(\vec{\phi}\wedge\vec{m}\right)/2+p\vec{\phi}\right)d\vec{S}\\&=\oiint_{S_m}\vec{\phi}\left[\sum_k\vec{e}_k\left(\vec{s}_k\vec{n}\right)+\left(\vec{m}\wedge\vec{n}\right)/2-\vec{n}p\right]dS-\oiint_{S_m}\left[\mu_L^*\vec{J}_L+\mu_I^*\vec{J}_I\right]\vec{n}\,dS\end{aligned} \qquad (6.15)$$

The first integral introduces the notion of a *surface force* $\vec{F}_S$, where $\vec{n}$ corresponds to the unit vector normal to the surface under consideration, and the second integral shows the appearance of *energy fluxes associated with the diffusion fluxes of vacancies and interstitials*

$$\begin{cases}\vec{F}_S=\sum_k\vec{e}_k\left(\vec{s}_k\vec{n}\right)+\frac{1}{2}\left(\vec{m}\wedge\vec{n}\right)-\vec{n}p\\ \vec{J}_w^L=\mu_L^*\vec{J}_L=\left[\mu_L-\frac{1}{2}m\left(\vec{\phi}_L^2-2\Delta\vec{\varphi}_L^2\right)\right]nC_L\Delta\vec{\varphi}_L\\ \vec{J}_w^I=\mu_I^*\vec{J}_I=\left[\mu_I+\frac{1}{2}m\vec{\phi}_I^2\right]nC_I\Delta\vec{\varphi}_I\end{cases} \qquad (6.16)$$

The last two relations show that the flux of work associated with the flux of diffusion of vacancies and interstitials contain both a flux of potential energy (the chemical potentials μ_L and μ_I

of vacancies and interstitials) and fluxes of kinetic energy.

The first relation could be calculated from the other mechanical potentials and would simply lead to the following expressions

$$\vec{F}_S = \sum_k \vec{e}_k \left(\vec{\sigma}_k \vec{n}\right) + \frac{1}{2}\left(\vec{m} \wedge \vec{n}\right) \tag{6.17}$$

$$\vec{F}_S = \sum_k \vec{e}_k \left(\vec{\Sigma}_k \vec{n}\right) \tag{6.18}$$

The three equations for the surface force allow us to specify the *limit conditions* external to the solid. At the limits of the medium, we can link the external forces by unit surface $\vec{F}_S$ to the stress tensors $\vec{\Sigma}_i$, $\vec{\sigma}_i$ or $\vec{s}_i$, the torque vector $\vec{m}$ and the scalar of pressure p.

From these expressions of the surface force $\vec{F}_S$ we easily deduce the physical interpretation, completely classical, of the stress potentials $\vec{\Sigma}_i$, $\vec{\sigma}_i$ or $\vec{s}_i$. Dimensionally they are forces by unit surface: the component Σ_{ij} of tensor $\vec{\Sigma}_i$ is equal to the *force per unit surface applied in direction i on a unit surface perpendicular to direction j*. Schematically, we can represent the whole set of forces and components on a unit cube as in figure 6.1.

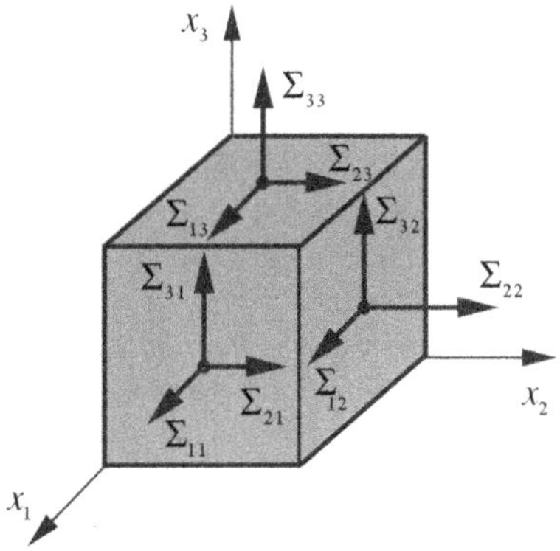

Figure 6.1 *- physical interpretation of the components of the stress tensor* $\vec{\Sigma}_i$

The component of the stress tensor $\vec{\Sigma}_i$ have the dimension of a stress, a force per unit surface, which is why we called $\vec{\Sigma}_i$ the *stress tensor* conjugated to $\vec{\beta}_i$. The same interpretation can be given for tensors $\vec{\sigma}_i$ and $\vec{s}_i$, meaning the stress tensors conjugated to $\vec{\varepsilon}_i$ and $\vec{\alpha}_i$ respectively.

With regards to vector $\vec{m}$, it can be interpreted as in the schematic figure 6.2. The vector $\vec{m}$ represents in fact the local torque (moment) applied to the continuous medium by the surface forces Σ_{ij}. $\vec{m}$ is the torque vector conjugated to the vector $\vec{\omega}$ of rotations, and so we will call $\vec{m}$ the *torsor of moments* or *the torque vector* conjugated to $\vec{\omega}$.

The scalar p is defined as the average of the three stresses of compression Σ_{ii} or σ_{ii}, with a changed sign. This scalar has the dimension of a hydrostatic pressure and we will call it the *scalar of pressure* conjugated to τ.

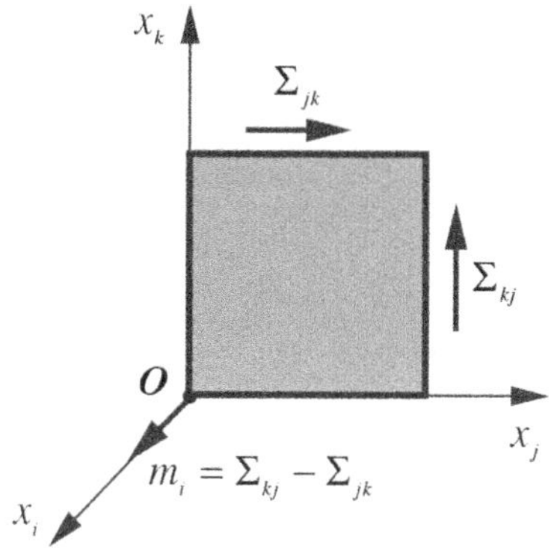

***Figure 9.3** - physical interpretation of the components of the torque vector or torsor of moment $\vec{m}$*

Finally equations *(6.16)* and *(6.18)* allow us to deduce the *pass-through conditions* at the interface of the two medias (see figure 3.9) for the stress potentials $\vec{\Sigma}_i$, $\vec{\sigma}_i$ or $\vec{s}_i$, for the torque vector $\vec{m}$ and for the scalar pressure p, by requiring that the forces be equal across the surface, with the three following equations

$$\sum_k \vec{e}_k\left(\vec{s}_k^{(1)}\vec{n}\right)+\frac{1}{2}\left(\vec{m}^{(1)}\wedge\vec{n}\right)-\vec{n}p^{(1)}=\sum_k \vec{e}_k\left(\vec{s}_k^{(2)}\vec{n}\right)+\frac{1}{2}\left(\vec{m}^{(2)}\wedge\vec{n}\right)-\vec{n}p^{(2)} \qquad (6.19)$$

$$\sum_k \vec{e}_k\left(\vec{\sigma}_k^{(1)}\vec{n}\right)+\frac{1}{2}\left(\vec{m}^{(1)}\wedge\vec{n}\right)=\sum_k \vec{e}_k\left(\vec{\sigma}_k^{(2)}\vec{n}\right)+\frac{1}{2}\left(\vec{m}^{(2)}\wedge\vec{n}\right) \qquad (6.20)$$

$$\sum_k \vec{e}_k\left(\vec{\Sigma}_k^{(1)}\vec{n}\right)=\sum_k \vec{e}_k\left(\vec{\Sigma}_k^{(2)}\vec{n}\right) \qquad (6.21)$$

The equations of heat for the lattice

The comparison of the scalar expressions within brackets leads us to the *equation of heat*, or *thermal equation*, of an elastic, anelastic and plastic media, with self-diffusion, which allows us, in principle, to compute the thermal evolution of the lattice

$$\begin{aligned} nT\frac{ds}{dt} &= -\mu_L^* S_L - \mu_I^* S_I - \left(e_{cin} + u - \mu_L^* C_L - \mu_I^* C_I\right)S_n \\ &-\left(\overrightarrow{\mathrm{grad}}\,\mu_I^* + m\frac{d\vec{\phi}_I}{dt} - m\vec{g}\right)\vec{J}_I - \left(\overrightarrow{\mathrm{grad}}\,\mu_L^* - m\frac{d\left(\vec{\phi}_L - 2\Delta\vec{\varphi}_L\right)}{dt} + m\vec{g}\right)\vec{J}_L \\ &+\vec{s}_k^{dis}\frac{d\vec{\beta}_k^{an}}{dt} + \vec{m}^{dis}\frac{d\vec{\omega}^{an}}{dt} + \vec{s}_k\frac{d\vec{\beta}_k^{pl}}{dt} + \vec{m}\frac{d\vec{\omega}^{pl}}{dt} - p\frac{d\tau^{pl}}{dt} - \mathrm{div}\,\vec{J}_q \end{aligned} \qquad (6.22)$$

As a matter of fact the way *(6.22)* is written is sub-optimal because it does not separate the processes of creation of vacancies and interstitials by creation-annihilation of pairs on the one hand and by plasticity on the other hand. To handle those, it is useful to rewrite this equation by using definition *(5.64) of* $d\tau^{pl} / dt$, and then definitions *(5.93)* for S_L and S_I, as well as the definition *(5.92)* of S_n. We then have an equation of heat that depends solely on S_{I-L}, S_L^{pl} and S_I^{pl}

$$nT\frac{ds}{dt} = -\left(\mu_L^* + \mu_I^*\right)S_{I-L} - \left(\mu_L^* + h^*\right)S_L^{pl} - \left(\mu_I^* - h^*\right)S_I^{pl}$$
$$+\left(-\overrightarrow{\text{grad}}\mu_L^* + m\frac{d}{dt}\left(\vec{\phi}_L - 2\Delta\vec{\varphi}_L\right) - m\vec{g}\right)\vec{J}_L + \left(-\overrightarrow{\text{grad}}\mu_I^* - m\frac{d}{dt}\left(\vec{\phi}_I\right) + m\vec{g}\right)\vec{J}_I \qquad (6.23)$$
$$+\vec{s}_k^{dis}\frac{d\vec{\beta}_k^{an}}{dt} + \vec{m}^{dis}\frac{d\vec{\omega}^{an}}{dt} + \vec{s}_k\frac{d\vec{\beta}_k^{pl}}{dt} + \vec{m}\frac{d\vec{\omega}^{pl}}{dt} - \text{div}\,\vec{J}_q$$

in which we have introduced energy h^* which is nothing more than the *thermal energy* taken from the lattice for the creation of a lattice site, given by

$$h^* = u + pv + e_{cin} - \mu_L^* C_L - \mu_I^* C_I$$
$$= f + Ts + pv + e_{cin} - \mu_L^* C_L - \mu_I^* C_I \qquad (6.24)$$

The denomination h^* given to this energy is simply due to the fact that it takes into account the *enthalpy* $h = u + pv = f + Ts + pv$ per site of lattice. This energy can be written using the definitions *(6.2)*

$$h^* = u + pv + m\vec{\phi}^2 / 2 - \mu_L C_L - \mu_I C_I$$
$$= f + Ts + pv + m\vec{\phi}^2 / 2 - \mu_L C_L - \mu_I C_I \qquad (6.25)$$

The sources of heat

In heat equation *(6.23)*, we have several *sources of heat:*

- a source of heat $S_q^{formation(I-L)}$ which corresponds to *thermal energy taken from or given to the lattice during the creation or annihilation of a vacancy-interstitial pair*

$$S_q^{formation\,(I-L)} = -\left(\mu_L^* + \mu_I^*\right)S_{I-L} \qquad (6.26)$$

- two sources of heat $S_q^{diffusion(L)}$ and $S_q^{diffusion(I)}$ which correspond to the *thermal energies involved in the diffusion of vacancies and interstitials respectively*

$$\begin{cases} S_q^{diffusion\,(L)} = \left(-\overrightarrow{\text{grad}}\mu_L^* + m\dfrac{d}{dt}\left(\vec{\phi}_L - 2\Delta\vec{\varphi}_L\right) - m\vec{g}\right)\vec{J}_L \\ S_q^{diffusion\,(I)} = \left(-\overrightarrow{\text{grad}}\mu_I^* - m\dfrac{d}{dt}\left(\vec{\phi}_I\right) + m\vec{g}\right)\vec{J}_I \end{cases} \qquad (6.27)$$

- a source of heat induced by the *dissipative anelastic phenomena*

$$S_q^{anélasticité} = \vec{s}_k^{dis}\frac{d\vec{\beta}_k^{an}}{dt} + \vec{m}^{dis}\frac{d\vec{\omega}^{an}}{dt} \qquad (6.28)$$

- a source of heat induced by the *dissipative plastic phenomena*, which can be written in two different forms using relation *(5.94)*

$$S_q^{plasticité} = -\left(\mu_L^* + h^*\right)S_L^{pl} - \left(\mu_I^* - h^*\right)S_I^{pl} + \vec{s}_k\frac{d\vec{\beta}_k^{pl}}{dt} + \vec{m}\frac{d\vec{\omega}^{pl}}{dt}$$
$$= -\mu_L^* S_L^{pl} - \mu_I^* S_I^{pl} - h^* S_n + \vec{s}_k\frac{d\vec{\beta}_k^{pl}}{dt} + \vec{m}\frac{d\vec{\omega}^{pl}}{dt} \qquad (6.29)$$

which shows that the thermal energies taken from the lattice respectively are worth μ_L^* for the creation of a vacancy, μ_I^* for the creation of a self-interstitial and h^* for the creation of a lattice site.

The source of entropy

If thermal equation *(6.23)* is compared with expression *(4.29)* of the second principle, in which the sources of lattice sites S_n is not null, we have the expression of *volume source of entropy* in an elastic, anelastic, plastic and self-diffusing lattice

$$\begin{aligned} S_e = & -\frac{1}{T}\left(\mu_L^* + \mu_I^*\right) S_{I-L} - \frac{1}{T}\left(\mu_L^* + g^*\right) S_L^{pl} - \frac{1}{T}\left(\mu_I^* - g^*\right) S_I^{pl} \\ & + \frac{1}{T}\left(- \overrightarrow{\mathrm{grad}\mu_L^*} + m\frac{d}{dt}\left(\vec{\phi}_L - 2\Delta\vec{\varphi}_L\right) - m\vec{g} \right)\vec{J}_L + \frac{1}{T}\left(- \overrightarrow{\mathrm{grad}\mu_I^*} - m\frac{d}{dt}\left(\vec{\phi}_I\right) + m\vec{g} \right)\vec{J}_I \\ & + \frac{1}{T}\left(\vec{s}_k^{\,dis}\frac{d\vec{\beta}_k^{an}}{dt} + \vec{m}^{dis}\frac{d\vec{\omega}^{an}}{dt} + \vec{s}_k\frac{d\vec{\beta}_k^{pl}}{dt} + \vec{m}\frac{d\vec{\omega}^{pl}}{dt} \right) + \vec{J}_q\,\overrightarrow{\mathrm{grad}}\left(\frac{1}{T}\right) \end{aligned} \qquad (6.30)$$

in which g^* is given by expression

$$\begin{aligned} g^* &= f + pv + e_{cin} - \mu_L^* C_L - \mu_I^* C_I \\ &= f + pv + m\vec{\phi}^2/2 - \mu_L C_L - \mu_I C_I \end{aligned} \qquad (6.31)$$

which contains among others the *free enthalpy* $g = u + pv - Ts = f + pv$ per lattice site.

6.2 - Phenomenological relations of a lattice

The function and equations of state

We already mentioned the state equations in section 5.1. However, the expressions we found there were derived from the internal energy u of the solid and thus depended on entropy s. It is in general a lot more convenient to use the temperature T as an independent variable, instead of entropy s, and we can introduce the state function of *free energy* f per site of lattice by applying the following *Legendre transform*

$$f = u - Ts \quad \Rightarrow \quad df = du - Tds - sdT \qquad (6.32)$$

In this expression of the differential of free energy f, we can introduce the expression for the differential of internal energy u obtained in *(5.83)*

$$ndf = \vec{s}_k d\vec{\alpha}_k^{el} + m_k\omega_k^{el} - pd\tau^{el} + \vec{s}_k^{cons} d\vec{\alpha}_k^{an} + m_k^{cons}\omega_k^{an} + n\mu_L dC_L + n\mu_I dC_I - nsdT \qquad (6.33)$$

The free energy $f = u - Ts$ becomes a function of the following state variables

$$f = f\left(\alpha_{ij}^{el}, \alpha_{ij}^{an}, \omega_k^{el}, \omega_k^{an}, \tau^{el}, C_L, C_I, T\right) \qquad (6.34)$$

An important realization can be made here which concerns this state function, namely that it does not depend on the plastic distortions of the lattice.

As we explained in section 5.1, the functions and state equations can take different forms depending on which of the three topological representations we use $\{\vec{\beta}_i\}$, $\{\vec{\varepsilon}_i, \vec{\omega}\}$ or $\{\vec{\alpha}_i, \vec{\omega}, \tau\}$, which each have their conjugate stress potentials, namely $\{\vec{\Sigma}_i\}$, $\{\vec{\sigma}_i, \vec{m}\}$ and $\{\vec{s}_i, \vec{m}, p\}$ respectively. For example, in the expression *(6.34)* for free energy f, we have the various equations of state for a lattice in the topological representation $\{\vec{\alpha}_i, \vec{\omega}, \tau\}$ which are directly deduced:

- the *elastic equations of state*:

$$\left\{\begin{aligned} s_{ij} &= \frac{n}{2}\left(\frac{\partial f}{\partial \alpha_{ij}^{él}} + \frac{\partial f}{\partial \alpha_{ji}^{él}}\right) - \frac{n}{3}\delta_{ij}\sum_k \frac{\partial f}{\partial \alpha_{kk}^{él}} = s_{ij}(\alpha_{lm}^{él}, \omega_n^{él}, \tau^{él}, \alpha_{lm}^{an}, \omega_n^{an}, C_L, C_I, T) \\ m_k &= n\frac{\partial f}{\partial \omega_k^{él}} = m_k(\alpha_{lm}^{él}, \omega_n^{él}, \tau^{él}, \alpha_{lm}^{an}, \omega_n^{an}, C_L, C_I, T) \\ p &= -n\frac{\partial f}{\partial \tau^{él}} = p(\alpha_{lm}^{él}, \omega_n^{él}, \tau^{él}, \alpha_{lm}^{an}, \omega_n^{an}, C_L, C_I, T) \end{aligned}\right. \tag{6.35}$$

- the *anelastic equations of state*:

$$\left\{\begin{aligned} s_{ij}^{dis} &= \frac{n}{2}\left(\frac{\partial f}{\partial \alpha_{ij}^{an}} + \frac{\partial f}{\partial \alpha_{ji}^{an}}\right) - \frac{n}{3}\delta_{ij}\sum_k \frac{\partial f}{\partial \alpha_{kk}^{an}} = s_{ij}^{dis}(\alpha_{lm}^{él}, \omega_n^{él}, \tau^{él}, \alpha_{lm}^{an}, \omega_n^{an}, C_L, C_I, T) \\ m_k^{dis} &= n\frac{\partial f}{\partial \omega_k^{an}} = m_k^{dis}(\alpha_{lm}^{él}, \omega_n^{él}, \tau^{él}, \alpha_{lm}^{an}, \omega_n^{an}, C_L, C_I, T) \end{aligned}\right. \tag{6.36}$$

- and finally the *equations of state for the entropy and the chemical potentials* for vacancies and interstitials:

$$\left\{\begin{aligned} s &= -\frac{\partial f}{\partial T} = s(\alpha_{lm}^{él}, \omega_n^{él}, \tau^{él}, \alpha_{lm}^{an}, \omega_n^{an}, C_L, C_I, T) \\ \mu_L &= \frac{\partial f}{\partial C_L} = \mu_L(\alpha_{lm}^{él}, \omega_n^{él}, \tau^{él}, \alpha_{lm}^{an}, \omega_n^{an}, C_L, C_I, T) \\ \mu_I &= \frac{\partial f}{\partial C_I} = \mu_I(\alpha_{lm}^{él}, \omega_n^{él}, \tau^{él}, \alpha_{lm}^{an}, \omega_n^{an}, C_L, C_I, T) \end{aligned}\right. \tag{6.37}$$

The dissipative equations of thermo-conduction and self-diffusion

The three terms of the source of entropy *(6.30)* depend on the thermodynamic fluxes and are linked to the following dissipative processes:

- *the process of thermo-conduction:*

$$S_e^{thermoconduction} = \vec{J}_q \, \overrightarrow{\text{grad}} \frac{1}{T} = \vec{J}_q \bullet \vec{X}_q \tag{6.38}$$

- *the process of vacancy diffusion:*

$$S_e^{auto-diffusion\,(L)} = \vec{J}_L \frac{1}{T}\left(-\overrightarrow{\text{grad}}\,\mu_L^* + m\frac{d}{dt}\left(\vec{\phi}_L - 2\Delta\vec{\varphi}_L\right) - m\vec{g}\right) = \vec{J}_L \bullet \vec{X}_L \tag{6.39}$$

- *the process of diffusion of self-interstitials:*

$$S_e^{auto-diffusion\,(I)} = \vec{J}_I \frac{1}{T}\left(-\overrightarrow{\text{grad}}\,\mu_I^* - m\frac{d}{dt}\left(\vec{\phi}_I\right) + m\vec{g}\right) = \vec{J}_I \bullet \vec{X}_I \tag{6.40}$$

These processes can be written as the product of vectorial fluxes by the vectors of thermodynamical forces which are their conjugates. The forces $\vec{X}_L$ and $\vec{X}_I$ conjugated to the flux of diffusion $\vec{J}_L$ and $\vec{J}_I$ contains several terms: a gravitational force, an inertial force and a generalized chemical force, this last one corresponds to the gradient of the chemical potential and the kinetic energy of both species of point defects. According to the *thermodynamics of irreversible processes*, there can exist a coupling between the diverse fluxes and thermodynamical forces which appear in the expression of the source of entropy, as long as these fluxes and forces are of the same tensorial order. In the present case, concerning the *flux of heat by thermo-conduction* $\vec{J}_q$ and the *flux of particles by self-diffusion* $\vec{J}_L$ and $\vec{J}_I$, there must exist *phenomenological dissipation equations of thermo-conduction and self-diffusion*, which depend strongly on the local state of the lattice and therefore on the values of quantities n, T, C_L, C_I

$$\begin{cases} \vec{J}_q = \vec{J}_q(\vec{X}_q, \vec{X}_L, \vec{X}_I, n, T, C_L, C_I, \dots) \\ \vec{J}_L = \vec{J}_L(\vec{X}_q, \vec{X}_L, \vec{X}_I, n, T, C_L, C_I, \dots) \\ \vec{J}_I = \vec{J}_I(\vec{X}_q, \vec{X}_L, \vec{X}_I, n, T, C_L, C_I, \dots) \end{cases} \tag{6.41}$$

To insure the positivity of the sources of entropy, *(6.38) to (6.40)* must depend in a linear fashion on the thermodynamic forces $\vec{X}_q$, $\vec{X}_L$ and $\vec{X}_I$, defined by relations

$$\begin{cases} \vec{X}_q = \overrightarrow{\text{grad}}\, \dfrac{1}{T} \\ \vec{X}_L = \dfrac{1}{T}\left(- \overrightarrow{\text{grad}}\ \mu_L^* + m\dfrac{d}{dt}\left(\vec{\phi}_L - 2\Delta\vec{\varphi}_L\right) - m\vec{g} \right) \\ \vec{X}_I = \dfrac{1}{T}\left(- \overrightarrow{\text{grad}}\ \mu_I^* - m\dfrac{d}{dt}\left(\vec{\phi}_I\right) + m\vec{g} \right) \end{cases} \tag{6.42}$$

The equation of creation-annihilation of vacancy-interstitial pairs

The first term of the source of entropy uses the product of the source S_{I-L} of pairs of vacancy-interstitial, meaning the number of pairs created or annihilated per unit volume by unit of time inside the lattice, by an entropic term $-(\mu_L^* + \mu_I^*)/T$ which is its conjugate, meaning the *entropy of formation* of vacancy-interstitial pair

$$S_e^{création-annihilation\ (I-L)} = S_{I-L} \bullet \left[-\frac{1}{T}(\mu_L^* + \mu_I^*) \right] \tag{6.43}$$

This entropy of formation is directly linked to the thermal energy (heat) extracted from the lattice to form a vacancy-interstitial pair, as shown by the source of heat $S_q^{formation\ (I-L)}$ obtained in *(6.26)*. *The energy of formation of a vacancy-interstitial pair* is consequently equal to the sum of chemical energy, meaning the chemical potentials μ_L and μ_I, and the kinetic energy of a pair, namely e_{cin}^L and e_{cin}^I as the comparison of relations *(5.108)* and *(6.2)* shows

$$\mu_L^* + \mu_I^* = \mu_L + \mu_I + e_{cin}^L + e_{cin}^I \tag{6.44}$$

The kinetic energies e_{cin}^L and e_{cin}^I need a longer discussion. To that end it is useful to write them using the relations *(6.2)* in which they were defined

$$\begin{cases} e^L_{cin} = -\frac{1}{2}m\left(\vec{\phi}^2_L - 2\Delta\vec{\varphi}^2_L\right) = -\frac{1}{2}m\vec{\phi}^2 - m\vec{\phi}\Delta\vec{\varphi}_L + \frac{1}{2}m\Delta\vec{\varphi}^2_L \\ e^I_{cin} = \frac{1}{2}m\vec{\phi}^2_I = \frac{1}{2}m\vec{\phi}^2 + m\vec{\phi}\Delta\vec{\varphi}_I + \frac{1}{2}m\Delta\vec{\varphi}^2_I \end{cases} \tag{6.45}$$

These energies are composed in a kinetic energy of entrainment proportional to $\vec{\phi}^2$, which is negative since the vacancy it represents is missing kinetic energy at the site under consideration, and an internal kinetic energy (Δe^L_{cin} and Δe^I_{cin}) linked to the relative movement of the point defect with respect to the lattice. The sum of energies e^L_{cin} and e^I_{cin} is then equal to

$$e^L_{cin} + e^I_{cin} = m\vec{\phi}\left(\Delta\vec{\varphi}_I - \Delta\vec{\varphi}_L\right) + \frac{1}{2}m\left(\Delta\vec{\varphi}^2_I + \Delta\vec{\varphi}^2_L\right) = \Delta e^L_{cin} + \Delta e^I_{cin} \tag{6.46}$$

with

$$\begin{cases} \Delta e^L_{cin} = \frac{1}{2}m\Delta\vec{\varphi}^2_L - m\vec{\phi}\Delta\vec{\varphi}_L \\ \Delta e^I_{cin} = \frac{1}{2}m\Delta\vec{\varphi}^2_I + m\vec{\phi}\Delta\vec{\varphi}_I \end{cases} \tag{6.47}$$

The kinetic energy Δe^{I-L}_{cin} necessary for the creation of a pair, is given by the thermal energy of the lattice and is therefore equal to

$$\Delta e^{I-L}_{cin} = \Delta e^L_{cin} + \Delta e^I_{cin} = e^L_{cin} + e^I_{cin} \tag{6.48}$$

so that the *energy of formation of a vacancy-interstitial pair* is composed of the chemical energy $\mu_L + \mu_I$ and the kinetic energy Δe^{I-L}_{cin}

$$\mu^*_L + \mu^*_I = \mu_L + \mu_I + \Delta e^{I-L}_{cin} \tag{6.49}$$

It is clear that the rate S_{I-L} of creation of pairs is strongly dependent on the threshold energy of formation. As a consequence there must exist a *phenomenological equation of creation and annihilation of vacancy-interstitial pair,* which depends on $\mu^*_L + \mu^*_I$. This equation will be written in a symbolic form in which we will include a strong dependence in the intensive local thermodynamic parameters n, T, C_L, C_I

$$S_{I-L} = S_{I-L}\left[\left(\mu^*_L + \mu^*_I\right), n, T, C_L, C_I, \ldots\right] \tag{6.50}$$

The dissipative equations of plasticity

The second and third term for the source of entropy make use of the product of the scalar sources of vacancies and interstitials, due to the creation or annihilation of sites of the lattice during the process of plastic volume expansion, by a term of formation of entropy

$$\begin{aligned} S^{site-L}_e &= S^{pl}_L \bullet \left[-\left(\mu^*_L + g^*\right)/T\right] \\ S^{site-I}_e &= S^{pl}_I \bullet \left[-\left(\mu^*_I - g^*\right)/T\right] \end{aligned} \tag{6.51}$$

This term of entropy of formation is conjugated to the sources of vacancies and interstitials by plasticity and can be written under the following form, by using relation *(6.45)* for e^L_{cin} and e^I_{cin}, and also using relation *(6.31)* giving g^*

$$
\begin{aligned}
&-\frac{1}{T}\left(\underbrace{\mu_L + e^L_{cin}}_{\substack{\text{energy of formation}\\ \text{of a vacancy}}} \underbrace{+g + e_{cin} - \mu_L^* C_L - \mu_I^* C_I}_{\substack{\text{energy of creation}\\ \text{of an additionnal lattice site}}}\right) = -\frac{1}{T}\left(\underbrace{\mu_L + e^L_{cin}}_{\substack{\text{energy of formation}\\ \text{of a vacancy}}} \underbrace{+g + \frac{m\vec{\phi}^2}{2} - \mu_L C_L - \mu_I C_I}_{\substack{\text{energy of creation}\\ \text{of an additionnal lattice site}}}\right)\\
&-\frac{1}{T}\left(\underbrace{\mu_I + e^I_{cin}}_{\substack{\text{energy of formation}\\ \text{of an interstitial}}} \underbrace{-g - e_{cin} + \mu_L^* C_L + \mu_I^* C_I}_{\substack{\text{energy of annihilation}\\ \text{of a lattice site}}}\right) = -\frac{1}{T}\left(\underbrace{\mu_I + e^I_{cin}}_{\substack{\text{energy of formation}\\ \text{of an interstitial}}} \underbrace{-g - \frac{m\vec{\phi}^2}{2} + \mu_L C_L + \mu_I C_I}_{\substack{\text{energy of annihilation}\\ \text{of a lattice site}}}\right)
\end{aligned}
\tag{6.52}
$$

We can see that this entropies of formation use two very important quantities: *the energy of formation of a vacancy by the creation of an additional site of lattice*, and the *energy of formation of an interstitial by annihilation of a site of lattice*. The creation or annihilation of a lattice site during a plastic deformation will satisfy a kinetic equation that must depend on the energies of formation of vacancies or interstitials, as well as physical conditions such as volume of lattice unit cell v, temperature T, atomic concentrations of vacancies and interstitials, etc. There must exist, as a consequence, *phenomenological equations of creation-annihilation of lattice sites*, which will translate the plasticity of the lattice under volume expansion and which will be symbolically written as such

$$
\begin{cases}
\dfrac{d\tau^{pl}}{dt} = \dfrac{S_n}{n} = \dfrac{1}{n}\left(S_L{}^{pl} - S_I{}^{pl}\right)\\
S_L{}^{pl} = S_L{}^{pl}\left[\left(\mu_L^* + g^*\right), v, T, C_L, C_I, \ldots\right]\\
S_I{}^{pl} = S_I{}^{pl}\left[\left(\mu_I^* - g^*\right), v, T, C_L, C_I, \ldots\right]
\end{cases}
\tag{6.53}
$$

As regards the dissipative terms associated with plasticity by shear and rotation, it leads to a source of entropy that we can write with the help of *(6.30) as*

$$
S_e^{pl} = \frac{1}{T}\left(\vec{s}_k \frac{d\vec{\beta}_k^{pl}}{dt} + \vec{m}\frac{d\vec{\omega}^{pl}}{dt}\right)
\tag{6.54}
$$

In this relation, the mechanical potential fields $\vec{s}_i$ and $\vec{m}$ are perfectly known, as it leads to the stress field of elastic shear $\vec{s}_i$ and the torsor of elastic moments $\vec{m}$ deduced from the state equations *(6.35)*. As a consequence the velocity of plastic deformations by shear and rotation must satisfy the *dissipative phenomenological equations of plasticity*

$$
\begin{cases}
\dfrac{d\vec{\alpha}_i^{pl}}{dt} = \dfrac{d\vec{\alpha}_i^{pl}}{dt}\left(\vec{s}_m, v, T, \ldots\right)\\
\dfrac{d\vec{\omega}^{pl}}{dt} = \dfrac{d\vec{\omega}^{pl}}{dt}\left(\vec{m}, v, T, \ldots\right)
\end{cases}
\tag{6.55}
$$

Writing it in this way reflects the fact that the terms of sources of entropy due to the plastic deformations by shear and rotation can be considered as the product of a *generalized flux* ($d\vec{\alpha}_i^{pl} / dt$ or $d\vec{\omega}^{pl} / dt$) by a *"generalized force"*, which is here represented by the shear stress tensor $\vec{s}_i$ or the torque torsor $\vec{m}$.

The dissipative equations of anelasticity

For the dissipative terms associated with anelasticity by shear and rotation, the source of entropy can be written, thanks to *(6.30)*

$$S_e^{an} = \frac{1}{T}\left(\vec{s}_k^{dis}\frac{d\vec{\beta}_k^{an}}{dt} + \vec{m}^{dis}\frac{d\vec{\omega}^{an}}{dt}\right) \tag{6.55}$$

The equations of state *(6.35)* show that the total potential fields $\vec{s}_i$ and $\vec{m}$ are the conjugates of the field of elastic shear deformations $\vec{\alpha}_i^{el}$ and elastic rotation $\vec{\omega}^{el}$ while the conservative potential fields $\vec{s}_i^{cons}$ and $\vec{m}^{cons}$ are the conjugates of the fields of anelastic shear deformation $\vec{\alpha}_i^{an}$ and anelastic rotation $\vec{\omega}^{an}$. By writing the decomposition of the energy of anelastic distortion in terms of a source of work and a source of heat, we had to decompose the potential elastic fields $\vec{s}_i$ and $\vec{m}$ in a conservative part and a dissipative part, in the following way

$$\begin{cases} \vec{s}_i = \vec{s}_i^{cons} + \vec{s}_i^{dis} \\ \vec{m} = \vec{m}^{cons} + \vec{m}^{dis} \end{cases} \tag{6.56}$$

In these relations, the fields $\vec{s}_i$, $\vec{s}_i^{cons}$, $\vec{m}$ and $\vec{m}^{cons}$ are given by the state equations. We only have the dissipative components $\vec{s}_i^{dis}$ and $\vec{m}^{dis}$ left to discuss.

As a matter of fact, the velocity of anelastic shear distortion $d\vec{\alpha}_i^{an} / dt$ and of anelastic rotation $d\vec{\omega}^{an} / dt$ are responsible for the appearance of intrinsic friction forces which give rise to dissipative stress fields $\vec{s}_i^{dis}$ and $\vec{m}^{dis}$. As a consequence, it is reasonable to admit that these dissipative fields are first and foremost function of the velocities of anelastic distortions by writing

$$\begin{cases} \vec{s}_i^{dis} = \vec{s}_i^{dis}\left(\dfrac{d\vec{\alpha}_m^{an}}{dt}, v, T, \ldots\right) \\ \vec{m}^{dis} = \vec{m}^{dis}\left(\dfrac{d\vec{\omega}^{an}}{dt}, v, T, \ldots\right) \end{cases} \tag{6.57}$$

This leads us to write, by taking into account the relations that exist between the various fields of mechanical potential, the *phenomenological dissipative equations of anelasticity*, under the following form

$$\begin{cases} \vec{s}_i = \vec{s}_i^{cons}\left(\vec{\alpha}_m^{an}, v, T, \ldots\right) + \vec{s}_i^{dis}\left(\dfrac{d\vec{\alpha}_m^{an}}{dt}, v, T, \ldots\right) \\ \vec{m} = \vec{m}^{cons}\left(\vec{\omega}^{an}, v, T, \ldots\right) + \vec{m}^{dis}\left(\dfrac{d\vec{\omega}^{an}}{dt}, v, T, \ldots\right) \end{cases} \tag{6.58}$$

The dissipative equations we just wrote, written under this form, are nothing more than mathematical differential equations that give us the anelastic responses $\vec{\alpha}_i^{an}$ and $\vec{\omega}^{an}$ of the solid lattice, to the elastic stresses $\vec{s}_i$ et $\vec{m}$.

6.3 - Energetic balance

We carry the following multiplication operations on the three equations of geometro-kinetic *(5.65)* and on the equation of newtonian dynamics *(6.11)*

$$\vec{s}_k \frac{d\vec{\beta}_k}{dt} = \vec{s}_k \overrightarrow{\operatorname{grad}} \phi_k \quad \text{and} \quad \vec{m}\frac{d\vec{\omega}}{dt} = \frac{1}{2}\vec{m}\overrightarrow{\operatorname{rot}}\vec{\phi} \quad \text{and} \quad -p\frac{d\tau}{dt} = -p\operatorname{div}\vec{\phi} \tag{6.59}$$

$$n\vec{\phi}\frac{d\vec{p}}{dt} = \rho\vec{g}\vec{\phi} + \phi_k \operatorname{div}\vec{s}_k - \frac{1}{2}\vec{\phi}\overrightarrow{\operatorname{rot}}\,\vec{m} - \vec{\phi}\,\overrightarrow{\operatorname{grad}}\,p + nm\vec{\phi}\vec{\phi}_I\frac{dC_I}{dt} - nm\vec{\phi}\vec{\phi}_L\frac{dC_L}{dt} \tag{6.60}$$

by summing all these equations, we have the *equation of energetic balance* inside the lattice

$$n\vec{\phi}\left(\frac{d\vec{p}}{dt} - m\vec{\phi}_I\frac{dC_I}{dt} + m\vec{\phi}_L\frac{dC_L}{dt}\right) + \vec{s}_k\frac{d\vec{\beta}_k}{dt} + \vec{m}\frac{d\vec{\omega}}{dt} - p\frac{d\tau}{dt} = \rho\vec{g}\vec{\phi} - \operatorname{div}\left[-\phi_k\vec{s}_k - \frac{1}{2}\left(\vec{\phi}\wedge\vec{m}\right) + p\vec{\phi}\right] \tag{6.61}$$

Each term of this equation can be easily interpreted in terms of a variation of energy densities inside the lattice:

$n\vec{\phi}\left(\frac{d\vec{p}}{dt} - m\vec{\phi}_I\frac{dC_I}{dt} + m\vec{\phi}_L\frac{dC_L}{dt}\right)$ is the variation of kinetic energy density,

$\vec{s}_k\ d\vec{\beta}_k / dt$ is the variation in density of shear elastic energy,

$\vec{m}\ d\vec{\omega} / dt$ is the variation in density of potential energy of global rotation of the lattice, if it exists,

$-p\ d\tau / dt$ is the variation of density of elastic energy of volume expansion,

$\rho\vec{g}\vec{\phi}$ is the power per unit volume given by the external field force of gravitation,

$-\phi_k\vec{s}_k - \left(\vec{\phi}\wedge\vec{m}\right)/2 + p\vec{\phi}$ is the flux of work of distorsion $\vec{J}_w$, which is analog of the Poynting vector.

With regards to the equations of energetic balance we can get from the other topological representations of the equation of thermo-kinetic, they are even simpler

$$n\vec{\phi}\left(\frac{d\vec{p}}{dt} - m\vec{\phi}_I\frac{dC_I}{dt} + m\vec{\phi}_L\frac{dC_L}{dt}\right) + \vec{\sigma}_k\frac{d\vec{\beta}_k}{dt} + \vec{m}\frac{d\vec{\omega}}{dt} = \rho\vec{g}\vec{\phi} - \operatorname{div}\left[-\phi_k\vec{\sigma}_k - \frac{1}{2}\left(\vec{\phi}\wedge\vec{m}\right)\right] \tag{6.62}$$

$$n\vec{\phi}\left(\frac{d\vec{p}}{dt} - m\vec{\phi}_I\frac{dC_I}{dt} + m\vec{\phi}_L\frac{dC_L}{dt}\right) + \vec{\Sigma}_k\frac{d\vec{\beta}_k}{dt} = \rho\vec{g}\vec{\phi} - \operatorname{div}\left(-\phi_k\vec{\Sigma}_k\right) \tag{6.63}$$

6.4 - Equations of spatiotemporal evolution of a lattice

It is now possible to combine all the results we have obtained in this chapter and the previous chapters to write the complete equations that describe the spatio-temporal evolution of a solid lattice with self-diffusion and phenomenological elasticity, anelasticity and plasticity. As seen in table 6.1, this system of equations is quite complex, in particular due to the high number of phenomenological equations of state and phenomenological equations of dissipation one needs for a complete description of all the possible behaviors and phenomena of this lattice.

The computation of the evolution of the lattice involves the resolution of this system of *fundamental equations*:

- *the topological equations*, which translate the galilean geometro-kinetic inside the solid, which allows us to calculate the global or local distortion fields from an absolute velocity field $\vec{\phi}$ or a relative velocity field $\vec{\varphi}$ inside the solid, from which we get the volume density n of sites of lattice. Furthermore the conditions of geometro-compatibility based on $\overrightarrow{\text{rot}}\,\vec{\beta}_i = \overrightarrow{\text{rot}}\,\vec{\beta}_i^{(\delta)} = 0$, are evidently valid for a lattice equipped with anelasticity and plasticity, because they assure that the local field $\vec{u}^{(\delta)}$, associated with the elastic, anelastic or plastic deformations, remains continuous. This is a necessary conditions to ensure the solidity of the lattice.
- *the dynamical equations*, translating the newtonian dynamic of the lattice, which lead us to the computation of the linear momentum $\vec{p}$, from which we extract velocities $\vec{\phi}$ and $\vec{\varphi}$, and the calculation of the density of inert mass ρ,
- *the equations of diffusion*, which translate the self-diffusion in the lattice, which lead us to the computation of the densities C_L and C_I of vacancies and self-interstitials,
- *the thermal equation*, which give us entropy s, from which we deduce the temperature T in the solid.

To this fundamental set of equations, which are the same for any type of lattice, we add the set of phenomenological equations which are particular to each type of lattice:

- *the equations of state*, which describe the phenomenological behavior of the lattice with regards to its elastic, anelastic, self-diffusive and thermal properties,
- *the dissipative equations*, which translate the phenomenology of dissipation by thermo-conduction, by self-diffusion, by creation-annihilation of pairs of vacancy-interstitial, by anelasticity and by plasticity of the lattice under consideration.

To the fundamental equations and the phenomenological ones necessary to describe the evolution of the lattice, we can add *additional equations*, such as the equations of mass continuity, flux of work and surface forces, sources of entropy and energetic balance. As a matter of fact, since these equations are deduced from the fundamental equations, they are not necessary to solving the problem of the evolution of the lattice. Nevertheless they shine an interesting light on the problem as far as fluxes of work, surfaces forces and exchanges of energy within the solid are concerned.

Table 6.1 - Fundamental equations of evolution of self-diffusive, elastic, anelastic and plastic solids

Topological equations

$$\begin{cases} \dfrac{d\vec{\beta}_i}{dt} = \overrightarrow{\text{grad}}\,\phi_i & (1) \\ \dfrac{d\vec{\omega}}{dt} = \dfrac{1}{2}\overrightarrow{\text{rot}}\,\vec{\phi} & (2) \\ \dfrac{d\tau}{dt} = \text{div}\,\vec{\phi} & (3) \end{cases} \qquad \begin{cases} \overrightarrow{\text{rot}}\,\vec{\beta}_i = 0 & (4) \\ \text{div}\,\vec{\omega} = 0 & (5) \\ d/dt = \partial/\partial t + (\vec{\varphi}\vec{\nabla}) & (6) \\ \vec{\varphi} = \vec{\phi} - \vec{\phi}_O(t) - \dot{\vec{\omega}}_O(t)\wedge\vec{r} & (7) \end{cases}$$

$$\begin{cases} \vec{\beta}_i = \vec{\beta}_i^{(\delta)} + \vec{e}_i\wedge\vec{\omega}_O(t) = \vec{\beta}_i^{el} + \vec{\beta}_i^{an} + \vec{\beta}_i^{pl} + \vec{e}_i\wedge\vec{\omega}_O(t) & (8) \\ \vec{\omega} = -\dfrac{1}{2}\sum_k \vec{e}_k\wedge\vec{\beta}_k = \vec{\omega}^{(\delta)} + \vec{\omega}_0(t) = \vec{\omega}^{el} + \vec{\omega}^{an} + \vec{\omega}^{pl} + \vec{\omega}_0(t) & (9) \\ \tau = \sum_k \vec{\beta}_k\vec{e}_k = \sum_k \vec{\beta}_k^{(\delta)}\vec{e}_k = \tau^{el} + \tau^{pl} \qquad (\tau^{an}\equiv 0 \text{ by hypothesis}) & (10) \\ \vec{\varepsilon}_i = \vec{\beta}_i - \vec{e}_i\wedge\vec{\omega} = \vec{\beta}_i^{(\delta)} - \vec{e}_i\wedge\vec{\omega}^{(\delta)} = \vec{\varepsilon}_i^{el} + \vec{\varepsilon}_i^{an} + \vec{\varepsilon}_i^{pl} & (11) \\ \vec{\alpha}_i = \vec{\varepsilon}_i - \dfrac{1}{3}\tau\vec{e}_i = \vec{\alpha}_i^{el} + \vec{\alpha}_i^{an} + \vec{\alpha}_i^{pl} & (12) \end{cases}$$

Dynamical equations

$$n\frac{d\vec{p}}{dt} = \rho\vec{g} + \sum_k \vec{e}_k\,\text{div}\,\vec{s}_k - \frac{1}{2}\overrightarrow{\text{rot}}\,\vec{m} - \overrightarrow{\text{grad}}p + nm\vec{\phi}_I\frac{dC_I}{dt} - nm\vec{\phi}_L\frac{dC_L}{dt} \qquad (13)$$

$$\begin{cases} n = 1/v = n_0\exp(-\tau^{el}) & (14) \\ \vec{p} = m\left(\vec{\phi} + C_I\vec{\phi}_I - C_L\vec{\phi}_L\right) = m\vec{\phi} + m\left(C_I - C_L\right)\vec{\phi} + \dfrac{m}{n}\left(\vec{J}_I - \vec{J}_L\right) & \\ \qquad = \left[\rho\vec{\phi} + m\left(\vec{J}_I - \vec{J}_L\right)\right]/n & (15) \\ \rho = mn\left(1 + C_I - C_L\right) & (16) \end{cases}$$

Diffusion equations

$$\begin{cases} n\dfrac{dC_L}{dt} = \left(S_{I\text{-}L} + S_L^{pl}\right) - C_L\left(S_L^{pl} - S_I^{pl}\right) - \text{div}\,\vec{J}_L & (17) \\ n\dfrac{dC_I}{dt} = \left(S_{I\text{-}L} + S_I^{pl}\right) - C_I\left(S_L^{pl} - S_I^{pl}\right) - \text{div}\,\vec{J}_I & (18) \end{cases}$$

$$\begin{cases} \vec{J}_L = nC_L\Delta\vec{\varphi}_L = nC_L\left(\vec{\phi}_L - \vec{\phi}\right) = nC_L\left(\vec{\varphi}_L - \vec{\varphi}\right) & (19) \\ \vec{J}_I = nC_I\Delta\vec{\varphi}_I = nC_I\left(\vec{\phi}_I - \vec{\phi}\right) = nC_I\left(\vec{\varphi}_I - \vec{\varphi}\right) & (20) \end{cases}$$

Thermal equations

$$nT\frac{ds}{dt} = -\left(\mu_L^* + \mu_I^*\right)S_{I-L} - \left(\mu_L^* + h^*\right)S_L^{pl} - \left(\mu_I^* - h^*\right)S_I^{pl} + T\vec{J}_L\vec{X}_L$$
$$+T\vec{J}_I\vec{X}_I + \vec{s}_k^{\,dis}\frac{d\vec{\beta}_k^{an}}{dt} + \vec{m}^{dis}\frac{d\vec{\omega}^{an}}{dt} + \vec{s}_k\frac{d\vec{\beta}_k^{pl}}{dt} + \vec{m}\frac{d\vec{\omega}^{pl}}{dt} - \operatorname{div}\vec{J}_q \quad (21)$$

$$\mu_L^* = \mu_L - \frac{1}{2}m\left(\vec{\phi}_L^{\,2} - 2\Delta\vec{\varphi}_L^2\right) \quad (22)$$

$$\mu_I^* = \mu_I + \frac{1}{2}m\vec{\phi}_I^{\,2} \quad (23)$$

$$\vec{X}_q = \overrightarrow{\operatorname{grad}}\frac{1}{T} \quad (24)$$

$$\vec{X}_L = \frac{1}{T}\left(-\overrightarrow{\operatorname{grad}}\ \mu_L^* + m\frac{d}{dt}\left(\vec{\phi}_L - 2\Delta\vec{\varphi}_L\right) - m\vec{g}\right) \quad (25)$$

$$\vec{X}_I = \frac{1}{T}\left(-\overrightarrow{\operatorname{grad}}\ \mu_I^* - m\frac{d}{dt}\left(\vec{\phi}_I\right) + m\vec{g}\right) \quad (26)$$

$$h^* = f + Ts + pv + \frac{1}{2}m\vec{\phi}^2 - \mu_L C_L - \mu_I C_I \quad (27)$$

Phenomenological equations of evolution of self-diffusive, elastic, anelastic and plastic solids

Functions and equations of state

$$f = f\left(\alpha_{ij}^{el}, \alpha_{ij}^{an}, \omega_k^{el}, \omega_k^{an}, \tau^{el}, C_L, C_I, T\right) \quad (28)$$

$$s_{ij} = \frac{n}{2}\left(\frac{\partial f}{\partial \alpha_{ij}^{el}} + \frac{\partial f}{\partial \alpha_{ji}^{el}}\right) - \frac{n}{3}\delta_{ij}\sum_k \frac{\partial f}{\partial \alpha_{kk}^{el}} = s_{ij}(\alpha_{lm}^{el}, \omega_n^{el}, \tau^{el}, \alpha_{lm}^{an}, \omega_n^{an}, C_L, C_I, T) \quad (29)$$

$$m_k = n\frac{\partial f}{\partial \omega_k^{el}} = m_k(\alpha_{lm}^{el}, \omega_n^{el}, \tau^{el}, \alpha_{lm}^{an}, \omega_n^{an}, C_L, C_I, T) \quad (30)$$

$$p = -n\frac{\partial f}{\partial \tau^{el}} = p(\alpha_{lm}^{el}, \omega_n^{el}, \tau^{el}, \alpha_{lm}^{an}, \omega_n^{an}, C_L, C_I, T) \quad (31)$$

$$s_{ij}^{dis} = \frac{n}{2}\left(\frac{\partial f}{\partial \alpha_{ij}^{an}} + \frac{\partial f}{\partial \alpha_{ji}^{an}}\right) - \frac{n}{3}\delta_{ij}\sum_k \frac{\partial f}{\partial \alpha_{kk}^{an}} = s_{ij}^{dis}(\alpha_{lm}^{el}, \omega_n^{el}, \tau^{el}, \alpha_{lm}^{an}, \omega_n^{an}, C_L, C_I, T) \quad (32)$$

$$m_k^{dis} = n\frac{\partial f}{\partial \omega_k^{an}} = m_k^{dis}(\alpha_{lm}^{el}, \omega_n^{el}, \tau^{el}, \alpha_{lm}^{an}, \omega_n^{an}, C_L, C_I, T) \quad (33)$$

$$\begin{cases} s = -\dfrac{\partial f}{\partial T} = s(\alpha_{lm}{}^{él}, \omega_n{}^{él}, \tau^{él}, \alpha_{lm}{}^{an}, \omega_n{}^{an}, C_L, C_I, T) & (34) \\ \mu_L = \dfrac{\partial f}{\partial C_L} = \mu_L(\alpha_{lm}{}^{él}, \omega_n{}^{él}, \tau^{él}, \alpha_{lm}{}^{an}, \omega_n{}^{an}, C_L, C_I, T) & (35) \\ \mu_I = \dfrac{\partial f}{\partial C_I} = \mu_I(\alpha_{lm}{}^{él}, \omega_n{}^{él}, \tau^{él}, \alpha_{lm}{}^{an}, \omega_n{}^{an}, C_L, C_I, T) & (36) \end{cases}$$

Equations of dissipation: self-diffusion and creation-annihilation of pairs

$$\begin{cases} \vec{J}_q = \vec{J}_q(\vec{X}_q, \vec{X}_L, \vec{X}_I, n, T, C_L, C_I, \ldots) & (37) \\ \vec{J}_L = \vec{J}_L(\vec{X}_q, \vec{X}_L, \vec{X}_I, n, T, C_L, C_I, \ldots) & (38) \\ \vec{J}_I = \vec{J}_I(\vec{X}_q, \vec{X}_L, \vec{X}_I, n, T, C_L, C_I, \ldots) & (39) \end{cases}$$

$$S_{I-L} = S_{I-L}\left(\mu_L^* + \mu_I^*, n, T, C_L, C_I, \ldots\right) \qquad (40)$$

Equations of dissipation: anelasticity

$$\begin{cases} \vec{s}_i = \vec{s}_i^{\,cons}\left(\vec{\alpha}_m^{an}, v, T, \ldots\right) + \vec{s}_i^{\,dis}\left(\dfrac{d\vec{\alpha}_m^{an}}{dt}, v, T, \ldots\right) & (41) \\ \vec{m} = \vec{m}^{cons}\left(\vec{\omega}^{an}, v, T, \ldots\right) + \vec{m}^{dis}\left(\dfrac{d\vec{\omega}^{an}}{dt}, v, T, \ldots\right) & (42) \end{cases}$$

Equations of dissipation: plasticity

$$\begin{cases} \dfrac{d\tau^{pl}}{dt} = \dfrac{S_n}{n} = \dfrac{1}{n}\left(S_L{}^{pl} - S_I{}^{pl}\right) & (43) \\ S_L{}^{pl} = S_L{}^{pl}\left[\left(\mu_L^* + g^*\right), v, T, C_L, C_I, \ldots\right] & (44) \\ S_I{}^{pl} = S_I{}^{pl}\left[\left(\mu_I^* - g^*\right), v, T, C_L, C_I, \ldots\right] & (45) \end{cases}$$

$$g^* = f + pv + m\vec{\phi}^2 / 2 - \mu_L C_L - \mu_I C_I \qquad (46)$$

$$\begin{cases} \dfrac{d\vec{\alpha}_i^{pl}}{dt} = \dfrac{d\vec{\alpha}_i^{pl}}{dt}\left(\vec{s}_m, v, T, \ldots\right) & (47) \\ \dfrac{d\vec{\omega}^{pl}}{dt} = \dfrac{d\vec{\omega}^{pl}}{dt}\left(\vec{m}, v, T, \ldots\right) & (48) \end{cases}$$

Additional equations of evolution

Continuity of mass

$$\frac{\partial \rho}{\partial t} = -\mathrm{div}\left[\rho\vec{\phi} + m\left(\vec{J}_I - \vec{J}_L\right)\right] = -\mathrm{div}\left(n\vec{p}\right) \qquad \text{in } Q\xi_1\xi_2\xi_3 \qquad (49)$$

Flux of work and surface force

$$\begin{cases} \vec{J}_w = \mu_L^{\ *}\vec{J}_L + \mu_I^{\ *}\vec{J}_I - \phi_k\vec{s}_k - \dfrac{1}{2}\left(\vec{\phi}\wedge\vec{m}\right) + p\vec{\phi} & (50) \\ \vec{F}_S = \displaystyle\sum_k \vec{e}_k\left(\vec{s}_k\vec{\boldsymbol{n}}\right) + \dfrac{1}{2}\left(\vec{m}\wedge\vec{\boldsymbol{n}}\right) - \vec{\boldsymbol{n}}p & (51) \end{cases}$$

Source of entropy

$$\begin{aligned} S_e = & -\frac{1}{T}\left(\mu_L^* + \mu_I^*\right)S_{I-L} - \frac{1}{T}\left(\mu_L^* + g^*\right)S_L^{pl} - \frac{1}{T}\left(\mu_I^* - g^*\right)S_I^{pl} \\ & + \vec{J}_L\vec{X}_L + \vec{J}_I\vec{X}_I + \frac{1}{T}\left(\vec{s}_k^{\,dis}\frac{d\vec{\beta}_k^{an}}{dt} + \vec{m}^{dis}\frac{d\vec{\omega}^{an}}{dt} + \vec{s}_k\frac{d\vec{\beta}_k^{pl}}{dt} + \vec{m}\frac{d\vec{\omega}^{pl}}{dt}\right) \\ & + \vec{J}_q\overrightarrow{\mathrm{grad}}\left(\frac{1}{T}\right) \end{aligned} \qquad (52)$$

Energetic balance

$$\begin{aligned} & n\vec{\phi}\left(\frac{d\vec{p}}{dt} - m\vec{\phi}_I\frac{dC_I}{dt} + m\vec{\phi}_L\frac{dC_L}{dt}\right) + \vec{s}_k\frac{d\vec{\beta}_k}{dt} + \vec{m}\frac{d\vec{\omega}}{dt} - p\frac{d\tau}{dt} \\ & \qquad = \rho\vec{g}\vec{\phi} - \mathrm{div}\left[-\phi_k\vec{s}_k - \frac{1}{2}\left(\vec{\phi}\wedge\vec{m}\right) + p\vec{\phi}\right] \end{aligned} \qquad (53)$$

PART B

Application: examples of phenomenologies of usual solids

State functions and equations of state of an isotropic solid

Elastic et thermal behaviours

Wave propagations and thermo-elastic relaxations

Transport phenomena and inertial relaxations

Phenomenologies of anelasticity and plasticity

Structural transitions of first-order and second-order

Chapter 7

Phenomenological examples of usual solids

In this chapter, we present some applications of the first part by some simple examples of phenomenologies encountered in common isotropic solids. At first, simplified state functions are introduced to describe the phenomenology of a *'linear' isotropic solid.* From that we deduce the state equations for that type of solid. Then we discuss some phenomenological behaviors associated with elasticity, anelasticity, structural transitions, plasticity and auto-diffusion in the usual solids.

7.1 - Functions and state equations of isotropic solids

For a solid to present isotropic elastic properties (which are independent of direction in space), its free energy f per site of lattice must be a scalar built from the invariants of the tensor of distortion $\vec{\beta}_i$. Given that the operators of divergence and rotational applied to a vector field $\vec{u}$ give us invariant quantities of this field, and hence do not depend on the system of coordinates we choose, the first two scalar invariants of tensor gradient β_{ij} are its trace τ and the square of the norm of its anti-symmetric part $\vec{\omega}$

$$\tau = -\operatorname{div}\vec{u} = \sum_i \vec{e}_i\vec{\beta}_i = \sum_i \vec{e}_i\vec{\varepsilon}_i = \textit{invariant} \tag{7.1}$$

$$\vec{\omega}^2 = \frac{1}{4}\left(\overrightarrow{\operatorname{rot}}\,\vec{u}\right)^2 = \frac{1}{4}\left(\sum_i \vec{e}_i \wedge \vec{\beta}_i\right)^2 = \textit{invariant} \tag{7.2}$$

From the tensor gradient β_{ij}, other invariant scalars can be deduced such as the sum of the minors (concept in linear algebra) of the tensors of distortion of order 2.

$$\sum_i \vec{e}_i\left(\vec{\beta}_j \wedge \vec{\beta}_k\right) = \textit{invariant} \;\; ; \;\; \sum_i \vec{e}_i\left(\vec{\varepsilon}_j \wedge \vec{\varepsilon}_k\right) = \textit{invariant} \;\; ; \;\; \sum_i \vec{e}_i\left(\vec{\alpha}_j \wedge \vec{\alpha}_k\right) = \textit{invariant} \tag{7.3}$$

or the sum of the squares of the components of the tensors of distortion of order 2.

$$\sum_i \vec{\beta}_i^2 = \textit{invariant} \;\; ; \;\; \sum_i \vec{\varepsilon}_i^2 = \textit{invariant} \;\; ; \;\; \sum_i \vec{\alpha}_i^2 = \textit{invariant} \tag{7.4}$$

The state function of elasticity of an isotropic solid

The expression of free energy by lattice site of an isotropic solid must be a function of the following invariants

$$f^{el} = f^{el}\left(\tau, \sum_i \vec{\varepsilon}_i^2, T\right) \quad \text{or} \quad f^{el} = f^{el}\left(\tau, \sum_i \vec{\alpha}_i^2, T\right) \tag{7.5}$$

It must be possible to correctly express it by separating the volume expansion and the shear by a development such as

$$f^{el} = -k_0(T)\tau + k_1(T)\tau^2 + k_2(T)\sum_i \vec{\alpha}_i^2 + k_3(T)\tau\sum_i \vec{\alpha}_i^2 + k_4(T)\tau^3 + ... \qquad (7.6)$$

If the development is limited to the terms containing $k_0(T), k_1(T), k_2(T)$, we will speak of a *"linear" isotropic elastic media* which we will henceforth call *perfect solid.* We deduce from that that the free elastic energy f^{el} averaged by site of lattice is

$$f^{el} = -k_0(T)\tau + k_1(T)\tau^2 + k_2(T)\sum_i \vec{\alpha}_i^2 \qquad (7.7)$$

The state function associated with the linear anelasticity of a solid

The macroscopic phenomenology of anelasticity of a solid lattice of particles can have many different physical microscopic causes. We can name the short distance defects of the lattice structure (substitutional or interstitial point defects, dislocations or disclinations, grain boundaries in polycrystalline materials or interphase boundaries in polyphase crystals, etc.) and even name some phase transitions. At the macroscopic scale of the continuous description of the lattice, the statistical manifestation of the collective movements of the whole set of structural defects is translated by the existence of a phenomenological equations of dissipative anelasticity.

The most straightforward way to introduce the phenomenological manifestation of anelasticity in a solid lattice is to first limits oneself to an isotropic lattice and suppose that *(i)* the anelasticity of the lattice can be fully described by the shear tensor, which also means we do not have terms from the anelastic rotations in the state function of free energy, *(ii)* that the terms of elasticity do not couple with anelasticity and self-diffusion, and *(iii)* the term of "free energy" by site of lattice linked to the anelasticity is a simple quadratic function of the components of the tensor of anelastic shear, multiplied by a coefficient k_{an} which could in theory depend on temperature T as well as the elastic volume expansion τ^{el} of the lattice. Under these conditions, the free energy associated with the anelasticity of the lattice can be written as

$$f^{an} = \frac{1}{2}k_{an}(T,\tau^{el})\sum_i (\vec{\alpha}_i^{an})^2 \qquad (7.8)$$

As a matter of fact, the anelastic term f^{an} we added to the state function f is nothing more than the average potential energy stored by site of lattice by the restoring forces acting on the structural defects of anelasticity. The quadratic function are equivalent to suppose that these forces are *linear in the tensor of anelastic shear.*

The state function associated with a solid with weak self-diffusion

For a solid lattice of volume density n of sites and containing n_L and n_I , respectively vacancies and interstitials by unit volume, the free energy will be computable if the concentration of point defects is weak enough that we can neglect the interaction of these defects, and that we also neglect the impact of these defects on the elastic and thermic properties of the lattice. We will then talk about a *weakly diffusive solid.* The free energy F by unit of volume of the solid must contain a F^{dp} component due to the presence of *point defects (PD).* If we neglect the interactions between point defects, then the function F^{dp} can be assumed to be proportional to the number n_L et n_I of point defects per unit of volume

$$F^{dp} = n_L \varepsilon_L(\tau^{el}) - T n_L s_L(\tau^{el}) + n_I \varepsilon_I(\tau^{el}) - T n_I s_I(\tau^{el}) - T\Delta S^{conf} \tag{7.9}$$

in which $\varepsilon_L(\tau^{el})$ and $\varepsilon_I(\tau^{el})$ are the *potential energies* of a vacancy and interstitial respectively, which must first depend on the elastic volume expansion τ^{el} of the lattice. $s_L(\tau^{el})$ and $s_I(\tau^{el})$ are the *entropies* associated with a vacancy or an interstitial and must also depend on τ^{el}, and ΔS^{conf} is the *entropy of configuration*, linked to the disorder introduced inside the lattice by the presence of the n_L and n_I point defects among the n sites per unit volume of the lattice.

The entropy of configuration ΔS^{conf}, by definition, equals $k \ln \Omega^{conf}$, where k is the *constant of Boltzmann* and Ω^{conf} is a number *counting the different microscopic configurations, which are known as 'complexions' of the system, that the n_L and n_I vacancies and interstitial can have.* For a network with a density n of substitutional sites, the density of interstitials is given by $z_I n$, where z_I is the number of interstitial sites associated with each substitutional site, which self evidently depends on the structure of the lattice under consideration. Define that the number of vacancy sites is n_L so that the number of interstitial site that are occupied is given by $n - n_L$ and that the number of unoccupied interstitial sites is $z_I n - n_I$, with these definitions, the number of complexions is written combinatorially

$$\Omega^{conf} = \frac{n!(z_I n)!}{(n-n_L)!n_L!(z_I n - n_I)!n_I!} \tag{7.10}$$

Using the *Stirling's approximation formula*, the entropy of configuration can be written

$$\Delta S^{conf} = k \ln \Omega^{conf} \cong -k\left(n_L \ln\frac{n_L}{n} + n_I \ln\frac{n_I}{z_I n}\right) \tag{7.11}$$

From these relations, the average free energy per site of the lattice due to the presence of point defects can be deduced by dividing F^{dp} by n and by introducing the atomic concentrations of point defects

$$f^{dp} = C_L\left[\varepsilon_L(\tau^{el}) - T s_L(\tau^{el})\right] + C_I\left[\varepsilon_I(\tau^{el}) - T s_I(\tau^{el})\right] + kT\left(C_L \ln C_L + C_I \ln\frac{C_I}{z_I}\right) \tag{7.12}$$

The equations of state of a perfect solid

By basing ourselves on the various components of the free energy f, and by introducing another thermic term $f_0(T)$ which contains all *the effects due to the temperature* of the lattice, and by making the following simplifying hypothesis

Hypothesis: $k_1, k_2, k_{an}, \varepsilon_L, s_L, \varepsilon_I$ *and* s_I *are constants* (7.13)

we have the following expression for the free energy $f \cong f_0(T) + f^{el} + f^{an} + f^{dp}$

$$\begin{aligned} f \cong\ & f_0(T) - k_0(T)\tau^{el} + k_1\left(\tau^{el}\right)^2 + k_2 \sum\left(\vec{\alpha}_i^{el}\right)^2 + \frac{1}{2}k_{an}\sum_i\left(\vec{\alpha}_i^{an}\right)^2 \\ & + C_L\left[\varepsilon_L - T s_L\right] + C_I\left[\varepsilon_I - T s_I\right] + kT\left(C_L \ln C_L + C_I \ln\frac{C_I}{z_I}\right) \end{aligned} \tag{7.14}$$

The equations of state of the lattice can be written in this case, by recalling that $n = n_0 \exp\left(-\tau^{él}\right)$

$$\left\{\begin{array}{l} \vec{s}_i = n \dfrac{\partial f}{\partial \alpha_{ik}^{él}} \vec{e}_k = 2nk_2 \vec{\alpha}_i^{él} \\ p = -n \dfrac{\partial f}{\partial \tau^{él}} = n\left(k_0(T) - 2k_1\tau^{él}\right) \text{ in the perfect solid} \\ \vec{s}_i^{\,cons} = n \dfrac{\partial f}{\partial \alpha_{ij}^{an}} \vec{e}_j = nk_{an} \vec{\alpha}_i^{an} \end{array}\right. \tag{7.15}$$

$$\left\{ s = -\frac{\partial f}{\partial T} = -\frac{\partial f_0(T)}{\partial T} + \frac{\partial k_0(T)}{\partial T}\tau^{él} + C_L s_L + C_I s_I - kC_L \ln C_L - kC_I \ln \frac{C_I}{z_I} \right. \tag{7.16}$$

$$\left\{\begin{array}{l} \mu_L = \dfrac{\partial f}{\partial C_L} = \varepsilon_L - Ts_L + kT\left(\ln C_L + 1\right) \\ \mu_I = \dfrac{\partial f}{\partial C_I} = \varepsilon_I - Ts_I + kT\left(\ln \dfrac{C_I}{z_I} + 1\right) \end{array}\right. \tag{7.17}$$

These state equations are very simple. Amid others, the shear stress $\vec{s}_i$ and $\vec{s}_i^{\,cons}$ depend linearly on the shear tensors $\vec{\alpha}_i^{él}$ and $\vec{\alpha}_i^{an}$. There is another remark to be made: the pressure p in the lattice depends on a term $k_0(T)$ associated with temperature and a term $-2k_1\tau^{él}$ which is a linear function of the volume expansion of the lattice.

7.2 - Elastic moduli of isotropic solids

The Lamé parameters μ and λ at small distortions of the solid

In general, the equations of state of elasticity of an isotropic solid are given in the representation $\{\vec{\varepsilon}_i, \vec{\sigma}_i\}$. It is very simple to write in this representation by using the relation $\vec{\sigma}_i = \vec{s}_i - p\vec{e}_i$, as well as pressure $p = nk_0(T) - 2nk_1\tau^{él}$

$$\vec{\sigma}_i = n\left(2k_2\vec{\alpha}_i - k_0(T)\vec{e}_i + 2k_1\tau\vec{e}_i\right) \tag{7.18}$$

With relations $\vec{\alpha}_i = \vec{\varepsilon}_i - \tau\vec{e}_i / 3$ and $\tau = \sum \varepsilon_{kk}$, we also have

$$\vec{\sigma}_i = n\left[2k_2\vec{\varepsilon}_i + 2\left(k_1 - \frac{k_2}{3}\right)\sum_k \varepsilon_{kk}\vec{e}_i - k_0(T)\vec{e}_i\right] \tag{7.19}$$

We can introduce the new elastic coefficients with the following relations

$$\left\{\begin{array}{l} \mu = nk_2(T) \\ \lambda = 2n\left(k_1(T) - \dfrac{k_2(T)}{3}\right) \end{array}\right. \tag{7.20}$$

These coefficients μ and λ are called the *Lamé parameters* of the perfect isotropic solid, and they allow us to write the elastic state equations under the form

$$\vec{\sigma}_i = 2\mu\vec{\varepsilon}_i + \vec{e}_i\left(\lambda\sum_k \varepsilon_{kk} - nk_0(T)\right) \tag{7.21}$$

This form has the distinct advantage of allowing the inversion of the state equation to express tensor $\vec{\varepsilon}_i$ as a function of tensor $\vec{\sigma}_i$

$$\vec{\varepsilon}_i = \frac{1}{2\mu}\vec{\sigma}_i + \vec{e}_i \frac{1}{2\mu+3\lambda}\left(nk_0(T) - \frac{\lambda}{2\mu}\sum_k \sigma_{kk} \right) \quad (7.22)$$

The Young's modulus E and the Poisson's ratio ν under simple traction

Let's imagine a parallelepipedic solid which is submitted to an *uniaxial tensile testing* (figure 7.1): a normal tensile stress σ_{33}, different than zero, is imposed on the faces perpendicular to the axis Ox_3, while the faces perpendicular to axis Ox_1 and Ox_2 are free of stress ($\sigma_{11}=0$ and $\sigma_{22}=0$). The state equation *(7.22)* allows us to calculate the components ε_{ii} of the tensor of deformation

$$\varepsilon_{11} = \varepsilon_{22} = \frac{1}{2\mu+3\lambda}\left(nk_0(T) - \frac{\lambda}{2\mu}\sigma_{33} \right) \quad \text{et} \quad \varepsilon_{33} = \frac{1}{2\mu+3\lambda}\left(nk_0(T) + \frac{\mu+\lambda}{\mu}\sigma_{33} \right) \quad (7.23)$$

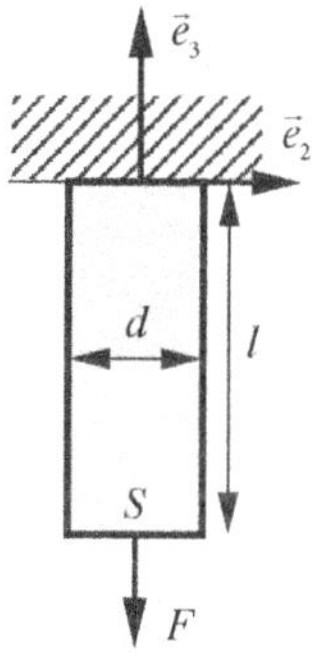

Figure 7.1 - *uniaxial tensile testing*

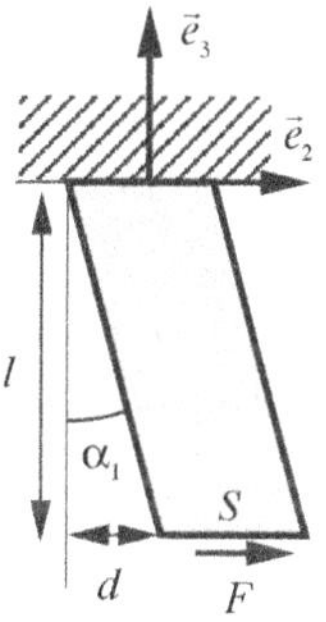

Figure 7.2 - *pure shear strain*

By definition, the *Young's modulus* E is defined by the *uniaxial tensile testing* (figure 7.1) as the ratio of the increase in the stress $\Delta\sigma_{33}$ along the tensile axis and the increase of the deformation $\Delta\varepsilon_{33}$ along the same axis, at a given temperature T

$$E = \frac{\Delta\sigma_{33}}{\Delta\varepsilon_{33}} = \frac{\mu(2\mu+3\lambda)}{\mu+\lambda} = n\frac{18k_1(T)k_2(T)}{6k_1(T)+k_2(T)} \quad (7.24)$$

With regards to the *Poisson's ratio* ν, it is defined as the ratio between the narrowing $-\Delta\varepsilon_{11}$ or $-\Delta\varepsilon_{22}$ along the transversal axis Ox_1 and Ox_2 and the stretching $\Delta\varepsilon_{33}$ along axis Ox_3 of traction, at a given temperature T

$$\nu = -\frac{\Delta\varepsilon_{11}}{\Delta\varepsilon_{33}} = -\frac{\Delta\varepsilon_{22}}{\Delta\varepsilon_{33}} = \frac{\lambda}{2(\mu+\lambda)} = \frac{3k_1(T)-k_2(T)}{6k_1(T)+k_2(T)} \quad (7.25)$$

The shear modulus G during pure shear strain

In the case of a deformation of the solid by *pure shear*, a parallelepipedic solid is deformed by shear in the plan i and j by applying tangential stress $\sigma_{ij} = \sigma_{ji}$ on the perpendicular planes to the axes Ox_i and Ox_j (figure 7.2). The state equation *(7.30)* allows us to calculate the components $\varepsilon_{ij} = \varepsilon_{ji}$ of the tensor of deformation as

$$\sigma_{ij} = 2\mu\varepsilon_{ij} = G\varepsilon_{ij} \tag{7.26}$$

The modulus G which appears here is called *shear modulus* or *modulus of rigidity* of the isotropic solid and its value is

$$G = 2\mu = 2nk_2 \tag{7.27}$$

The relations between the elastic moduli

In general moduli E, v and G are shown in the tables of elastic moduli of isotropic materials. From knowing E, v and G for a given material, we find the Lamé parameters μ and λ of this material by

$$\mu = \frac{E}{2(1-v)} = \frac{G}{2} \quad \text{and} \quad \lambda = \frac{Ev}{(1+v)(1-2v)} \tag{7.28}$$

As regards the elastic coefficients k_1 and k_2 of a rigid material, they can be deduced either from the μ and λ parameters, either from the moduli E, v and G

$$K_1 = \frac{\lambda}{2} + \frac{\mu}{3} = \frac{E}{6(1-2v)} \quad \text{et} \quad \lambda = \frac{Ev}{(1+v)(1-2v)} \tag{7.29}$$

The Poisson's ratio v of an isotropic material is very interesting, as it permits us to make the link between the k_1 coefficient associated to the compressibility of the media and the k_2 coefficient, linked to the shear-ability of the media. Indeed

$$\frac{k_1}{k_2} = \frac{1+v}{3(1-2v)} \tag{7.30}$$

For example, in metals, the Poisson's ratio is in general close to 1/3, which implies that the ratio k_1 / k_2 is approximatively 4/3.

7.3 - Thermal behavior of isotropic solids

In general, experiments show that, in solids at *high temperature*, the coefficient $k_0(T)$ as well as the thermal term $f_0(T)$ of the state function are simple functions of temperature. Indeed, in the usual solids, we observe a *specific heat* c_V measured by site of lattice, which at high temperature is approximatively a constant with value $c_V = 3k$ *(Dulong-Petit law)*, where k is the *Boltzmann constant*

$$c_V = T\frac{\partial s}{\partial T} = T\left[-\frac{\partial^2 f_0(T)}{\partial T^2} + \frac{\partial^2 k_0(T)}{\partial T^2}\tau^{el}\right] \cong cste \tag{7.31}$$

This relation can be easily satisfied if we make the hypothesis that the dependence on temperature of the elastic coefficient $k_0(T)$ is linear

$$T\frac{\partial^2 k_0(T)}{\partial T^2}\tau^{el} \cong 0 \;\Rightarrow\; k_0(T) \cong \gamma_0 c_V T \tag{7.32}$$

This hypothesis then implies for $f_0(T)$ that

$$-T\frac{\partial^2 f_0(T)}{\partial T^2} \cong c_V \;\Rightarrow\; f_0(T) \cong u_0 - Ts_0 + c_V T(1-\ln T) \tag{7.33}$$

The coefficient $k_0(T)$ depends more or less linearly on T via a constant $\gamma_0 c_V$, or γ_0 which is called the *Grünheisen parameter*, and whose numerical value is between 1,2 and 2,6 for most of the usual isotropic solids. We also observe, in the usual solids, that the coefficients k_1 and k_2 do not depend much on temperature, and can most of the time be treated as constants. These thermal behavior can be obtained from statistical physics, which shows the existence of the Grünheisen parameter γ_0 and that it is linked to the *anharmonic properties of the lattice.*

At constant pressure or with constant volume expansion, as in the case of weak distortions of the lattice, the introduction of *(7.32)* in *(7.15)* shows that the coefficient $k_0(T)$ is directly linked to the *phenomenon of thermic volume expansion* of the lattice

$$\tau^{el}(T)\Big|_{p_{cste}} \cong \frac{1}{2nk_1}\left(\gamma_0 n c_V T - p_{cste}\right) \quad \text{or} \quad p(T)\Big|_{\tau^{el}_{cste}} \cong n\left(\gamma_0 c_V T - 2k_1\tau^{el}_{cste}\right) \tag{7.34}$$

Let's imagine that a parallelepipedic solid is not submitted to any axial tensile stress, so that $\sigma_{ii}=0$, and we change its temperature. Because of coefficient $k_0(T)\cong\gamma_0 c_V T$, the uniaxial strain ε_{ii} (7.23) then becomes dependent on temperature, a phenomenon we call *linear thermal expansion of the solid*

$$\varepsilon_{11}(T)\Big|_{\sigma_{33}=0} = \varepsilon_{22}(T)\Big|_{\sigma_{33}=0} = \varepsilon_{33}(T)\Big|_{\sigma_{33}=0} = \frac{\gamma_0 n c_V T}{2\mu+3\lambda} \tag{7.35}$$

7.4 - Transport phenomena in isotropic solids

Let's consider the phenomenological equations of dissipative thermal conduction and self-diffusion *(6.41)*. In the case where a solid is isotropic and is not too removed from the thermodynamic equilibrium, and additionally if the concentrations of point defects remain small, we can linearize the vectorial dependences of $\vec{J}_q$, $\vec{J}_L$ and $\vec{J}_I$ on the thermodynamic forces $\vec{X}_q$, $\vec{X}_L$ and $\vec{X}_I$ by introducing *kinetic coefficients* of the type $L_{qL}(\tau,T,C_L,C_I)$, which allows us to write the equation of heat transport and the chemical species in the form

$$\begin{cases} \vec{J}_q = L_{qq}\vec{X}_q + L_{qL}\vec{X}_L + L_{qI}\vec{X}_I \\ \vec{J}_L = L_{Lq}\vec{X}_q + L_{LL}\vec{X}_L + L_{LI}\vec{X}_I \\ \vec{J}_I = L_{Iq}\vec{X}_q + L_{IL}\vec{X}_L + L_{II}\vec{X}_I \end{cases} \tag{7.36}$$

The source of entropy $S_e^{transport}$ associated to transport is written

$$S_e^{transport} = \sum_x \vec{X}_x\vec{J}_x \tag{7.37}$$

which we expand as

$$\begin{aligned} S_e^{transport} &= L_{qq}\vec{X}_q^2 + L_{LL}\vec{X}_L^2 + L_{II}\vec{X}_I^2 \\ &\quad + \left(L_{qL}+L_{Lq}\right)\vec{X}_q\vec{X}_L + \left(L_{qI}+L_{Iq}\right)\vec{X}_q\vec{X}_I + \left(L_{LI}+L_{IL}\right)\vec{X}_L\vec{X}_I \end{aligned} \tag{7.38}$$

and which can be transformed by expressing the scalar product in product of the components of the vectors

$$S_e^{transport} = \sum_{k=1}^{3} \left[L_{qq} X_{qk}^2 + L_{LL} X_{Lk}^2 + L_{II} X_{Ik}^2 + \left(L_{qL} + L_{Lq} \right) X_{qk} X_{Lk} + \left(L_{qI} + L_{Iq} \right) X_{qk} X_{Ik} + \left(L_{LI} + L_{IL} \right) X_{Lk} X_{Ik} \right] \quad (7.39)$$

The source of entropy is made up of the sum of three quadratic functions X_{qi}, X_{Li} and X_{Ii}. To insure that whatever the values of X_{qi}, X_{Li} and X_{Ii}, we have the positivity of the source of entropy as required by the second principle, it must be that these three functions are individually positive in the 3D space of X_{qi}, X_{Li} et X_{Ii}, in such a way that the three quadratic functions can be transformed as the sum of square with positive coefficients by an orthogonal transformation of the variables in 3D space. Mathematically speaking, this is only possible if the following matrix with kinetic coefficients, composed of real numbers

$$\begin{pmatrix} L_{qq} & L_{qL} & L_{qI} \\ L_{Lq} & L_{LL} & L_{LI} \\ L_{Iq} & L_{IL} & L_{II} \end{pmatrix} \quad (7.40)$$

is a symmetric matrix, which can be diagonalized with eigenvalues that are positive. We impose then the following conditions, which are known as the *Onsager relations*,

$$L_{qL} = L_{Lq} \quad ; \quad L_{qI} = L_{Iq} \quad ; \quad L_{LI} = L_{IL} \quad (7.41)$$

The coefficients of thermal conductivity and diffusion

In the case of an isotropic solid with no coupling between the flux of particle transport and heat, or if these couplings are negligible, the equations of transport can be rewritten by introducing a *thermal conductivity coefficient* χ *and two coefficients of diffusion* D_L and D_I for the vacancies and the interstitials respectively

$$\begin{cases} \vec{J}_q = L_{qq} \vec{X}_q = L_{qq} \, \overrightarrow{\text{grad}} \dfrac{1}{T} = -\dfrac{L_{qq}}{T^2} \, \overrightarrow{\text{grad}} T = -\chi \, \overrightarrow{\text{grad}} T \\ \vec{J}_L = L_{LL} \vec{X}_L = n C_L \dfrac{D_L}{k} \vec{X}_L \\ \vec{J}_I = L_{II} \vec{X}_I = n C_I \dfrac{D_I}{k} \vec{X}_I \end{cases} \quad (7.42)$$

If C_L and C_I are small, the previous coefficients do not depend on them, so that we have a dependency on the volume expansion and the temperature (of Boltzmann type for the diffusion coefficients)

$$\chi \cong \chi(\tau, T) \quad ; \quad D_L \cong D_{L0} \exp\left[-E_L^{diff}(\tau) / kT \right] \quad ; \quad D_I \cong D_{I0} \exp\left[-E_I^{diff}(\tau) / kT \right] \quad (7.43)$$

7.5 - Propagation of waves and thermoelastic relaxation

The thermo-plastic equations of the perfect solid

The exact computation of the spatiotemporal evolution of a perfect solid is a rather complex problem, because to all the local deformations of the lattice is associated a local perturbation of the temperature of the lattice. This coupling between deformations and temperature comes from the entropy expression of the solid. By neglecting the effects of the field of gravitation $\rho \vec{g}$, and assuming that there are no point defects, nor anelasticity or plasticity in the lattice, and by using relations *(3.5) and (3.10)*, one obtains for a given compatible solid

$$\begin{cases} \sum_k \vec{e}_k \operatorname{div}\vec{\alpha}_k = -\overrightarrow{\mathrm{rot}}\,\vec{\omega} + \frac{2}{3}\overrightarrow{\mathrm{grad}}\,\tau - \cancel{\sum_k\left(\vec{e}_k \wedge \overrightarrow{\mathrm{rot}}\,\vec{\beta}_k\right)} = -\overrightarrow{\mathrm{rot}}\,\vec{\omega} + \frac{2}{3}\overrightarrow{\mathrm{grad}}\,\tau \\ \sum_k\left(\overrightarrow{\mathrm{grad}}\,\tau\ \vec{\alpha}_k\right)\vec{e}_k = \overrightarrow{\mathrm{grad}}\,\tau \wedge \cancel{\sum_k\left(\vec{e}_k \wedge \vec{\alpha}_k\right)} + \sum_k\left(\vec{e}_k\,\overrightarrow{\mathrm{grad}}\,\tau\right)\vec{\alpha}_k = \sum_k\left(\vec{e}_k\,\overrightarrow{\mathrm{grad}}\,\tau\right)\vec{\alpha}_k \end{cases} \tag{7.44}$$

The equation of Newton *(6.11)* can be written by using the state equations *(7.15)* for $\vec{\alpha}_i$ and τ and by using the fact that $k_0(T) \cong \gamma_0 c_V T$ under the form

$$\frac{d\vec{p}}{dt} = -2k_2\,\overrightarrow{\mathrm{rot}}\,\vec{\omega} - 2k_2\sum_k\left(\vec{e}_k\,\overrightarrow{\mathrm{grad}}\,\tau\right)\vec{\alpha}_k + \left[\gamma_0 c_V T + \frac{4}{3}k_2 + 2k_1(1-\tau)\right]\overrightarrow{\mathrm{grad}}\,\tau - \gamma_0 c_V\,\overrightarrow{\mathrm{grad}}\,T \tag{7.45}$$

By using relation *(7.15)* for entropy and by introducing the expressions $k_0(T) \cong \gamma_0 c_V T$ and $f_0(T) \cong u_0 - Ts_0 + c_V T(1-\ln T)$, we obtain the expression of entropy at high temperature inside a perfect lattice

$$s = s_0 + c_V \ln T + \gamma_0 c_V \tau^{el} \tag{7.46}$$

By introducing that expression of entropy as well as the flux of heat $\vec{J}_q = -\chi\,\overrightarrow{\mathrm{grad}}\,T$ in the heat equation *(6.22)*, we obtain the following equation for the *evolution of temperature*

$$\frac{dT}{dt} = -\gamma_0 T\frac{d\tau}{dt} + \frac{\chi}{nc_V}\operatorname{div}\overrightarrow{\mathrm{grad}}\,T = -\gamma_0 T\frac{d\tau}{dt} + \frac{\chi}{nc_V}\Delta T \tag{7.47}$$

The equations *(7.45)* and *(7.47)* are called the *thermo-elastic equations* of the solid.

The propagation of thermo-elastic longitudinal waves

In an isotropic lattice without point defects and without anelasticity, first homogeneous in volume expansion ($\tau = \tau_0$), isothermal ($T = T_0$), non sheared and immobile, we introduce a *longitudinal perturbation* in a local frame $Ox_1x_2x_3$ under the form of a velocity field $\vec{\phi} = \vec{\phi}^{(p)} = \phi_j^{(p)}(x_j,t)\vec{e}_j$ parallel to the axis Ox_j and which changes along the axis Ox_j. Via the geometro-kinetic equations for τ and $\vec{\alpha}_i$, this perturbation of the velocity field implies perturbations $\tau^{(p)}$, $\vec{\omega}^{(p)}$ and $\vec{\alpha}_i^{(p)}$ of distortions along axis Ox_j

$$\begin{cases} \dfrac{d\tau^{(p)}}{dt} = \operatorname{div}\vec{\phi}^{(p)} & \Rightarrow \quad \tau = \tau_0 + \tau^{(p)}(x_j,t) \\ \dfrac{d\vec{\omega}^{(p)}}{dt} = \dfrac{1}{2}\overrightarrow{\mathrm{rot}}\,\vec{\phi}^{(p)} = 0 & \Rightarrow \quad \vec{\omega}^{(p)} = 0 \\ \dfrac{d\vec{\alpha}_i^{(p)}}{dt} = -\dfrac{1}{3}\vec{e}_i\operatorname{div}\vec{\phi}^{(p)} = -\dfrac{1}{3}\dfrac{d\tau^{(p)}}{dt}\vec{e}_i & \Rightarrow \quad \vec{\alpha}_i^{(p)} = -\dfrac{1}{3}\vec{e}_i\tau^{(p)}(x_j,t) \\ \dfrac{d\vec{\alpha}_j^{(p)}}{dt} = \overrightarrow{\mathrm{grad}}\,\phi_j^{(p)} - \dfrac{1}{3}\vec{e}_j\operatorname{div}\vec{\phi}^{(p)} = \dfrac{2}{3}\dfrac{d\tau^{(p)}}{dt}\vec{e}_j & \Rightarrow \quad \vec{\alpha}_j^{(p)} = +\dfrac{2}{3}\vec{e}_j\tau^{(p)}(x_j,t) \\ \dfrac{d\vec{\alpha}_k^{(p)}}{dt} = -\dfrac{1}{3}\vec{e}_k\operatorname{div}\vec{\phi}^{(p)} = -\dfrac{1}{3}\dfrac{d\tau^{(p)}}{dt}\vec{e}_k & \Rightarrow \quad \vec{\alpha}_k^{(p)} = -\dfrac{1}{3}\vec{e}_k\tau^{(p)}(x_j,t) \end{cases} \tag{7.48}$$

By considering the heat equation *(7.56)*, it is clear that there exists a perturbation of the temperature field along the Ox_j axis

$$T = T_0 + T^{(p)}(x_j, t) \tag{7.49}$$

By introducing these perturbations in the equations *(7.45)* and *(7.47)*, and by taking into account the first relation *(7.48)*, we obtain the following set of equations, which we have linearized for weak longitudinal perturbations

$$\begin{cases} \dfrac{\partial^2 \tau^{(p)}}{\partial t^2} \cong \dfrac{1}{m}\left[\gamma_0 c_V T_0 + \dfrac{4}{3}k_2 + 2k_1(1-\tau_0)\right]\Delta\tau^{(p)} - \dfrac{\gamma_0 c_V}{m}\Delta T^{(p)} \\ \dfrac{\partial T^{(p)}}{\partial t} \cong -\gamma_0 T_0 \dfrac{\partial \tau^{(p)}}{\partial t} + \dfrac{\chi}{n c_V}\Delta T^{(p)} \qquad \left(n = n_0 e^{-\tau_0}\right) \end{cases} \tag{7.50}$$

Possible solutions to this set of equations are the *high frequency, isothermal, dampened, longitudinal waves* as well as *the low frequency, adiabatic, non dampened, longitudinal waves*, with respective phase velocities c_{iso} and c_{adia}

$$\begin{cases} c_{iso} = \sqrt{\dfrac{1}{m}\left[\gamma_0 c_V T_0 + \dfrac{4}{3}k_2 + 2k_1(1-\tau_0)\right]} \\ c_{adia} = \sqrt{\dfrac{1}{m}\left[\gamma_0 c_V T_0 + \dfrac{4}{3}k_2 + 2k_1(1-\tau_0)\right] + \dfrac{\gamma_0^2 T_0 c_V}{m}} \end{cases} \tag{7.51}$$

The presence of the term $2k_1(1-\tau_0)$ implies that the velocity of wave propagation depends on the state of volume expansion τ_0 of the lattice.

Logarithmic decrement and the Debye relaxation due to thermo-elasticity

The longitudinal wave described by equations *(7.50)* presents a *spatial logarithmic decrement* Δ_{log}, defined as the logarithm of the ratio of two amplitudes of the wave measured at two points separated by a distance Δx_j which is equal to the wavelength $\lambda = c_\varphi / f$, and evaluates to

$$\Delta_{log} = \delta = \frac{1}{\pi}\ln\frac{\tau^{(p)}(x_j, t)}{\tau^{(p)}(x_j + \lambda, t)} = \Delta_{rel}\,\frac{\omega\tau_{rel}}{1+(\omega\tau_{rel})^2} \tag{7.52}$$

with $\Delta_{rel} = \dfrac{\gamma_0^2 T_0 c_V}{m c_{iso} c_{adia}}$ et $\tau_{rel} = \dfrac{\chi}{n c_V c_{iso} c_{adia}}$ (7.53)

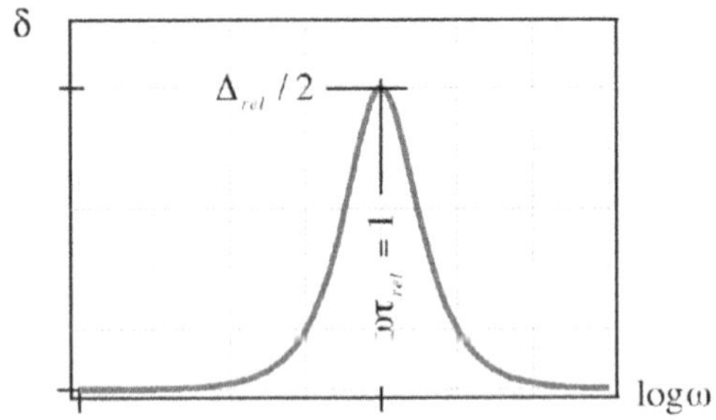

Figure 7.3 *- Loss factor and Debye relaxation due to thermo-elasticity*

The dependency of the *logarithmic decrement* Δ_{log} on the frequency, is also called the *loss factor* δ of the lattice. It is typical of a *relaxation phenomenon*. There appears a characteristic time τ_{rel}, called the *thermo-elastic relaxation time*. The logarithmic decrement Δ_{log} exhibits a *Debye relaxation peak* (figure 7.3) which finds its common origin in the thermal expansion and the process of the thermo-conduction. Indeed, the *relaxation amplitude* $\Delta_{rel} = \gamma_0^2 T_0 c_V / (mc_{iso} c_{adia})$ is proportional to γ_0^2 and depends as a consequence directly on the thermal expansion of the solid, while the relaxation time $\tau_{rel} = \chi / (nc_V c_{iso} c_{adia})$ is proportional to χ and depends then essentially to the process of thermo-conduction

The propagation of invariant transversal waves

If we consider deformations of the solid by pure shear, with constant volume expansion τ_0, the equation of Newton *(7.45)* simplifies into the form

$$\frac{d\vec{p}}{dt} = -2k_2 \overrightarrow{\mathrm{rot}}\,\vec{\omega} \tag{7.54}$$

By introducing transversal perturbations of the velocity field $\vec{\phi}^{(p)}$ under the form of a velocity field $\vec{\phi}^{(p)} = \phi_k^{(p)}(x_j, t)\vec{e}_k$ varying along axis Ox_j, we show with the geometro-kinetic equations that this perturbation of the velocity field implies the following perturbations for the volume expansion and the rotational vector along the axis Ox_j

$$\begin{cases} \dfrac{d\tau^{(p)}}{dt} = \mathrm{div}\,\vec{\phi}^{(p)} = 0 & \Rightarrow \quad \tau = \tau_0 = cste \\ \dfrac{d\vec{\omega}^{(p)}}{dt} = \dfrac{1}{2}\overrightarrow{\mathrm{rot}}\,\vec{\phi}^{(p)} = \dfrac{1}{2}\dfrac{\partial \phi_k^{(p)}}{\partial x_j}\vec{e}_i & \Rightarrow \quad \vec{\omega}^{(p)} = \omega_i^{(p)}(x_j, t)\vec{e}_i \end{cases} \tag{7.55}$$

it also implies the following perturbations for the shear tensor along the Ox_j axis, which are directly linked to the perturbations of rotation

$$\begin{cases} \dfrac{d\vec{\alpha}_i^{(p)}}{dt} = 0 & \Rightarrow \quad \vec{\alpha}_i^{(p)} = 0 \\ \dfrac{d\vec{\alpha}_j^{(p)}}{dt} = \dfrac{1}{2}\dfrac{\partial \phi_k^{(p)}}{\partial x_j}\vec{e}_k & \Rightarrow \quad \vec{\alpha}_j^{(p)} = \omega_i^{(p)}(x_j, t)\vec{e}_k \\ \dfrac{d\vec{\alpha}_k^{(p)}}{dt} = \dfrac{1}{2}\dfrac{\partial \phi_k^{(p)}}{\partial x_j}\vec{e}_j & \Rightarrow \quad \vec{\alpha}_k^{(p)} = \omega_i^{(p)}(x_j, t)\vec{e}_j \end{cases} \tag{7.56}$$

The momentum is linked to the velocity field $\vec{\phi}^{(p)}$ via relation $\vec{p}^{(p)} = m\vec{\phi}^{(p)}$ and the Newton equation can be written

$$\frac{d\vec{p}^{(p)}}{dt} = -2k_2 \overrightarrow{\mathrm{rot}}\,\vec{\omega}^{(p)} \quad \Rightarrow \quad \frac{d\vec{\phi}^{(p)}}{dt} = -2\frac{k_2}{m}\overrightarrow{\mathrm{rot}}\,\vec{\omega}^{(p)} \tag{7.57}$$

Because the perturbations $\vec{\phi}^{(p)}(\vec{r}, t) = \phi_k^{(p)}(x_j, t)\vec{e}_k$ only have one component along the Ox_k axis, and since all the propagating values vary only along the Ox_j axis, the material derivative d/dt can be replaced by the partial derivative $\partial/\partial t$ with respect to time everywhere in the local frame $Ox_1x_2x_3$. This allows us to obtain the following set of scalar differential equations, which control the transversal perturbation in the local frame of the perfect solid

$$\begin{cases} \dfrac{\partial \omega_i^{(p)}}{\partial t} = \dfrac{1}{2}\dfrac{\partial \phi_k^{(p)}}{\partial x_j} \\ \dfrac{\partial \phi_k^{(p)}}{\partial t} = 2\dfrac{k_2}{m}\dfrac{\partial \omega_i^{(p)}}{\partial x_j} \end{cases} \Rightarrow \begin{cases} \dfrac{\partial^2 \omega_i^{(p)}}{\partial t^2} = \dfrac{k_2}{m}\dfrac{\partial^2 \omega_i^{(p)}}{\partial x_j^2} \\ \dfrac{\partial^2 \phi_k^{(p)}}{\partial t^2} = \dfrac{k_2}{m}\dfrac{\partial^2 \phi_k^{(p)}}{\partial x_j^2} \end{cases} \tag{7.58}$$

Thus, no matter the amplitude of a transversal wave propagating in a perfect solid, it moves about in the local frame $Ox_1x_2x_3$ with a *wave velocity* equal to

$$c_t = \sqrt{\frac{k_2}{m}} \tag{7.59}$$

Therefore, in a perfect isotropic solid, the intrinsic velocity of transversal waves is an «*invariant quantity*», meaning one that does not depend on the volume expansion τ_0 of the lattice, on the condition that it be homogenous and constant. It should be also noted that, if the static volume expansion τ of the solid is not homogenous, meaning if $\overrightarrow{\text{grad}}\,\tau \neq 0$, the propagation of the transverse perturbations is impacted by the presence of a coupling term $-2k_2\sum\left(\vec{e}_k \overrightarrow{\text{grad}}\,\tau\right)\vec{\alpha}_k$ between the fields τ and $\vec{\alpha}_i$ in the Newton equation *(7.45)*. The propagation of transversal waves depends in this case on the gradient of the volume expansion τ, and the resolution of this propagation problem cannot do without the explicit use of the shear field $\vec{\alpha}_i$!

7.6 - Equations of transport and inertial relaxation

The equations of transport for point defects *(7.42)* can be transformed by introducing the fluxes $\vec{J}_L$ et $\vec{J}_I$ *(5.78)* and the forces $\vec{X}_L$ and $\vec{X}_I$ *(6.42)*

$$\begin{cases} \Delta\vec{\varphi}_L = \vec{\phi}_L - \vec{\phi} = \dfrac{D_L(\tau,T)}{kT}\left(-\overrightarrow{\text{grad}}\;\mu_L^* + m\dfrac{d}{dt}\left(\vec{\phi}_L - 2\Delta\vec{\varphi}_L\right) - m\vec{g}\right) \\ \Delta\vec{\varphi}_I = \vec{\phi}_I - \vec{\phi} = \dfrac{D_I(\tau,T)}{kT}\left(-\overrightarrow{\text{grad}}\;\mu_I^* - m\dfrac{d}{dt}\left(\vec{\phi}_I\right) + m\vec{g}\right) \end{cases} \tag{7.60}$$

The new equations link the relative velocities of the point defects with respect to the lattice to the chemical forces and the mechanical forces, via the coefficients D/kT. Such equations bear the name of *Einstein's diffusion relations*. It is possible to transform these equations to explicitly show the *equations of movement* of the vacancies and interstitials with respect to the lattice, along a path of the lattice, by decomposing the absolute velocity of the vacancies and regrouping the terms in a different way. We obtain

$$\begin{cases} m\dfrac{d\Delta\vec{\varphi}_L}{dt} + B_L(\tau,T)\Delta\vec{\varphi}_L = +m\dfrac{d\vec{\phi}}{dt} - m\vec{g} - \overrightarrow{\text{grad}}\;\mu_L^* \\ m\dfrac{d\Delta\vec{\varphi}_I}{dt} + B_I(\tau,T)\Delta\vec{\varphi}_I = -m\dfrac{d\vec{\phi}}{dt} + m\vec{g} - \overrightarrow{\text{grad}}\;\mu_I^* \end{cases} \tag{7.61}$$

in which $B_L(\tau,T)$ and $B_I(\tau,T)$ are the *coefficients of viscous friction* of the vacancies and interstitials respectively, and are linked to the coefficients of diffusion by

$$B_L(\tau,T) = \frac{kT}{D_L(\tau,T)} \quad \textit{and} \quad B_I(\tau,T) = \frac{kT}{D_I(\tau,T)} \tag{7.62}$$

In the equations of movement *(7.61)*, the relative dampened movement of point defects with respect to the lattice is controlled by three different applied forces: a drag-along force due to the acceleration $md\vec{\phi}/dt$ of the lattice, a force $m\vec{g}$ due to gravity, a chemical and kinetic force equal to the gradient of μ_L^* or of μ_I^*.

The processes of inertial relaxation of point defects

If a solid lattice whose point defects are in thermodynamical equilibrium is submitted to a vibration, characterized by an absolute oscillating velocity field $\vec{\phi}$, the movement of the point defects inside the solid is described by the equations of movement *(7.61)*, in which the terms $-\overrightarrow{\text{grad}}\ \mu_L^*$ and $-\overrightarrow{\text{grad}}\ \mu_I^*$ can be considered null since the point defects are at thermodynamical equilibrium. By also neglecting the gravity field, and by imposing in a direction $\vec{n}$ a vibration $\vec{\phi}(t)=\vec{n}\phi_0\cos\omega t$ with angular frequency ω to the isotropic lattice, the field of absolute velocity of the solid can be written, for convenience, by using the formalism of *rotating vectors in the complex plane* (we will underscore complex quantities)

$$\underline{\vec{\phi}}(t)=\vec{n}\phi_0\,\mathrm{e}^{i\omega t} \tag{7.63}$$

According to equations *(7.61)*, the point defects are going to start oscillating but with amplitudes and phase shift angles δ_L and δ_I, which is different for the vacancies and the interstitials

$$\begin{cases} \underline{\Delta\vec{\varphi}_L}(t)=\vec{n}\Delta\varphi_{L0}\phi_0\,\mathrm{e}^{i(\omega t-\delta_L)}=\underline{\alpha_L}\ \underline{\vec{\phi}}(t) \\ \underline{\Delta\vec{\varphi}_I}(t)=\vec{n}\Delta\varphi_{I0}\phi_0\,\mathrm{e}^{i(\omega t-\delta_I)}=\underline{\alpha_I}\ \underline{\vec{\phi}}(t) \end{cases} \tag{7.64}$$

By introducing the complex coefficients $\underline{\alpha_L}$ and $\underline{\alpha_I}$, the equations of movement *(7.61)* allow us to deduce the values of these coefficients, which are associated to the oscillations of vacancies and interstitials by the following relations

$$\underline{\alpha_L}=-\frac{m\omega}{iB_L-m\omega}=\frac{(\tau_L\omega)^2+i\tau_L\omega}{1+(\tau_L\omega)^2} \quad\text{and}\quad \underline{\alpha_I}=\frac{m\omega}{iB_I-m\omega}=-\frac{(\tau_I\omega)^2+i\tau_I\omega}{1+(\tau_I\omega)^2} \tag{7.65}$$

in which τ_L and τ_I are the *relaxation times for vacancies and interstitials*

$$\tau_L=\frac{m}{B_L}=\frac{mD_L}{kT} \quad\text{and}\quad \tau_I=\frac{m}{B_I}=\frac{mD_I}{kT} \tag{7.66}$$

The real fields of the relative velocities of point defects with respect to the lattice are then deduced as the real part of $\underline{\Delta\vec{\varphi}_L}(t)$ and $\underline{\Delta\vec{\varphi}_I}(t)$. The average momentum $\vec{p}=m(\vec{\phi}-C_L\vec{\phi}_L+C_I\vec{\phi}_I)$ per site of lattice is deduced in the complex representation, which manifests a *complex apparent mass* $\underline{m}$

$$\underline{\vec{p}}(t)=m\left[1-C_L\frac{1+2(\tau_L\omega)^2+i\tau_L\omega}{1+(\tau_L\omega)^2}+C_I\frac{1-i\tau_I\omega}{1+(\tau_I\omega)^2}\right]\underline{\vec{\phi}}(t)=\underline{m}\underline{\vec{\phi}}(t) \tag{7.67}$$

The average momentum $\vec{p}$ per site of lattice in the presence of an oscillatory movement of the lattice then has two contributions: a contribution which is in phase with the movement of the lattice and an other one which is out of phase by $\pi/2$. There appears in fact two *relaxation phenomena* on the average momentum $\vec{p}(t)$ per site of lattice, which are due respectively to the movement of vacancies and interstitials inside the lattice. These relaxation processes depend strongly on the frequency ω of the oscillatory movement and give rise to open cycles when we plot $\vec{pn}$ as a function of $\vec{\phi n}$, as shown in figure 7.4, in which the effects along the vertical axis have been greatly exaggerated

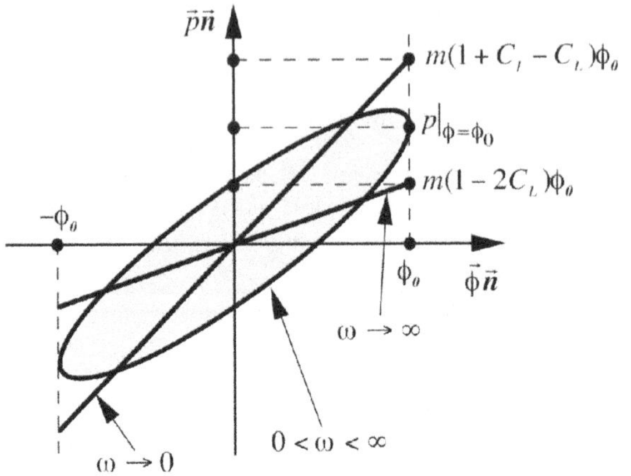

Figure 7.4 - *relaxation cycle of* $\vec{pn}$ as a function of $\vec{\phi n}$

7.7 - Creation-annihilation of a vacancy-interstitial pair

The phenomenological equation *(6.50)* giving the volume source S_{I-L} of pairs of vacancies and interstitials in the lattice can be deduced thanks to a very simplified model of creation-annihilation. Let's assume that the energetic diagram, as a function of distance d separating a vacancy from an interstitial, presents a maximum for a distance which will be called the *activation distance* d^{act} (figure 7.5). For distances d different than zero, but inferior to d^{act}, the vacancy and interstitial do not have a real existence. There are only thermal fluctuations of the position of the particle around the site it occupies. For distances d superior to d^{act}, the vacancy and interstitial become real point defects of the lattice, that have in common an increase in the pair energy $\Delta\varepsilon_0$ with respect to the initial energy of the particle on its initial lattice site. The creation of a pair is due to a thermal fluctuation which momentarily is equal or superior to $\Delta\varepsilon^{act}$. This is why $\Delta\varepsilon^{act}$ will be called *activation energy* of a vacancy-interstitial pair. The annihilation of a vacancy-interstitial pair, meaning the reverse path in the energetic diagram, also needed a thermal fluctuation but of energy superior or equal to $\Delta\varepsilon^{act}-\Delta\varepsilon_0$. The annihilation is thus a thermally activated process which has an *activation energy of annihilation* equal to $\Delta\varepsilon^{act}-\Delta\varepsilon_0$.

The value of the activation energy $\Delta\varepsilon^{act}$ of creation of a pair of vacancy-interstitial is evidently linked to the nature of the lattice under consideration, it must also depend strongly on the hydrostatic pressure p or the local volume expansion τ of the lattice so that

$$\Delta\varepsilon^{act}=\Delta\varepsilon^{act}(\tau) \qquad (7.68)$$

With regards to the activation energy $\Delta\varepsilon^{act}-\Delta\varepsilon_0$ of annihilation of a vacancy-interstitial pair, it depends on $\Delta\varepsilon^{act}(\tau)$, but also on the increase $\Delta\varepsilon_0$ of the energy of the vacancy-interstitial pair with respect to the energy of the particle in its original site. This energy increase $\Delta\varepsilon_0$ is nothing but the sum of *potential energies* $\varepsilon_L(\tau)$ and $\varepsilon_I(\tau)$ of the vacancy and interstitial, introduced previously in the state function Δf^{dp}, and the *increase in kinetic energy* Δe_{cin}^{I-L} defined by the relation *(6.48)*

$$\Delta\varepsilon_0 = \varepsilon_L(\tau) + \varepsilon_I(\tau) + \Delta e_{cin}^{I-L} \tag{7.69}$$

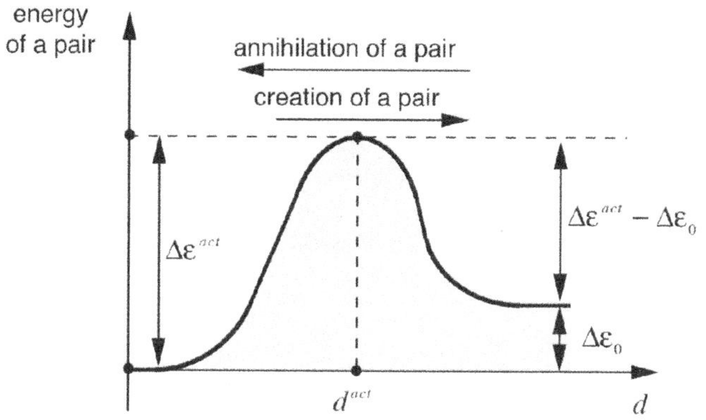

Figure 7.5 *- Energy diagram representing the creation-annihilation of a vacancy-interstitial pair as a function of distance* d

The terms of creation and annihilation of pairs can thus be written

$$S_{I-L}^{création} = n(1-C_L)v_c \exp\left(-\frac{\Delta\varepsilon^{act}}{kT}\right) \tag{7.70}$$

$$S_{I-L}^{annihilation} = nC_L C_I v_a \exp\left(-\frac{\Delta\varepsilon^{act} - \Delta\varepsilon_0}{kT}\right) \tag{7.71}$$

Indeed, as S_{I-L} represents a volume source term, S_{I-L} must be proportional to the volume density of sites n. In the case of creation, only occupied substitutional sites can contribute to the formation of a pair, hence the proportionality in $(1-C_L)$. In the case of annihilation, it is clear that the number of annihilations must be proportional to both C_L and C_I of vacancies and interstitials, and thus to the product $C_L C_I$.

In the two expressions we have a frequency of attack $v_c(\tau)$ or $v_a(\tau)$, which corresponds to the frequency at which favorable conditions to the creation or annihilation of a pair occur. This frequency is essentially a function of the local thermal fluctuations and must also depend on τ.

Finally, each of the expressions is multiplied by a *Boltzmann term* corresponding to the probability of the appearance of a thermal fluctuation with an energy equal or superior to the activation energies $\Delta\varepsilon^{act}$ or $\Delta\varepsilon^{act} - \Delta\varepsilon_0$, which allows us to 'jump' above the energy peak shown in figure 7.5. With this elementary model, the term S_{I-L} which is a global source or sink of pairs can be written by remembering that concentration C_L is supposed weak

$$S_{I-L} = S_{I-L}^{création} - S_{I-L}^{annihilation} = n\left[\left(1-C_L\right)\nu_c - C_L C_I \nu_a \exp\left(\frac{\Delta\varepsilon_0}{kT}\right)\right]\exp\left(-\frac{\Delta\varepsilon^{act}}{kT}\right)$$
$$\cong n\left[\nu_c - C_L C_I \nu_a \exp\left(\frac{\varepsilon_L(\tau)+\varepsilon_I(\tau)+\Delta e_{cin}^{I-L}}{kT}\right)\right]\exp\left(-\frac{\Delta\varepsilon^{act}}{kT}\right) \quad (7.72)$$

The global term of source S_{I-L} is null when the argument between brackets is null, meaning when the product $C_L C_I$ is approximatively equal to

$$C_L C_I \cong \frac{\nu_c}{\nu_a}\exp\left(-\frac{\Delta\varepsilon_0}{kT}\right) = \frac{\nu_c}{\nu_a}\exp\left(-\frac{\varepsilon_L(\tau)+\varepsilon_I(\tau)}{kT}\right)\exp\left(-\frac{\Delta e_{cin}^{I-L}}{kT}\right) \quad (7.73)$$

This condition is nothing more than the *relation of kinetic equilibrium* of the mechanism of creation and annihilation of pairs of vacancy-interstitial. For a temperature T and a volume expansion τ of the solid at thermodynamical equilibrium, it is possible to calculate the *thermodynamical equilibrium concentrations* of vacancies and interstitials from the state function f, simply by looking for the values of C_L and C_I which minimize f. It follows as a consequence the following values for $C_L^{équilibre}$ and $C_I^{équilibre}$

$$\begin{cases} \dfrac{\partial f}{\partial C_L} = \mu_L = 0 \quad \Rightarrow \quad C_L^{équilibre} = \exp\left(\dfrac{s_L(\tau)}{k} - 1\right)\exp\left(-\dfrac{\varepsilon_L(\tau)}{kT}\right) \\ \dfrac{\partial f}{\partial C_I} = \mu_I = 0 \quad \Rightarrow \quad C_I^{équilibre} = z_I \exp\left(\dfrac{s_I(\tau)}{k} - 1\right)\exp\left(-\dfrac{\varepsilon_I(\tau)}{kT}\right) \end{cases} \quad (7.74)$$

These two conditions of thermodynamical equilibrium allows us to calculate the *thermodynamic equilibrium relation* of the mechanism of creation-annihilation of pairs, under the form

$$C_L^{équilibre} C_I^{équilibre} = z_I \exp\left(\frac{s_L(\tau)+s_I(\tau)}{k} - 2\right)\exp\left(-\frac{\varepsilon_L(\tau)+\varepsilon_I(\tau)}{kT}\right) \quad (7.75)$$

As the relations of *kinetic equilibrium (7.73)* and *thermodynamical equilibrium (7.75)* must be equal at thermodynamical equilibrium of the solid, it is necessary that the two following equations be true at thermodynamical equilibrium

$$\begin{cases} \dfrac{\nu_c}{\nu_a} = z_I \exp\left(\dfrac{s_L(\tau)+s_I(\tau)}{k} - 2\right) \\ \dfrac{\Delta e_{cin}^{I-L}}{kT} = 0 \end{cases} \quad (7.76)$$

The second relation implies that at thermo-dynamical equilibrium, the kinetic energies of the vacancies and interstitials thermalize, which is expressed by the following equations, with the help of *(6.48)*

$$\Delta e_{cin}^{I-L} = \Delta e_{cin}^{L} + \Delta e_{cin}^{I} = e_{cin}^{L} + e_{cin}^{I} = 0 \quad (7.77)$$

This translates in term of relative velocities of vacancies and interstitials into the relation

$$\Delta e_{cin}^{I-L} = m\vec{\phi}\left(\Delta\vec{\varphi}_I - \Delta\vec{\varphi}_L\right) + \frac{1}{2}m\left(\Delta\vec{\varphi}_I^2 + \Delta\vec{\varphi}_L^2\right) = 0 \quad \textit{(at thermodynamical equilibrium)} \quad (7.78)$$

With regards to the first equation, it implies the existence of a strong relationship between the activation frequencies $\nu_c(\tau)$ and $\nu_a(\tau)$ and the entropies $s_L(\tau)$ and $s_I(\tau)$ of vacancies and interstitials

$$s_L(\tau)+s_I(\tau)=k\ln\left(\frac{e^2\nu_c(\tau)}{z_I\nu_a(\tau)}\right)=k\ln\Omega^{vibration} \tag{7.79}$$

This shows that the entropies $s_L(\tau)$ and $s_I(\tau)$ measure in fact the variations of $\Omega^{vibration}$, the number of possible *microscopic vibrational configurations of the lattice* in the neighborhood of vacancies and interstitials. From this relation is deduced, thanks to the state equations *(7.15)* for μ_L and μ_I, another relation giving $\nu_a(\tau)$ as a function of $\nu_c(\tau)$ and the chemical potentials

$$\nu_a(\tau)=\frac{\nu_c(\tau)}{z_I}\exp\left(2-\frac{s_L(\tau)+s_I(\tau)}{k}\right)=\frac{\nu_c(\tau)}{C_LC_I}\exp\left(\frac{\mu_L+\mu_I}{kT}\right)\exp\left(-\frac{\varepsilon_L(\tau)+\varepsilon_I(\tau)}{kT}\right) \tag{7.80}$$

This last relation allows us to express the global term S_{I-L} of sources or sinks of pairs *(7.72)* with the following form

$$S_{I-L}\cong n\nu_c\exp\left(-\frac{\Delta\varepsilon^{act}(\tau)}{kT}\right)\left[1-\exp\left(\frac{\mu_L^*+\mu_I^*}{kT}\right)\right] \tag{7.81}$$

which corresponds perfectly to the form *(6.50)* previously hypothesized for the phenomenological equation of creation-annihilation of a vacancy-interstitial pair.

7.8 - Phenomenology of anelasticity

The phenomenological equations of anelasticity

The global shear $\vec{\alpha}_i$ of an anelastic solid are given by the sum of the elastic and anelastic shears of a lattice

$$\vec{\alpha}_i=\vec{\alpha}_i^{el}+\vec{\alpha}_i^{an} \tag{7.82}$$

In this constitutive relation, the tensor of elastic shear and the conservative part of the tensor of anelastic shear are deduced from *(7.15)*

$$\begin{cases} \vec{s}_i=2nk_2(T)\vec{\alpha}_i^{el}=K_{el}\vec{\alpha}_i^{el} & \Rightarrow \quad K_{el}=K_{el}(n,T)=2nk_2(T) \\ \vec{s}_i^{\,cons}=nk_{an}(T,\tau^{el})\vec{\alpha}_i^{an}=K_{an}\vec{\alpha}_i^{an} & \Rightarrow \quad K_{an}=K_{an}(n,T)=nk_{an}(T,\tau^{el}) \end{cases} \tag{7.83}$$

It is the dissipative part of the phenomenological equation of anelasticity *(6.58)* which must essentially translate the dynamics of the structural defects responsible for anelasticity. It thus takes different forms depending on the obstacles and interaction that counter the movement of defects in the structure. Four typical cases of this dynamic will be reviewed now.

The process of viscous friction and the phenomenon of anelastic relaxation

Since the dissipative part of the phenomenological equation of anelasticity must essentially depend on the velocities of anelastic distortions, the simplest form we can imagine between $\vec{s}_i^{\,dis}$ and $d\vec{\alpha}_i^{an}/dt$ is simple proportionality. We introduce a *viscous friction coefficient* $B_{an}(n,T)$ which links $\vec{s}_i^{\,dis}$ with $d\vec{\alpha}_i^{an}/dt$ and which can depend on temperature T and the density of lattice sites

$$\vec{s}_i^{\,dis} = B_{an}(n,T)\frac{d\vec{\alpha}_i^{an}}{dt} \tag{7.84}$$

Therefore, the phenomenological equation of dissipative anelasticity *(6.58)* becomes a differential equation of first order in $\vec{\alpha}_i^{an}$, which allows us to link the behavior of anelastic shear $\vec{\alpha}_i^{an}$ to the spatiotemporal evolution of the field of elastic stress $\vec{s}_i$

$$\vec{s}_i = \vec{s}_i^{\,cons} + \vec{s}_i^{\,dis} = K_{an}(n,T)\,\vec{\alpha}_i^{an} + B_{an}(n,T)\,\frac{d\vec{\alpha}_i^{an}}{dt} \tag{7.85}$$

The elastic and anelastic shears of the lattice can be represented by a rheological model (figure 7.6a). In fact, the rheological model of the anelastic solid described above corresponds to put in parallel a spring of constant K_{an} with a viscous damper B_{an} in order to represent the anelastic part of the shears described by the equation of anelasticity. To this we add, in series, a spring of constant K_2 to represent the elastic part of the shears described by the equation of state *(15.5)*. In this case, it is clear that the dynamic structural defects responsible for anelasticity is controlled by a mechanism leading to *viscous friction* .

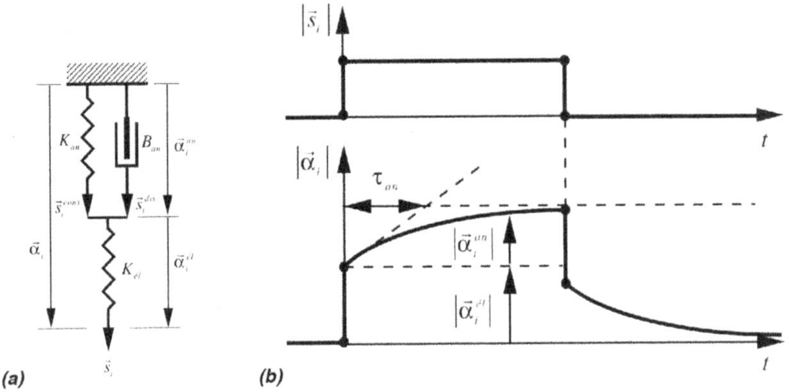

(a) (b)

Figure 7.6 *- **(a)** Rheological model of viscous friction*
*(**b**) instantaneous elastic response and relaxation*

With such rheological model, the response of the total shear $\vec{\alpha}_i$ to a solicitation by jump of the shear stress $\vec{s}_i$ has an instantaneous elastic part and a delayed anelastic portion (figure 7.6b), along with a *relaxation time* τ_{an} equal to

$$\tau_{an} = \frac{B_{an}}{K_{an}} \tag{7.86}$$

If the lattice is loaded by an oscillatory shear stress field $\vec{s}_i\,(t) = \vec{n}s_{i0}\cos\omega t$ oscillating in a certain direction $\vec{n}$, the anelastic response will then present a *relaxation phenomenon* at a frequency $\omega = 1/\tau_{an}$ similar to that reported in Figure 7.3. The temporal response of the overall shear field $\vec{\alpha}_i(t)$ to this cyclical shear stress $\vec{s}_i(t)$ itself has an amplitude and opening which correspond to the frequency of oscillation (fig. 7.7).

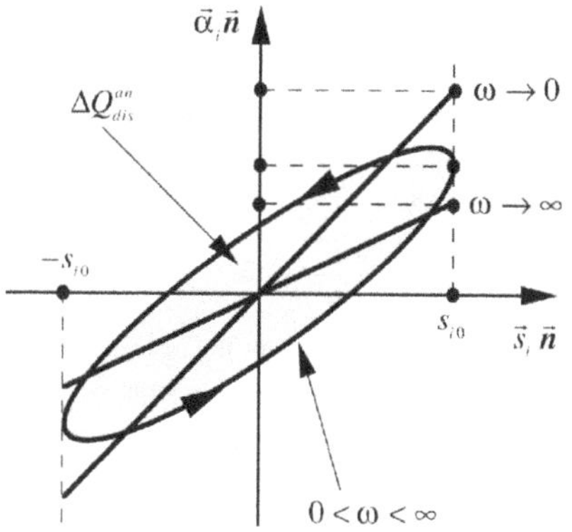

***Figure 7.7** - Hysteresis cycle in presence of viscous friction*

If the temperature T and the volume expansion τ^{el} of the medium can be considered as constant over time and space, and only in this case, equations *(7.83)* and *(7.85)* can be combined so as to write a single differential equation connecting directly together the shear strain tensor $\vec{\alpha}_i$ and the shear stress tensor $\vec{s}_i$. In this relationship the two components $\vec{\alpha}_i^{el}$ and $\vec{\alpha}_i^{an}$ *do not appear anymore*

$$B_{an}\frac{d\vec{\alpha}_i}{dt}+K_{an}\vec{\alpha}_i=\frac{B_{an}}{K_{el}}\frac{d\vec{s}_i}{dt}+\left(1+\frac{K_{an}}{K_{el}}\right)\vec{s}_i \qquad (7.87)$$

Existence of an inertial term and the phenomenon of anelastic resonance

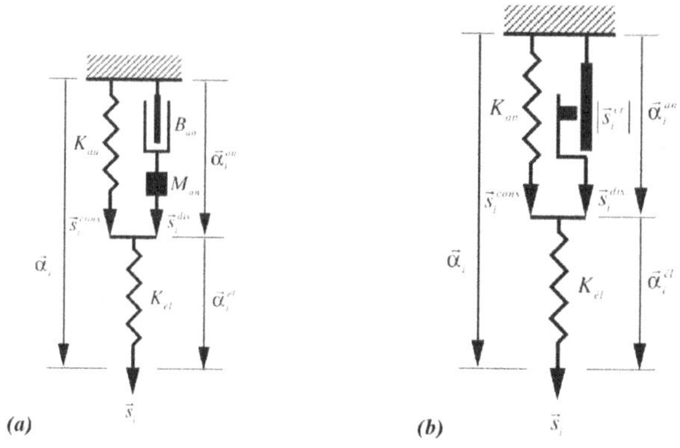

***Figure 7.8** - rheological models **(a)** with viscous friction and inertial mass*
***(b)** with a dry friction pad*

It is still possible that structural defects responsible for anelasticity have their own inertial mass, the effect of which is significant vis-à-vis the viscous friction forces. In this case, the dissipative portion of the phenomenological equation of anelasticity must contain this inertia effect in the form of a term proportional to the second derivative in the tensor of anelastic shears, multiplied by a *coefficient of inertia* $M_{an}(n,T)$ representing the inertial mass involved and which may also depend on the network site density

$$\vec{s}_i^{dis} = M_{an}(n,T)\,\frac{d^2\vec{\alpha}_i^{an}}{dt^2} + B_{an}(n,T)\,\frac{d\vec{\alpha}_i^{an}}{dt} \qquad (7.88)$$

The presence of the inertial mass transforms the dissipative anelasticity phenomenological equation *(6.58)* in a differential equation of second order in $\vec{\alpha}_i^{an}$, which can be represented by a rheological model of an anelastic solid containing an inertial mass (fig. 7.8a)

$$\vec{s}_i = \vec{s}_i^{cons} + \vec{s}_i^{dis} = K_{an}(n,T)\,\vec{\alpha}_i^{an} + B_{an}(n,T)\,\frac{d\vec{\alpha}_i^{an}}{dt} + M_{an}(n,T)\,\frac{d^2\vec{\alpha}_i^{an}}{dt^2} \qquad (7.89)$$

The equation of anelasticity of the lattice *(7.89)* involves an *anelastic relaxation time* τ_{an} and a *resonance frequency* ω_0, given by the expressions

$$\tau_{an} = \frac{B_{an}}{K_{an}} \quad \text{and} \quad \omega_0 = \sqrt{\frac{K_{an}}{M_{an}}} \qquad (7.90)$$

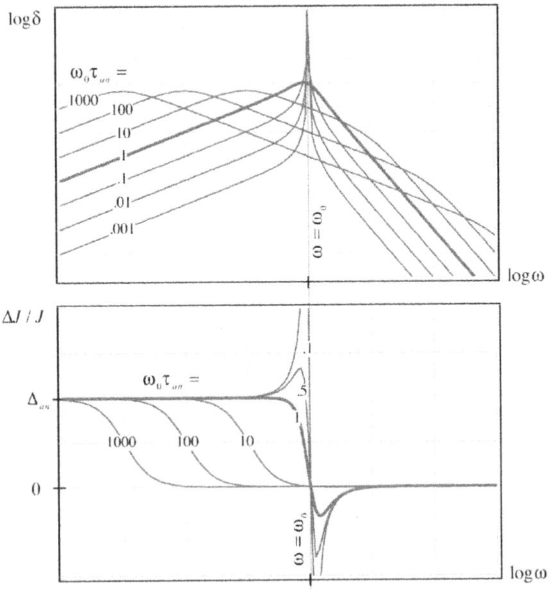

Figure 7.9 - *relaxation or resonance observed as a function of* $\omega_0\tau_{an}$

If the lattice is loaded by an oscillatory shear stress field $\vec{s}_i(t) = \vec{n}s_{i0}\cos\omega t$ in a certain direction $\vec{n}$, the anelastic response may present (fig. 7.9):

- a *phenomenon of relaxation* at frequency $\omega = 1/\tau_{an}$ if $\omega_0\tau_{an} > 1$, because the resonance frequency ω_0 is in this case higher than the relaxation frequency $1/\tau_{an}$,

- a *phenomenon of frequency resonance* at $\omega = \omega_0$ if $\omega_0 \tau_{an} \leq 1$, because the resonant frequency ω_0 is lower than the relaxation frequency $1/\tau_{an}$.

In the case where the temperature T and the volume expansion τ^{el} of the medium can be considered as constant over time and space, the equations *(7.83)* and *(7.86)* can be combined in a single differential relationship between $\vec{\alpha}_i$ and $\vec{s}_i$

$$M_{an}\frac{d^2\vec{\alpha}_i}{dt^2} + B_{an}\frac{d\vec{\alpha}_i}{dt} + K_{an}\vec{\alpha}_i = \frac{M_{an}}{K_{él}}\frac{d^2\vec{s}_i}{dt^2} + \frac{B_{an}}{K_{él}}\frac{d\vec{s}_i}{dt} + \left(1+\frac{K_{an}}{K_{él}}\right)\vec{s}_i \qquad (7.91)$$

The process of dry friction and the phenomenon of anelastic hysteresis

The two examples of dynamics presented above involve linear phenomena, including a linear coefficient of restoring force K_{an} and a linear viscous damping coefficient B_{an}. The phenomenological relationship of anelasticity is therefore a linear differential equation of first or second order, depending on whether or not there is an inertial mass M_{an}. But nothing prevents *a priori* the existence of strongly nonlinear anelasticity mechanisms. For example, the relationship describing the dissipative term of anelasticity could present a *dry friction* behavior. The easiest way to think of such behavior is to replace in the rheological model in Figure 7.6a, the viscous damper by a *dry friction pad* (fig. 7.8b) whose property is to be blocking while the absolute value $\left|\vec{s}_i^{dis}\right|$ of the stress $\vec{s}_i^{dis}$ is below a certain *critical stress,* $\left|\vec{s}_i^{cr}\right| = s_i^{cr}$, which can depend on the temperature T and the frequency ω of the stress $\vec{s}_i$, and become infinitely flexible when the absolute value $\left|\vec{s}_i^{dis}\right|$ of the stress $\vec{s}_i^{dis}$ becomes equal to the critical stress $\vec{s}_i^{cr}$. Mathematically, the phenomenological equation of dissipative anelasticity *(6.58)* is then replaced by a system of equations and inequalities

$$\begin{cases} \dfrac{d\vec{\alpha}_i^{an}}{dt} = 0 & si \quad \left|\vec{s}_i^{dis}\right| < \left|\vec{s}_i^{cr}\right| \\ K_{an}(n,T)\,\vec{\alpha}_i^{an} = \vec{s}_i - \vec{s}_i^{cr} & si \quad \vec{s}_i^{dis} = \vec{s}_i^{cr} \\ K_{an}(n,T)\,\vec{\alpha}_i^{an} = \vec{s}_i + \vec{s}_i^{cr} & si \quad \vec{s}_i^{dis} = -\vec{s}_i^{cr} \end{cases} \qquad (7.93)$$

In this case, the anelastic response to an oscillatory shear field $\vec{s}_i(t) = \vec{n}s_{i0}\cos\omega t$ in the direction $\vec{n}$ does not depend on the frequency ω, but only on the amplitude s_{i0} of the shear stress field. This response can be deduced from equations *(7.93)* and is shown in Figure 7.10. If the amplitude s_{i0} is smaller than the critical stress value of 'undocking', $s_i^{cr} = \left|\vec{s}_i^{cr}\right|$, there is no shear anelastic response ($\vec{\alpha}_i^{an} = 0$). On the other hand, when the amplitude s_{i0} exceeds the critical stress value of undocking, there appears a hysteresis cycle like the one shown in the right figure.

Thus, when the solid's anelasticity is controlled by a dry friction, the behavior of the tensor of anelastic shear $\vec{\alpha}_i^{an}$ explicitly depends only on the magnitude s_{i0} of the shear stress tensor.

One should note that if the shear stress $\vec{s}_i(t)$ is canceled after one cycle whose amplitude s_{i0} has exceeded the critical stress s_i^{cr}, there appears a *non-zero residual value* of the anelastic shear tensor $\vec{\alpha}_i^{an}$. This is sometimes called the phenomenon of "*micro-plasticity*". However, if the dry friction under consideration here is actually an approximation of a highly nonlinear function, which will always be the case if the anelastic process is thermally activated, then this residual value will gradually disappear over time as a result of the restoring anelastic force, so that over the medium term, the anelastic deformation will eventually be fully recovered.

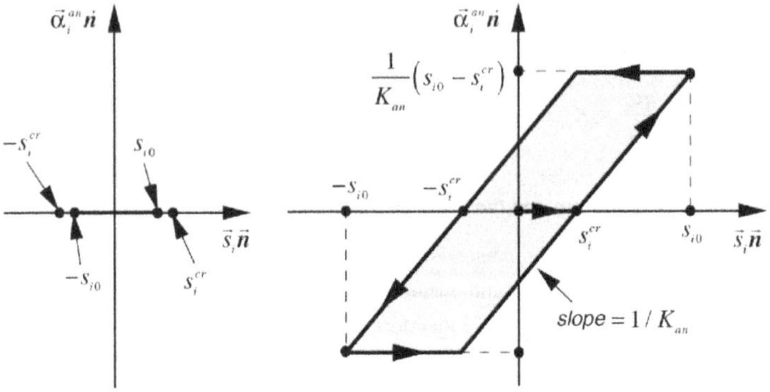

Figure 7.10 - *Hysteresis cycles* $\vec{\alpha}_i^{an}$ *in the presence of a dry friction*

7.9 - Displacive structural transitions of the first and second order

The existence of phenomena of structural phase transitions in structural nature of the lattice, including *martensitic phase transitions* are sources of anelasticity within the solid. In this section, we will develop for example two simple models for displacive martensitic transformations of the second and first order, meaning, structural transformations without diffusion leading to a homogeneous deformation of the crystal lattice such that it results in a shear deformation on a microscopic scale if the lattice is submitted to an external applied stress.

An example of displacive martensitic transition of second order

We can develop a very simplified imaginary model of martensitic phase transitions, by assuming a lattice with a unit volume slightly distorted by a given local shear angle ψ_0, representing a local shear $\alpha_{ij}^{(0)}$. In a given direction, successive atomic planes can have a positive ψ_0 shear angle and we will call it *variant* ***A***, or a negative shear angle $-\psi_0$ and we will then call it variant B (fig. 7.11).

Two successive plans of the lattice in a given direction then have two options of arrangements, namely two planes with the same local shear $\alpha_{ij}^{(0)}$ which we will call ***A-A*** connection, or with $-\alpha_{ij}^{(0)}$ called ***B-B*** connection, or two planes having opposed shears and we will call them ***A-B*** or ***B-A*** connection. Figure 7.11 illustrates such a sequence of successive variations.

If we now consider N successive planes in a given direction of the lattice, we can count the number n_A of layers of type ***A*** and the number n_B of variant ***B*** layers, with

$$n_B = N - n_A \tag{7.94}$$

On the basis of this counting, we can then find the angle ψ of the macroscopic shear

$$\tan\psi = \frac{l}{L} = \frac{\Delta a}{a}\frac{2n_A - N}{N} = \tan\psi_0\left(\frac{2n_A}{N} - 1\right) \tag{7.95}$$

The macroscopic shear α_{ij}^{an} of the solid associated with the angle ψ can be considered as an anelastic shear since it is not linked to the elasticity of the lattice

$$\alpha_{ij}^{an} = \alpha_{ij}^{(0)}\frac{\tan\psi}{\tan\psi_0} = \alpha_{ij}^{(0)}\left(\frac{2n_A}{N} - 1\right) \tag{7.96}$$

Suppose that the binding energies between atomic planes are slightly different depending on whether we have a ***A-A***, ***B-B***, ***A-B*** or ***B-A*** type bond. Assume further that ***A-A*** and ***B-B*** bonds are stronger than the bonds ***A-B*** and ***B-A***. In this case, it is clear that the bonding energies per unit cell, which are of course negative, are smaller for the pairs ***A-A*** and ***B-B*** than for the ***A-B*** and ***B-A*** pairs , so that

$$\varepsilon_{A\text{-}A} = \varepsilon_{B\text{-}B} < \varepsilon_{A\text{-}B} = \varepsilon_{B\text{-}A} < 0 \tag{7.97}$$

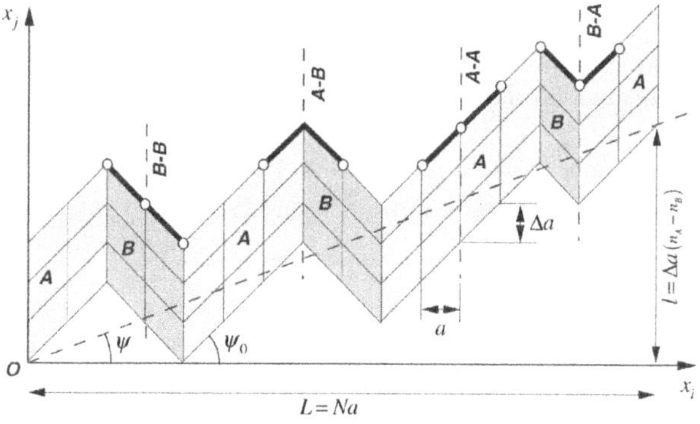

Figure 7.11 *- succession of lattice planes in a given direction with bonds* ***A-A*** *,* ***B-B*** *and* ***A-B***

The internal energy $U^{liaison}$ associated to the bonds of the N successive atomic planes in a given direction is written by enumerating the binding pairs

$$U^{liaison} = \varepsilon_{A\text{-}A} n_A \frac{n_A}{N} + \varepsilon_{B\text{-}B} n_B \frac{n_B}{N} + \varepsilon_{A\text{-}B} n_A \frac{n_B}{N} + \varepsilon_{B\text{-}A} n_B \frac{n_a}{N} < 0 \tag{7.98}$$

The configuration entropy $S^{liaison}$ associated with the bonds of the N successive atomic planes in a given direction is deduced from the number of possible configurations binding pairs

$$S^{liaison} = k\ln\Omega = k\ln\frac{N!}{n_A!\,n_B!} \tag{7.99}$$

We deduce the binding free energy of the N successive atomic planes with the relationship $F^{liaison} = U^{liaison} - TS^{liaison}$ in which we develop $U^{liaison}$ and $S^{liaison}$

$$F^{liaison} = N\varepsilon_{A\text{-}A}\left[\left(\frac{n_A}{N}\right)^2 + \left(\frac{n_B}{N}\right)^2\right] + 2N\varepsilon_{A\text{-}B}\left(\frac{n_A}{N}\right)\left(\frac{n_B}{N}\right) - kT\ln\frac{N!}{n_A!\,n_B!} \tag{7.100}$$

Using the relation *(7.94)* and Stirling's approximation, we obtain after some calculations the average free energy of the lattice bond $f^{liaison} = F^{liaison} / N$ per unit cell in the direction considered

$$f^{liaison} = 2\left(\varepsilon_{A\text{-}A} - \varepsilon_{A\text{-}B}\right)\frac{n_A}{N}\left(\frac{n_A}{N} - 1\right) + \varepsilon_{A\text{-}A} + kT\left[\frac{n_A}{N}\ln\left(\frac{n_A}{N}\right) + \left(1 - \frac{n_A}{N}\right)\ln\left(1 - \frac{n_A}{N}\right)\right] \quad (7.101)$$

This average free energy of bonds per unit cell of the lattice can be expressed by introducing the *order parameter* $\alpha_{ij}^{an} / \alpha_{ij}^{(0)}$ associated with the macroscopic shear of the solid lattice, whose values range between -1 and +1. It's expression is then

$$\frac{n_A}{N} = \frac{1}{2}\left(1 + \frac{\alpha_{ij}^{an}}{\alpha_{ij}^{(0)}}\right) \quad \text{and} \quad 1 - \frac{n_A}{N} = \frac{1}{2}\left(1 - \frac{\alpha_{ij}^{an}}{\alpha_{ij}^{(0)}}\right) \quad (7.102)$$

With this order parameter, the average free energy of bonds per unit cell of the lattice can be written as

$$\begin{aligned} f^{liaison} &= \varepsilon_{A\text{-}A} + \frac{1}{2}\left(\varepsilon_{A\text{-}A} - \varepsilon_{A\text{-}B}\right)\left[\left(\frac{\alpha_{ij}^{an}}{\alpha_{ij}^{(0)}}\right)^2 - 1\right] \\ &+ kT\left[\left(1 + \frac{\alpha_{ij}^{an}}{\alpha_{ij}^{(0)}}\right)\ln\left(1 + \frac{\alpha_{ij}^{an}}{\alpha_{ij}^{(0)}}\right) + \left(1 - \frac{\alpha_{ij}^{an}}{\alpha_{ij}^{(0)}}\right)\ln\left(1 - \frac{\alpha_{ij}^{an}}{\alpha_{ij}^{(0)}}\right) - 2\ln 2\right] \end{aligned} \quad (7.103)$$

At zero temperature, the free energy has a minimum for $\alpha_{ij}^{an} / \alpha_{ij}^{(0)} = \pm 1$. So it is the ***A-A*** or ***B-B*** configurations which are favored at low temperature, and free energy per unit cell is $\varepsilon_{A\text{-}A} < 0$

By introducing the energy difference $\Delta\varepsilon$ between ***A-B*** and ***A-A*** configurations

$$\Delta\varepsilon = \varepsilon_{A\text{-}B} - \varepsilon_{A\text{-}A} > 0 \quad (7.104)$$

it is possible to define the increase in anelastic free energy Δf^{an}

$$\Delta f^{an} = f^{liaison} - \varepsilon_{A\text{-}A} \quad (7.105)$$

that is associated with the macroscopic shear state α_{ij}^{an} of the lattice. We then obtain

$$\Delta f^{an} = -\frac{\Delta\varepsilon}{2}\left[\left(\frac{\alpha_{ij}^{an}}{\alpha_{ij}^{(0)}}\right)^2 - 1\right] + \frac{kT}{2}\left[\left(1 + \frac{\alpha_{ij}^{an}}{\alpha_{ij}^{(0)}}\right)\ln\left(1 + \frac{\alpha_{ij}^{an}}{\alpha_{ij}^{(0)}}\right) + \left(1 - \frac{\alpha_{ij}^{an}}{\alpha_{ij}^{(0)}}\right)\ln\left(1 - \frac{\alpha_{ij}^{an}}{\alpha_{ij}^{(0)}}\right) - 2\ln 2\right] \quad (7.106)$$

We can express the state function Δf^{an} as a function of the macroscopic shear α_{ij}^{an} for various temperatures (fig. 7.12), and there is phase transition behavior of the second order, which has a *critical transition temperature* T_c.

For temperatures $T > T_c$ the order parameter is 0 and shearing $\alpha_{ij}^{an} = 0$. The ***A-B*** and ***B-A*** bonds are favored by entropic effect. In other words, the *high-temperature phase* corresponds to a sequence of atomic planes ***A-B-A-B-A-B-...***

For temperatures $T < T_c$, there appears two values $\left(\alpha_{ij}^{an}\right)_1^{st}$ or $\left(\alpha_{ij}^{an}\right)_2^{st}$ of macroscopic shear which correspond to local minima of Δf^{an} and two values of macroscopic shear $\left(\alpha_{ij}^{an}\right)_1^{inst}$ or $\left(\alpha_{ij}^{an}\right)_2^{inst}$ which correspond to inflection points of Δf^{an}. For a given temperature, the alternatives ***A*** and ***B*** tend to cluster, with, in equilibrium, a phase containing an amount of variant ***B*** and having a shear and a phase containing a proportion $\left|\alpha_{ij}^{an}\right|_1^{st} / \alpha_{ij}^{(0)}$ of variant ***A*** and with a shear $\left(\alpha_{ij}^{an}\right)_2^{st} > 0$. Under zero stress, the proportion of these two phases, which are separated by anti-phase walls are equal, so that the overall macroscopic shear can only be zero. Gradually, as the temperature drops, the proportions at equilibrium $\left|\alpha_{ij}^{an}\right|_1^{st} / \alpha_{ij}^{(0)}$ and $\left|\alpha_{ij}^{an}\right|_2^{st} / \alpha_{ij}^{(0)}$ tend to 1, so it should appear at zero temperature *low temperature phase* corresponding to two pure variants ***A-A-A-A-A-...*** and ***B-B-B-B-...*** . separated by anti-phase walls.

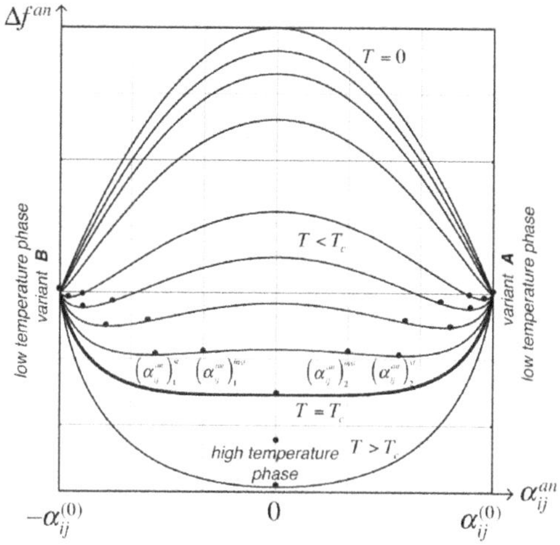

Figure 7.12 - *The anelastic free energy based on anelastic shear for different temperature values*

The *bifurcation diagram* (fig. 7.13) shows the evolution of the various phases with temperature, and the structural aspect of these phases. This second order displacive transformation presents a singular behavior of the specific heat at the critical *transition temperature* T_c. Indeed, the increase Δc_v^{tr} in the specific heat per unit cell due to the phase transition may be deduced from the internal energy $u^{liaison}$ expressed in terms of α_{ij}^{an} using the equation *(B.34)*

$$\Delta c_v^{tr} = \left.\frac{du^{liaison}}{dT}\right|_{v,\alpha_{ij}^{an}} = -\frac{\Delta\varepsilon}{2}\frac{d}{dT}\left(\frac{\alpha_{ij}^{an}}{\alpha_{ij}^{(0)}}\right)^2_{v,\alpha_{ij}^{an}} \qquad (7.107)$$

It follows immediately from this relationship that there is a singularity of the specific heat Δc_v^{tr} during the transition as Δc_v^{tr} is proportional to the derivative of the square of α_{ij}^{an} in the bifurcation diagram in figure 7.13.

The effect of shear applied to the second order displacive transformation

From the state function *(7.106)* of anelasticity, we can deduce the equations of state of the lattice for the quasi-static shear stress s_{ij}^{qs} and entropy s^{an}

$$\begin{cases} s_{ij}^{qs} = n\dfrac{\partial \Delta f^{an}}{\partial \alpha_{ij}^{an}} = \dfrac{n}{\alpha_{ij}^{(0)}}\left\{-\Delta\varepsilon\dfrac{\alpha_{ij}^{an}}{\alpha_{ij}^{(0)}} + \dfrac{kT}{2}\left[\ln\left(1+\dfrac{\alpha_{ij}^{an}}{\alpha_{ij}^{(0)}}\right) - \ln\left(1-\dfrac{\alpha_{ij}^{an}}{\alpha_{ij}^{(0)}}\right)\right]\right\} \\ s^{an} = -\dfrac{\partial \Delta f^{an}}{\partial T} = -\dfrac{k}{2}\left[\left(1+\dfrac{\alpha_{ij}^{an}}{\alpha_{ij}^{(0)}}\right)\ln\left(1+\dfrac{\alpha_{ij}^{an}}{\alpha_{ij}^{(0)}}\right) + \left(1-\dfrac{\alpha_{ij}^{an}}{\alpha_{ij}^{(0)}}\right)\ln\left(1-\dfrac{\alpha_{ij}^{an}}{\alpha_{ij}^{(0)}}\right) - 2\ln 2\right] \end{cases} \qquad (7.108)$$

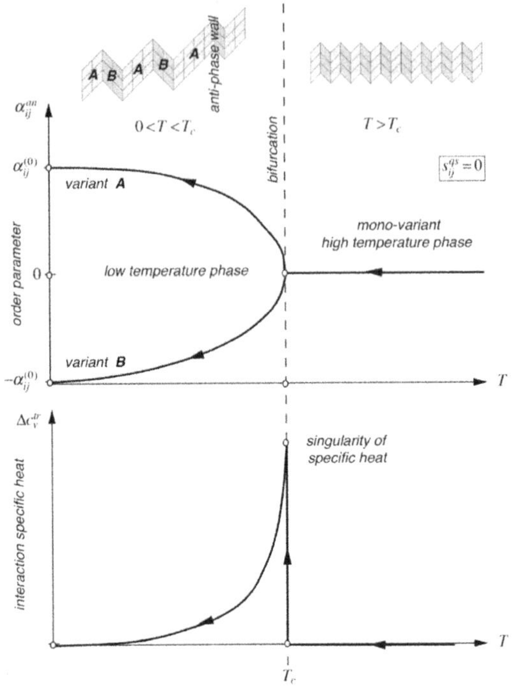

Figure 7.13 - *the bifurcation diagram of the phase transition, and the structural aspect of the phases at different temperatures*

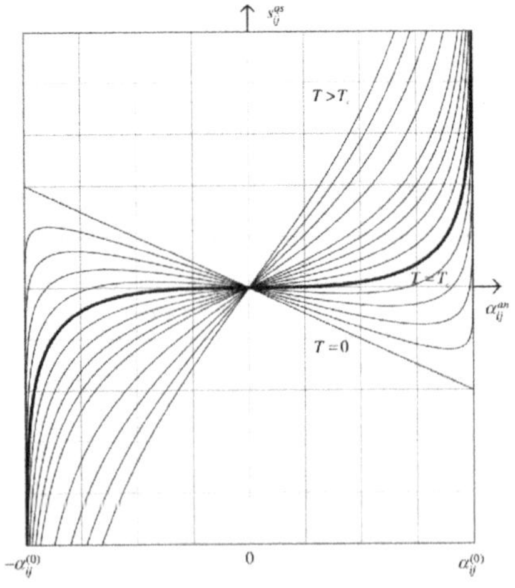

Figure 7.14 - *the anelastic stress-strain $s_{ij}^{qs}-\alpha_{ij}^{an}$ diagram*

The stress-strain diagram of anelasticity $s_{ij}^{qs} - \alpha_{ij}^{an}$ can be drawn for different temperatures (fig. 7.14). In this diagram, the shear values in the region $\left[(\alpha_{ij}^{an})_1^{inst}, (\alpha_{ij}^{an})_2^{inst}\right]$ of negative slope correspond to unstable values in the free energy diagram of figure 7.12, and therefore can not be achieved.

This implies that if an increasing or decreasing stress is applied to the solid in the domain $T < T_c$, there will appear a hysteresis in the graph $s_{ij} - \alpha_{ij}^{an}$,, corresponding to an abrupt change of a variant to the other alternative, induced by the external stress effect (fig. 7.15). During the transformation, the applied stress s_{ij} can be separated into quasi-static stress s_{ij}^{qs} derived from the free energy state function, and a dissipative stress s_{ij}^{dis}

$$s_{ij} = s_{ij}^{qs} + s_{ij}^{dis} \qquad (7.109)$$

The dissipative stress s_{ij}^{dis} is a source of energy dissipation during the transformation, equal to the area S reported in figure 7.15. This hysteresis effect will cause the appearance of a dissipative phenomenon of internal friction at temperatures lower than T_c, depending on both the temperature and the magnitude of the applied stress s_{ij}.

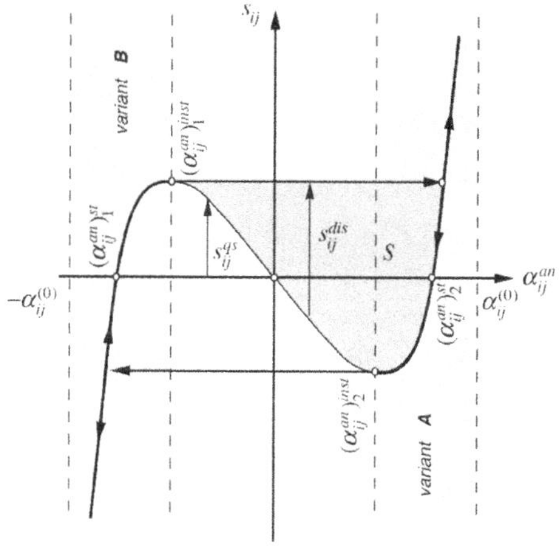

Figure 7.15 *- Hysteresis cycle* $s_{ij} - \alpha_{ij}^{an}$ *in the domain of temperature* $T < T_c$

Pseudo-plasticity and the irreversible shape memory effect

The combination of the effects of stress and temperature may reveal some surprising behavior of this type of solid.

If the stress is increased to a value $s_{ij(\max)}$ at a given temperature $T < T_c$, variant ***B*** will turn into variant ***A*** and there will appear a nonzero macroscopic shear in the solid (point $(\alpha_{ij}^{an})_1^{inst}$ in figure 7.15). If the stress is then released, the solid will retain a non-zero macroscopic shear, as if he had been plastically deformed (point $(\alpha_{ij}^{an})_2^{st}$ in figure 7.15). If this solid is then heated above the critical temperature, the macroscopic deformation disappears during the formation of the high-temperature phase. If the same solid is then cooled from the high temperature phase without being stressed, after passing the critical temperature, variants ***A*** and ***B*** will appear in equal proportion in the solid, so there will not be macroscopic shear in solid. This is why this low-temperature phenomenon of macroscopic deformation is called *pseudo-plasticity*.

On the other hand, if the solid is cooled from the high temperature phase with a fixed shear stress, the transition will show a macroscopic shear due to the fact that one of the variants will be favored over the other. And this macroscopic shear will again be kept in the low temperature phase if the stress is released. But the initial null macroscopic shear state is again recovered if the solid is heated in its high-temperature phase. This amazing effect is called *irreversible shape memory effect.*

An example of displacive martensitic transformation of first order

Let's consider the highly simplified imaginary model of martensitic phase transition described in the previous section (fig. 7.11), but this time let's assume that the binding energies between atomic planes depend not only on immediate first neighbors, but also the second planes of nearest neighbors. In this case, there are 16 different configurations of four planes of nearest neighbors of a given plane, which are shown in figure 7.16 .

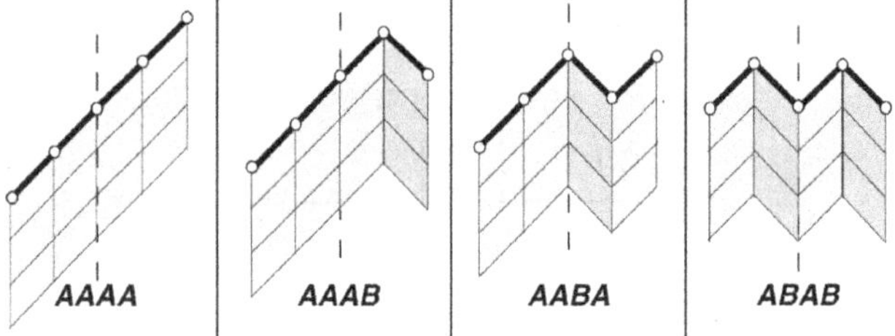

	0 bond **A-B**	1 bond **A-B**	2 bonds **A-B**	3 bonds **A-B**
Configurations	**AAAA** **BBBB**	**AAAB** **BAAA** **BBBA** **ABBB** **AABB** **BBAA**	**AABA** **ABAA** **BBAB** **BABB** **ABBA** **BAAB**	**ABAB** **BABA**
	ε_1	ε_3		ε_2

Figure 7.16 *- the 16 possible configurations of the 4 successive atomic planes*

For a set of *N* successive atomic planes, the probability that one of the planes be in one of the possible configurations, for example the probability P_{AABA} of being in the ***AABA*** configuration, is easily calculated

$$P_{AABA} = \frac{n_A}{N}\frac{n_A}{N}\frac{n_B}{N}\frac{n_A}{N} = \frac{n_A^3 n_B}{N^4} \qquad (7.110)$$

It is easily verified that the sum of the probabilities of all configurations is worth 1

$$\sum P_{XXX} = \frac{1}{N^4}\left(n_A^4 + n_B^4 + 6n_A^2 n_B^2 + 4n_A n_B^3 + 4n_A^3 n_B\right) = \frac{1}{N^4}\left(n_A + n_B\right)^4 = 1 \qquad (7.111)$$

Imagine further that both ***AAAA*** and ***BBBB*** configurations are very stable, with a binding energy $\varepsilon_{AAAA} = \varepsilon_{BBBB} = \varepsilon_1$, and that both ***ABAB*** and ***BABA*** configurations are also very stable, with a binding energy $\varepsilon_{ABAB} = \varepsilon_{BABA} = \varepsilon_2$, and that the 12 other configurations are less stable than the ones we just discussed and that, to simplify the problem, the all share the same binding energy ε_3.

We then have

$$\varepsilon_1 < \varepsilon_3 < 0 \quad \text{and} \quad \varepsilon_2 < \varepsilon_3 < 0 \tag{7.112}$$

The average internal energy $u^{liaison}$ per unit cell in a given direction is written using the probabilities of the different possible linking configurations. It follows

$$u^{liaison} = \left[\varepsilon_1 \left(n_A^4 + n_B^4 \right) + \varepsilon_2 \left(2 n_A^2 n_B^2 \right) + \varepsilon_3 \left(N^4 - n_A^4 - n_B^4 - 2 n_A^2 n_B^2 \right) \right] / N^4 \tag{7.113}$$

As to the average entropy $s^{liaison}$ per site of the lattice in a given direction, it is deduced from the number of possible configurations as

$$s^{liaison} = \frac{k}{N} \ln \Omega = \frac{k}{N} \ln \frac{N!}{n_A! n_B!} = -k \left[\frac{n_A}{N} \ln \frac{n_A}{N} + \left(1 - \frac{n_A}{N} \right) \ln \left(1 - \frac{n_A}{N} \right) \right] \tag{7.114}$$

We deduce the mean free energy per lattice unit in a given direction

$$\begin{aligned} f^{liaison} = u^{liaison} - T s^{liaison} &= \varepsilon_3 + (\varepsilon_1 - \varepsilon_3) \left[\left(\frac{n_A}{N} \right)^4 + \left(\frac{n_B}{N} \right)^4 \right] + 2(\varepsilon_2 - \varepsilon_3) \left(\frac{n_A}{N} \right)^2 \left(\frac{n_B}{N} \right)^2 \\ &+ kT \left[\frac{n_A}{N} \ln \frac{n_A}{N} + \left(1 - \frac{n_A}{N} \right) \ln \left(1 - \frac{n_A}{N} \right) \right] \end{aligned} \tag{7.115}$$

By introducing the following energy differences of the bond energy

$$\Delta\varepsilon_I = \varepsilon_3 - \varepsilon_1 > 0 \quad \text{and} \quad \Delta\varepsilon_{II} = \varepsilon_3 - \varepsilon_2 > 0 \tag{7.116}$$

It is possible to define the increase in anelastic free energy Δf^{an} with the relation

$$\Delta f^{an} = f^{liaison} - \varepsilon_3 \tag{7.117}$$

which is associated to the macroscopic state of shear α_{ij}^{an} via the *order parameter* $\alpha_{ij}^{an} / \alpha_{ij}^{(0)}$ thanks to relations *(7.94)*. We obtain

$$\begin{aligned} \Delta f^{an} = &-\frac{\Delta\varepsilon_I}{16} \left[\left(1 + \frac{\alpha_{ij}^{an}}{\alpha_{ij}^{(0)}} \right)^4 + \left(1 - \frac{\alpha_{ij}^{an}}{\alpha_{ij}^{(0)}} \right)^4 \right] - \frac{\Delta\varepsilon_{II}}{8} \left(1 + \frac{\alpha_{ij}^{an}}{\alpha_{ij}^{(0)}} \right)^2 \left(1 - \frac{\alpha_{ij}^{an}}{\alpha_{ij}^{(0)}} \right)^2 \\ &+ \frac{kT}{2} \left[\left(1 + \frac{\alpha_{ij}^{an}}{\alpha_{ij}^{(0)}} \right) \ln \left(1 + \frac{\alpha_{ij}^{an}}{\alpha_{ij}^{(0)}} \right) + \left(1 - \frac{\alpha_{ij}^{an}}{\alpha_{ij}^{(0)}} \right) \ln \left(1 - \frac{\alpha_{ij}^{an}}{\alpha_{ij}^{(0)}} \right) - 2 \ln 2 \right] \end{aligned} \tag{7.118}$$

One can see the state function Δf^{an} as a function of the macroscopic shear α_{ij}^{an} for different temperatures (fig. 7.17). There is a new behavior with respect to the displacive transition of second order described in the previous section.

At very low temperature, for $T < T_1$, the state function has two minima adjacently located to $\pm\alpha_{ij}^{(0)}$. We will call *phase I* this low temperature phase, which presents a ***A*** variant in the vicinity of $+\alpha_{ij}^{(0)}$ and a ***B*** variant in the vicinity of $-\alpha_{ij}^{(0)}$.

At high temperature $T > T_2$, the state function Δf^{an} no longer has a single minimum for $\alpha_{ij}^{an} = 0$. We will call *phase II* this mono-variant high temperature phase.

In the temperature range $[T_1, T_2]$, the state function Δf^{an} has two minima located in the vicinity of $\pm\alpha_{ij}^{(0)}$ and a minimum located at $\alpha_{ij}^{an} = 0$. This implies that, in this temperature range, there is coexistence of phases I and II.

The phase transition therefore occurs no longer at a fixed temperature as above, but proceeds gradually between the two *critical temperatures* T_1 and T_2, which is a characteristic of a *phase transition of the first order.*

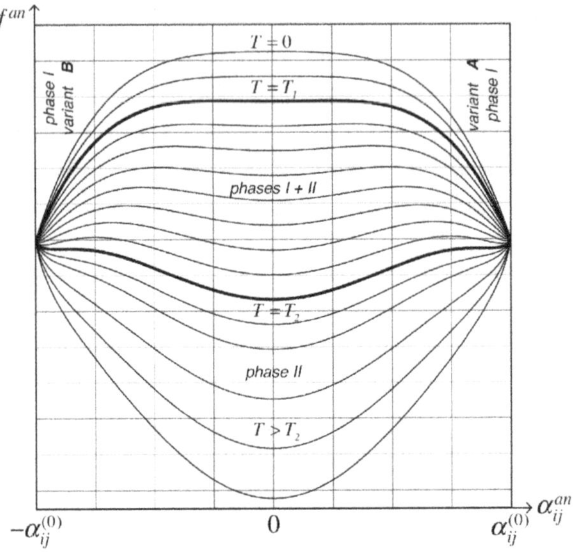

Figure 7.17 *- The anelastic free energy as a function of anelastic shear for different values of temperature*

The effect of the stress applied to first-order displacive transformation

From the state function *(7.118)* of the anelasticity of the lattice, we can deduce the equations of state of anelasticity of this lattice in the chosen direction, namely quasi-static shear stress s_{ij}^{qs} and entropy s^{an}

$$\left\{\begin{aligned} s_{ij}^{qs} &= n\frac{\partial \Delta f^{an}}{\partial \alpha_{ij}^{an}} = \frac{n}{\alpha_{ij}^{(0)}}\left\{-\frac{\Delta\varepsilon_I}{4}\left[\left(1+\frac{\alpha_{ij}^{an}}{\alpha_{ij}^{(0)}}\right)^3-\left(1-\frac{\alpha_{ij}^{an}}{\alpha_{ij}^{(0)}}\right)^3\right]+\frac{\Delta\varepsilon_{II}}{2}\left[1-\left(\frac{\alpha_{ij}^{an}}{\alpha_{ij}^{(0)}}\right)^2\right]\frac{\alpha_{ij}^{an}}{\alpha_{ij}^{(0)}}\right\} \\ &\qquad +\frac{kT}{2}\left[\ln\left(1+\frac{\alpha_{ij}^{an}}{\alpha_{ij}^{(0)}}\right)-\ln\left(1-\frac{\alpha_{ij}^{an}}{\alpha_{ij}^{(0)}}\right)\right] \\ s^{an} &= -\frac{\partial \Delta f^{an}}{\partial T} = -\frac{k}{2}\left[\left(1+\frac{\alpha_{ij}^{an}}{\alpha_{ij}^{(0)}}\right)\ln\left(1+\frac{\alpha_{ij}^{an}}{\alpha_{ij}^{(0)}}\right)+\left(1-\frac{\alpha_{ij}^{an}}{\alpha_{ij}^{(0)}}\right)\ln\left(1-\frac{\alpha_{ij}^{an}}{\alpha_{ij}^{(0)}}\right)-2\ln 2\right] \end{aligned}\right. \qquad (7.119)$$

The stress-strain diagram of transition anelasticity $s_{ij}^{qs} - \alpha_{ij}^{an}$ can be drawn for different temperatures (fig. 7.18).

In this diagram, the shear values in the negative slope areas correspond to values in the unstable free energy diagram of figure 7.17, and therefore cannot be achieved. By observing this chart in detail, it appears there is actually four characteristic temperatures T_1, T_1', T_2' and T_2, during the phase transformation, as illustrated by the four diagrams shown in the figure 7.19 .

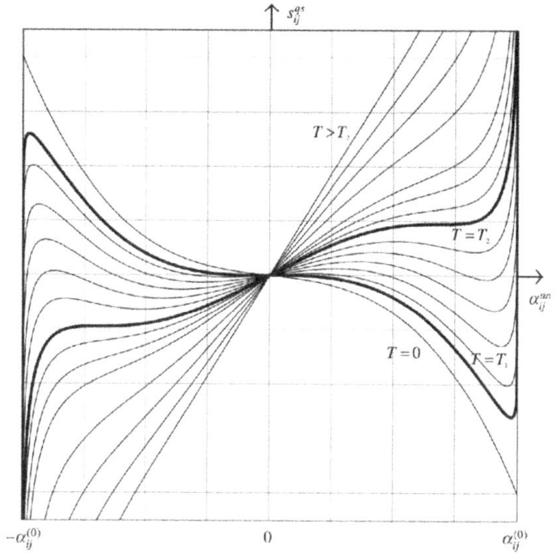

Figure 7.18 *- The stress-strain diagram $s_{ij}^{qs} - \alpha_{ij}^{an}$ of the anelasticity of transition with the transition temperatures T_1 and T_2*

These four characteristic temperatures, T_1, T_1', T_2' and T_2 are used to define four domains in which the dissipative behavior of anelastic transformation are different if the solid is subjected to an external shear stress s_{ij} (fig. 7.20).

Indeed, if an increasing or decreasing stress is applied to the solid in these different temperature ranges, there appears a phenomena of hysteresis in the diagram $s_{ij} - \alpha_{ij}^{an}$, corresponding to abrupt changes between the two variants of phase I and phase II. In the domain $T < T_1$, only the two variants of the phase I exist, and there are transformations between these two variants. In the domain $[T_1, T_1']$, the two variants in phase I are stable, and phase II is metastable. There are transformations between the two variants of phase I, but also irreversible transformations between the metastable phase II and the two variants of phase I. In the domains $[T_1', T_2']$ and $[T_2', T_2]$, phases I and II are stable, and it is only transformations between the variants of phase I and phase II. During these transformations, the applied stress s_{ij} can be separated into quasi-static strain s_{ij}^{qs} derived from the state function free energy, and dissipative stress s_{ij}^{dis}, according to equation (7.109). The dissipative stress s_{ij}^{dis} is again a source of energy dissipation during transformation, equal to the hatched areas S_1 and S_2 drawn in figure 7.20. Such hysteresis effects result in the onset of dissipative phenomena of internal friction in the different temperature ranges, dependent on both the temperature and the magnitude of the applied stress s_{ij}.

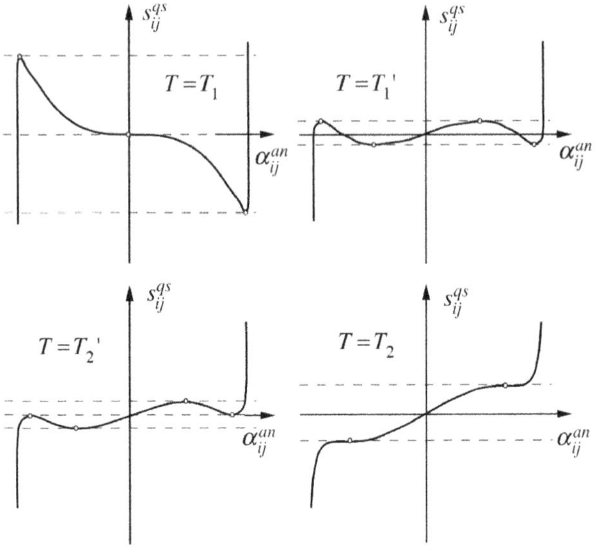

Figure 7.19 - *The diagrams $s_{ij}^{qs} - \alpha_{ij}^{an}$ for the characteristic temperatures T_1, T_1', T_2' and T_2.*

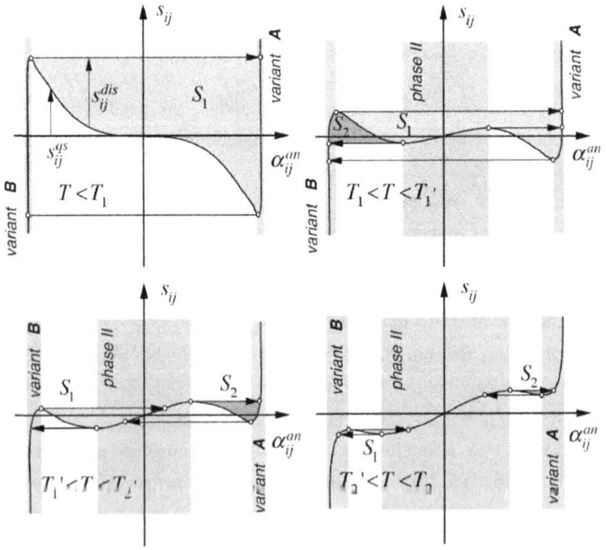

Figure 7.20 - *The cycle of hysteresis $s_{ij} - \alpha_{ij}^{an}$ in the four temperature domains*

Pseudo-plasticity , super-elasticity and the irreversible shape memory effect

In this more complicated case of a solid with first order martensitic transition, the combination of effects of stress and temperature will bring up many quite surprising behaviors of the solid, which we will discuss very briefly.
In the temperature ranges $T < T_2'$, if the stress is increased to a sufficient value $s_{ij(\max)}$, the solid will contain only variant ***A*** and there will appear a nonzero macroscopic shear in the solid. If the stress is then released, the solid will retain a non-zero macroscopic shear, as if he had been plastically deformed. If this solid is then heated above the critical temperature T_2', the solid turns into phase II and the macroscopic deformation disappears. If the same solid is then cooled without being stressed below the critical temperature, there will appear equal proportions of variants ***A*** and ***B*** of phase I in solid, so that there will no longer be macroscopic shear in the solid. This phenomenon of macroscopic deformation associated with the two variants of the low temperature phase I is called *pseudo-plasticity.*
In the domain $[T_2', T_2]$, the solid is in phase II without applied stress. If one applies sufficient stress, the solid will present a sudden major shear deformation for a small change in the applied stress, as if the solid had deformed plastically, because phase II becomes the variant ***A*** of phase I. And this intense deformation is completely reversible if the stress lowers again. This is called a *super-elastic* phenomenon.
If such solid is cooled from the high temperature stage II under a sufficient shearing stress, the passage of the transition temperature T_2 will bring about a macroscopic shearing due to the fact that one of the phase I variants will be preferred over the other. And macroscopic shear will be retained if the stress is released, provided that $T < T_2'$. The initial zero macroscopic shear state is again recovered if the solid is heated above the temperature T_2'. This phenomena is called the *irreversible shape memory effect.*
Materials with a martensitic phase transition of first order, and phenomenology such as the pseudo-plasticity, super-elasticity and irreversible shape memory effect, as well as other more complex phenomena, but also within the martensitic transformation realm, such as reversible shape memory effect, are well known and well studied. These include amongst the better known, the copper-aluminum-zinc alloys as well as the titanium-nickel alloys.

7.10 - Phenomenology of plasticity

The plasticity of a solid lattice of particles is controlled mainly by the long-distance movements of lattice defects such as dislocations, that move under the influence of a stress or strain field.
At the level of macroscopic description of said continuous solid, the statistical manifestation of movements involving structural defects results from the phenomenological equations of dissipative plasticity *(6.53)* and *(6.55)*.

Phenomenological approach to plasticity

As the dynamics of microscopic structural defects depends essentially on the obstacles they face in the lattice and the nature of the interactions they may have with these obstacles, the shape of the dissipative phenomenological equations of plasticity must therefore reflect this dynamic and can therefore have many different behaviors.

Typically, the plasticity mechanisms are by nature highly nonlinear . The intrinsic non-linearity of plasticity must obviously be captured in the dissipative phenomenological relationship of plasticity *(6.53)* and *(6.55)* .

By considering here, for example, only the plastic behavior of a strong shear, the phenomenological approach to the plasticity of this solid consist in finding an expression of the relation *(6.55)*, so that it best reflects all the plastic shear behavior observed in this solid. In other words, to test the material with the aid of various mechanical tests such as a creep test, a tensile test, a fatigue test, etc. , to select a function $\boldsymbol{F}$ in the following equation

$$\frac{d\vec{\alpha}_i^{pl}}{dt} = \boldsymbol{F}\left(\vec{s}_i, T, n, ...\right) \tag{7.120}$$

This function $\boldsymbol{F}$ will always be strongly nonlinear and often brutally growing from a certain critical value of the stress $\vec{s}_i^{cr}$ (fig. 7.21a). $\boldsymbol{F}$ will also usually depend heavily on the temperature T , because the plasticity phenomena are generally thermally activated and may also depend on the volume density n of the lattice . Having such a solid plastic behavior can be represented by a rheological model corresponding to a serial setup of a non-linear friction pad, representing plastic deformation and characterized by the function $\boldsymbol{F}\left(\vec{s}_i, T, n, ...\right)$, and a spring representing the elastic deformation and characterized by the elastic modulus K_2 (fig. 7.21b).

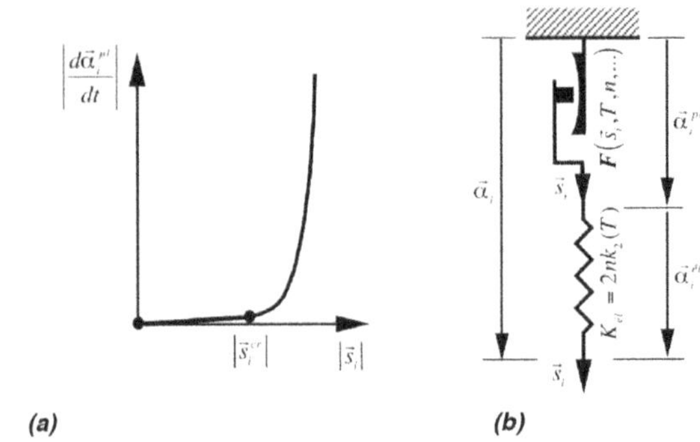

Figure 7.21 - ***(a)*** *typical behavior of the speed of plastic deformation with shear and* ***(b)*** *rheological model for the plastic deformation*

Such an approach may allow a more or less adequate description of the plastic behavior of a solid, including some cases of simple mechanical tests such as creep testing, tensile testing or testing of fatigue.

On the limits of the phenomenological approach to the study of plasticity

The phenomenological approach to the study of continuous plasticity based on the dissipative phenomenological plasticity relations such as relations *(6.53)*, *(6.55)* and *(7.120)* has its limits, as it usually faces several difficulties inherent to the plastic deformation process.

First of all, it is very rare for mechanical tests such as a creep test or a tensile test to be translated properly by a such a simple relationship as the relationships of *(7.120)*. In general, the behavior observed during mechanical testing are *very complex* (nonlinear creep, upper and lower yield points, diverse plateaus, instabilities as Portevin-Le Chatelier effect, hardening effects, etc.) and depend on many parameters, such as the nature of the solid lattice, its crystallinity, its purity, and the history of the thermo-mechanical treatments, etc. These behaviors actually reflect the complexity of the microscopic mechanisms involved during plastic deformation.

On the other hand, the presence of the necessary conditions of geometro-compatibility also imply that the deformations, including plastic deformation, cannot present spatial location, which is not always the case in real strong lattices such as metals, in which it may appear highly localized plastic deformations, such as slip lines for example.

Finally, what essentially distinguishes the phenomenology of solids endowed with anelasticity and plasticity from that of the purely elastic solids is the thermodynamic non-reversibility of anelastic and plastic responses, which results in the existence of sources of entropy associated with anelasticity and plasticity. In the case of the anelasticity, thermodynamic irreversibility leads to a recoverable but *non-immediate* response of anelastic deformation. In other words, the displacement field $\vec{u}$ associated with the anelasticity lags behind the stress field applied to the solid, but it can be recovered if the stress field is canceled. The anelasticity is therefore a Markov process which can relatively easily lead to a phenomenological equation such as the equation *(6.58)*. On the other hand, in the case of plastic deformation, the response of the plastic deformation to a stress field is unrecoverable: after plastic deformation, it inevitably appears a residual plastic displacement field, which can only be canceled by another plastic deformation. This behavior of the plastic deformation, due to the absence of a potential energy stored in the solid lattice by the plastic deformation, often leads to *non-markovian* behaviors of plasticity. In other words, the plastic deformation is a dissipative process which may also depend on the history of the solid, that is to say, of all the previous plastic processes it underwent. As a first approximation, the existence of this "memory" of the solid could be introduced in relation (7.120) by assuming an explicit dependence of the function $\boldsymbol{F}$ on the local state of plastic deformation $\vec{\alpha}_i^{pl}$ such as

$$\boldsymbol{F} = \boldsymbol{F}\left(\vec{s}_i, T, n, \vec{\alpha}_i^{pl}, \ldots\right) \tag{7.121}$$

In fact, the non-markovian behavior of plasticity of solids is extremely difficult, if not impossible, to translate, for the simple reason that it is directly associated to the microscopic structure of the solid, including the behaviors of the defects of this structure (such as dislocations, for example). However, there is no unique, bi-directional, relationship between the local plastic deformation state, characterized by the value of the plastic strain tensor, and the microscopic state of the lattice responsible for that plastic deformation. In other words, there is a multitude of structural defects arrangements leading to *same* local values of a given plastic deformation, which may evolve differently under the effect of a given constraint $\vec{s}_i(t)$.

It is concluded that only a microscopic approach may be able to properly describe the complexity of the plastic behavior of the solid lattice, as it is the microscopic behavior of the topological singularities of this lattice which is responsible for its plasticity. This study will be undertaken in the following chapters.

PART C

Dislocation charges and disclination charges in eulerian lattices

Density and flux of dislocation and disclination charges

Topological singularities associated to the charges

Flux dislocation charges and Orowan's relations

Equations of evolution and force of Peach and Koehler

Chapter 8

Density and flux of charges of dislocations and disclinations

The description of the plasticity of solid lattices with a plastic distortion tensor is very limited, especially due to the fact that there is no unique relationship between the local state of plastic deformation $\vec{\beta}_i^{pl}$ and the microscopic state of the structural defects of the lattice responsible for this plastic deformation. This is why the way of expressing the presence of distortions in a plastic lattice must be modified so that it is possible to take into account the microscopic state of structural defects of the lattice. A very elegant way to make this change is to introduce the concepts of *dislocation densities and corresponding flow of charges*, which are responsible for the plastic distortions of the solid, and *disclination densities* responsible for the plastic contortions of the solid.

8.1 - On the macroscopic concept of charges of plastic distortions

On the intuitive notion of plastic charges

The concept of *charges of plastic distortions* of the solid, which will now be called simply *dislocation charges*, is intuitively and visually easy to grasp, especially with the help of the famous "*Volterra tubes*"[1] (1907) and an analogy to electrical charges.

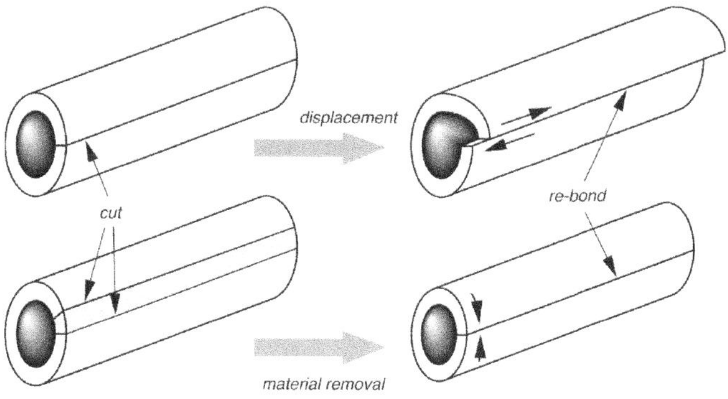

Figure 8.1 - «*Volterra tubes*»

[1] *V. Volterra, L'équilibre des corps élastiques, Ann. Ec. Norm. (3), XXIV, Paris, 1907*

Indeed, just imagine that, either you cut a solid environment and distort it by sliding the jaws, or alternatively you take away a part of the tube and you proceed again to re-bond the tube, as shown by the examples in figure 8.1. In both examples, the deformations undergone by the solid after re-bonding are irreversible and irretrievable, so of the plastic kind. On the other hand, it is intuitively clear that internal forces of stress are developed within the solid after re-bonding. They appeared in the elastic deformation that was imposed on the rest of the solid.
In fact, everything is exactly as if a localized topological discontinuity appeared in the center of the tube after re-bonding. This discontinuity creates an elastic distortion field in the macroscopically continuous medium that makes up the tube. And this field distortion, by its presence, is itself a source of a field-conjugated stresses, which can be called the *field of internal stresses*.
Mathematically, the discontinuity due to re-bonding should translate in terms of a local density of *"plastic charges"*, which are the sources of the elastic distortion field, and thus the internal stress field. In the same way as in electromagnetism the presence a local density of electric charge ρ is responsible for the appearance of an electric displacement field $\vec{D}$, as shown in the Maxwell equation

$$\rho = \operatorname{div}\vec{D} \tag{8.1}$$

and therefore a conjugate electric field $\vec{E}$, as $\vec{D} = \varepsilon_0 \vec{E}$.
The purpose of this chapter will be to mathematically translate the plasticity phenomena inside a solid, not only by introducing densities of plastic charges, but also the flow of these plastic charges, by analogy with the flow of electric charges $\vec{j}$ appearing in the following equation of electromagnetism (wherein $\vec{H}$ represents the magnetic field)

$$\vec{j} = -\frac{\partial \vec{D}}{\partial t} + \overrightarrow{\operatorname{rot}}\,\vec{H} \tag{8.2}$$

On the density and tensorial flow of plastic charges of the solid

By using the definition of the distortion tensor, as it had been obtained in chapter 5, the distortions are the sum of elastic, anelastic and plastic distortions

$$\vec{\beta}_i = \vec{\beta}_i^{\,el} + \vec{\beta}_i^{\,an} + \vec{\beta}_i^{\,pl} \tag{8.3}$$

A new notation for these distortions can be introduced, which allows to separate the contributions of plastic deformation from the contributions of elastic and inelastic deformations

$$\vec{\beta}_i \rightarrow \vec{\beta}_i + \vec{\beta}_i^{\,pl} \qquad \text{with} \qquad \vec{\beta}_i = \vec{\beta}_i^{\,el} + \vec{\beta}_i^{\,an} \tag{8.4}$$

The topological equations for distortions, which were written

$$\frac{d\vec{\beta}_i}{dt} = \overrightarrow{\operatorname{grad}}\,\phi_i \qquad \text{and} \qquad \overrightarrow{\operatorname{rot}}\,\vec{\beta}_i = 0 \tag{8.5}$$

can now be rewritten with the new notation as

$$\frac{d\vec{\beta}_i^{\,pl}}{dt} = -\frac{d\vec{\beta}_i}{dt} + \overrightarrow{\operatorname{grad}}\,\phi_i \qquad \text{and} \qquad -\overrightarrow{\operatorname{rot}}\,\vec{\beta}_i^{\,pl} = \overrightarrow{\operatorname{rot}}\,\vec{\beta}_i \tag{8.6}$$

This allows us to introduce, by analogy with the equations of electromagnetism, the concept of *tensorial charge density* $\vec{\lambda}_i$, the source of plastic distortions, by assuming *a priori* the definition of $\vec{\lambda}_i$ from the plastic distortion tensor $\vec{\beta}_i^{pl}$

$$\vec{\lambda}_i = -\overrightarrow{\text{rot}}\,\vec{\beta}_i^{pl} \tag{8.7}$$

We also introduce the concept of *tensorial flow of charges* $\vec{J}_i$, which is responsible for the temporal variation of the plastic distortion, by assuming *a priori* the following definition deriving from the temporal derivative $d\vec{\beta}_i^{pl} / dt$ of the plastic distortion tensor $\vec{\beta}_i^{pl}$

$$\vec{J}_i = \frac{d\vec{\beta}_i^{pl}}{dt} \tag{8.8}$$

It should be noted here that the concept of charge flow is defined as *a flow with respect to the lattice*, as $\vec{J}_i$ is deduced from the material derivative of $\vec{\beta}_i^{pl}$, that is to say to the temporal derivative taken *along the path of the solid lattice*.

The introduction of these new tensorial densities and charge flows is not gratuitous, they best meet the need to find a way of expressing the presence of plastic distortions. With it, it is possible to take into account the microscopic state of the structural defects of the solid lattice. We will verify *a posteriori*, when interpreting these tensors in the next chapter, that this approach is indeed appropriate. With this approach to the plasticity phenomena with tensors $\vec{J}_i$ and $\vec{\lambda}_i$, the topological equations describing geometro-kinetics and geometro-compatibility of an elastic, anelastic and plastic solid are now written as

$$\begin{cases} \vec{J}_i = -\dfrac{d\vec{\beta}_i}{dt} + \overrightarrow{\text{grad}}\,\phi_i \\ \vec{\lambda}_i = \overrightarrow{\text{rot}}\,\vec{\beta}_i \end{cases} \tag{8.9}$$

in which the distortion tensor $\vec{\beta}_i = \vec{\beta}_i^{\acute{e}l} + \vec{\beta}_i^{an}$ represents both distortions by elasticity and anelasticity.

From this new distortion tensor $\vec{\beta}_i$, we can deduce other distortion tensors always accepting the hypothesis *(5.55)* and decomposition relationships *(5.60)*

$$\begin{cases} \vec{\beta}_i = \vec{\beta}_i^{(\delta)} + \vec{e}_i \wedge \vec{\omega}_O(t) = \vec{\beta}_i^{\acute{e}l} + \vec{\beta}_i^{an} + \vec{e}_i \wedge \vec{\omega}_O(t) \\ \vec{\omega} = -\dfrac{1}{2}\sum_k \vec{e}_k \wedge \vec{\beta}_k = \vec{\omega}^{(\delta)} + \vec{\omega}_0(t) = \vec{\omega}^{\acute{e}l} + \vec{\omega}^{an} + \vec{\omega}_0(t) \\ \tau = \sum_k \vec{\beta}_k \vec{e}_k = \sum_k \vec{\beta}_k^{(\delta)} \vec{e}_k = \tau^{\acute{e}l} \qquad (\tau^{an} \equiv 0 \textit{ by hypothesis}) \\ \vec{\varepsilon}_i = \vec{\beta}_i - \vec{e}_i \wedge \vec{\omega} = \vec{\beta}_i^{(\delta)} - \vec{e}_i \wedge \vec{\omega}^{(\delta)} = \vec{\varepsilon}_i^{\acute{e}l} + \vec{\varepsilon}_i^{an} \\ \vec{\alpha}_i = \vec{\varepsilon}_i - \dfrac{1}{3}\tau \vec{e}_i = \vec{\alpha}_i^{\acute{e}l} + \vec{\alpha}_i^{an} \end{cases} \tag{8.10}$$

This new version of the topological distortion tensors and geometro-kinetic equations is in fact nothing else than a change in terminology for the plastic distortions based on an analogy with Maxwell's equations of electromagnetism. To exploit the potential that this formulation of topological equations unleashes will therefore be subject of this chapter.

On the tensorial density of charges of dislocation

Using again relations *(3.35)* and *(3.36)*, in the presence of plastic charges $\vec{\lambda}_i$

$$\vec{B} = \vec{u}_B - \vec{u}_A = \oint_C \delta\vec{u} = -\sum_k \vec{e}_k \iint_S \overrightarrow{\mathrm{rot}}\vec{\beta}_k \, d\vec{S} = -\sum_k \vec{e}_k \iint_S \vec{\lambda}_k \, d\vec{S} \neq 0 \qquad (8.11)$$

there appears a closure defect vector $\vec{B} = \vec{u}_B - \vec{u}_A \neq 0$ of the displacement field, around a closed contour C within the medium (fig. 3.5). The differential $\delta\vec{u}$ is no longer a total differential, but a *Pfaff differential form*, which only has meaning under local and differential form, as it represents a basic translation associated with elastic and anelastic deformations and rotations.
The integral on the closed contour C results in a quantity $\vec{B}$ called macroscopic *Burgers vector* [2] defined on the contour C, which corresponds to the macroscopic translation necessary to accommodate the environment in the presence of the charge density $\vec{\lambda}_i$, to ensure the compatibility of total deformations and rotations (the absence of voids and material recoveries in the solid).
The discontinuity $\vec{B}$ is called a *macroscopic dislocation* of the solid in the sense of Volterra, and therefore we will call it *density of dislocation charges*, the tensorial density of charges $\vec{\lambda}_i$ responsible for plastic distortions.

The appearance of macroscopic dislocations

Such macroscopic dislocation is carried out in a continuous solid by locally cutting the solid and moving in parallel the two jaws of the cut, before bonding them back together. This process is illustrated schematically in figure 8.2a using a tube material being cut along the plane $abcd$ and which we glue back together after a parallel shift along the direction of the cut. There appears a *one-dimensional topological singularity* of the localized distortion field on the axis cd. This macroscopic singularity, characterized by a translation vector $\vec{B}$ parallel to the singularity line is called *screw dislocation.*

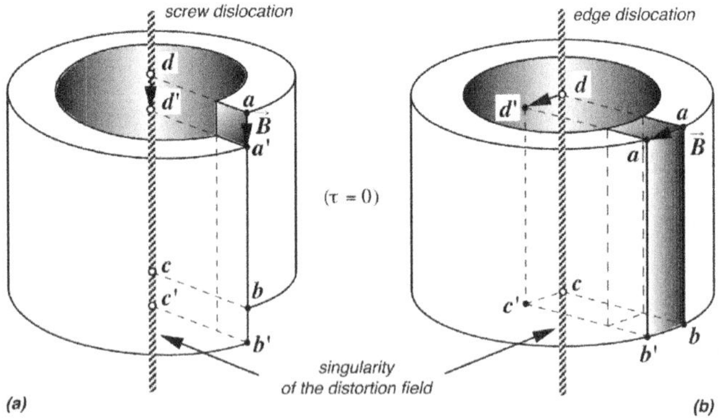

Figure 8.2 *- creation of a screw dislocation* **(a)** *and an edge dislocation* **(b)** *by cutting in gluing*

On the other hand, if we glue the two jaws after a parallel translation perpendicular to the plane of the cut, and with the addition or subtraction of a parallelepiped of material (fig. 8.2b), there

[2] *J. M. Burgers, Proc. Kon. Ned. Akad. Weten, vol.42, p. 293, 378, 1939*

appears another one-dimensional topological singularity of the distortion field, located on the axis cd. This macroscopic singularity, characterized by a translation vector $\vec{B}$ perpendicular to the singularity line is called *edge dislocation.* Another approach to achieve an edge dislocation, but without addition or subtraction of material is to glue the two jaws after a parallel translation in the cleavage plane, perpendicular to the direction of the cut, as shown in figure 8.3. Under the *sine qua non* precondition that the elastic volume expansion τ of the medium remained zero during the plastic deformation process, the Burgers vector $\vec{B}$ obtained by the integral *(8.11)* on a contour surrounding the singularity corresponds exactly to the macroscopic translation the jaw $abcd$ underwent.

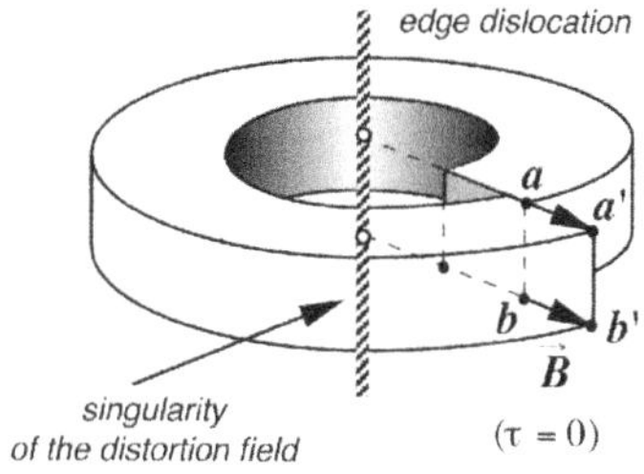

***Figure 8.3** - another creation of an edge dislocation by cutting and gluing*

Let's imagine that within a solid continuum first we cut a torus-shaped hole, as illustrated in the section shown in figure 8.4a, and second we cut the median plane in the center of the torus. The two jaws ab and $a'b'$ so formed can then be displaced with respect to one another, and then reattached.

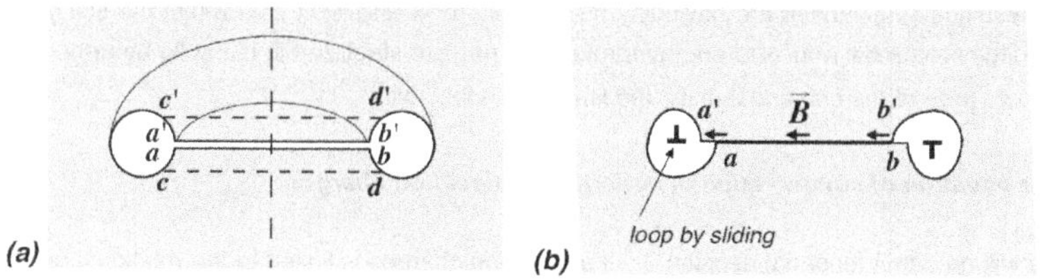

Figure 8.4** - cut of a torus and the median plane to form loops **(a)
*and creation of a dislocation loop by sliding of the jaws **(b)***

The first possible case is to move the two jaws parallel to the cleavage plane by a distance $\vec{B}$ as shown in figure 8.4b. After rejoining the middle is deformed by shear and the torus contains a macroscopic dislocation of *sliding loop type*, composed of edge, screw and mixed dislocation parts.

One can also insert additional material in the form of a thin disk of thickness $\vec{B}$ between the two jaws and by welding the disk to the two jaws (fig. 8.5A). This results in a deformation of the medium, obviously a curvature of the medium on each side of the torus. As for the core, it is the seat of a macroscopic dislocation of *prismatic loop type*. In this case, the prismatic loop is called interstitial because it contains additional material, and it is composed of a single edge dislocation which closes on itself.

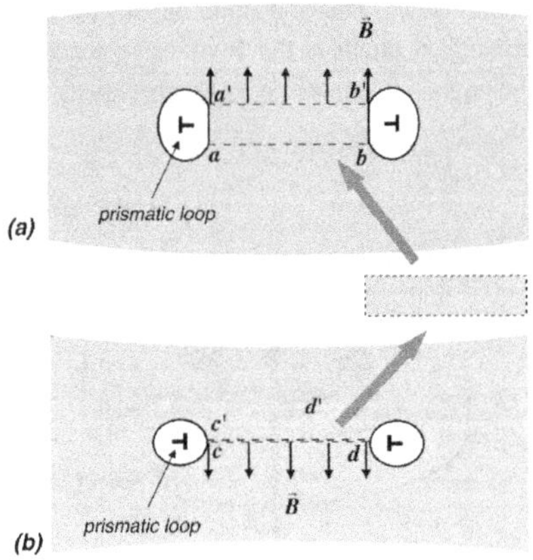

Figure 8.5 *- creation of prismatic loops of dislocation by addition* ***(a)*** *or subtraction* ***(b)*** *of thin disk of material*

A very similar case is obtained if, instead of adding a disk of material, we subtract a disc of material with thickness $\vec{B}$, as shown in figure 8.5b. We also obtain a macroscopic dislocation of *prismatic loop type*, but the loop is called *vacancy loop* because it lacks a certain amount of material. Within the core, there are also a single edge dislocation that closes on itself.

All resulting singularities are obviously responsible for a distortion field within the solid. Therefore, they require a non-zero energy formation. They are stabilized in the solid by re-bonding of the two jaws of the cut, and thus by the links within the solid.

The equation of conservation of density of dislocation charges

We will note that tensorial density $\vec{\lambda}_i$ of dislocation charges is linked to the rotational of $\vec{\beta}_i$, so that it satisfies the relation

$$\operatorname{div}\vec{\lambda}_i = 0 \qquad (8.12)$$

which we will call the *dislocation charge conservation equation*, which will be called later to play a significant role in the topological interpretation of dislocation charges.

The continuity equation connecting the density to the flow of dislocation charges

Topological equations *(8.9)* can be further modified by taking the curl of geometro-kinetic equation and the time derivative of the equation of geometro-compatibility

$$\overrightarrow{\mathrm{rot}}\,\vec{J}_i = -\overrightarrow{\mathrm{rot}}\,\frac{d\vec{\beta}_i}{dt} \quad \text{and} \quad \frac{d\vec{\lambda}_i}{dt} = \frac{d}{dt}\left(\overrightarrow{\mathrm{rot}}\,\vec{\beta}_i\right) \tag{8.13}$$

From these two relations we obtain

$$\frac{d\vec{\lambda}_i}{dt} = \left[\frac{d}{dt}\left(\overrightarrow{\mathrm{rot}}\,\vec{\beta}_i\right) - \overrightarrow{\mathrm{rot}}\left(\frac{d\vec{\beta}_i}{dt}\right)\right] - \overrightarrow{\mathrm{rot}}\,\vec{J}_i \tag{8.14}$$

which is nothing else than the *continuity equation* for the density of dislocation charges. In fact, in the term in brackets, which is the space-time commutator of $\vec{\beta}_i$, there is a *source term of dislocation charges*. In the event that this commutator is zero, equation *(8.14)* then connects directly the change in the charge density $\vec{\lambda}_i$ along the trajectory to the curl of the flux of charges $\vec{J}_i$ in relation to the lattice, which is characteristic of a continuity equation, in the case of a *tensorial charge density*. It is concluded that the tensorial flux $\vec{J}_i$ is nothing but the flow of dislocation charges.

The scalar density of charges of rotation

Equation *(8.9)* for the compatibility of the distortion tensor $\vec{\beta}_i$, namely $\vec{\lambda}_i = \overrightarrow{\mathrm{rot}}\,\vec{\beta}_i$ allows us to deduce immediately, by calculating the trace, a compatibility equation for the rotation vector $\vec{\omega}$

$$\frac{1}{2}\sum_k \vec{e}_k \vec{\lambda}_k = \frac{1}{2}\sum_k \vec{e}_k\,\overrightarrow{\mathrm{rot}}\,\vec{\beta}_k = \mathrm{div}\,\vec{\omega} \tag{8.15}$$

By introducing a *scalar density charge* λ, which is responsible for the plastic rotation within the solid and defined from the trace of the tensor of density of dislocation charges $\vec{\lambda}_i$

$$\lambda = -\,\mathrm{div}\,\vec{\omega}^{pl} = \frac{1}{2}\sum_k \vec{e}_k \vec{\lambda}_k \tag{8.16}$$

the compatibility equation of the vector $\vec{\omega}$ for the elastic and anelastic rotations becomes

$$\lambda = \frac{1}{2}\sum_k \vec{e}_k \vec{\lambda}_k = \mathrm{div}\,\vec{\omega} \tag{8.17}$$

It is noted that this scalar charge density allows us to write

$$\oiint_S \omega_\perp\, dS = \oiint_S \vec{\omega}\, d\vec{S} = \iiint_V \mathrm{div}\,\vec{\omega}\, dV = \iiint_V \lambda\, dV \neq 0 \tag{8.18}$$

Thus, the localized presence of a scalar charge density λ gives us a divergence of the rotation field around the charge. That's why we will give to the scalar λ the name of *rotational charge density*.

Note that the equation relating the rotation vector $\vec{\omega}$ to the density of charges of rotation λ is of the same tensorial order and presents a strong analogy with Maxwell's equations $\rho = \mathrm{div}\,\vec{D}$ relating the electric displacement field $\vec{D}$ to the electric charge density ρ.

The vectorial flow of torsion charges

Decomposition properties of the distortion tensor can also be used to derive the geometro-kinetic equation for the rotation vector $\vec{\omega}$. The operation of only taking the anti-symmetric part of the geometro-kinetic equation *(8.9)* leads to the relation

$$-\frac{1}{2}\sum_k \vec{e}_k \wedge \vec{J}_k = -\frac{1}{2}\sum_k \vec{e}_k \wedge \left(-\frac{d\vec{\beta}_k}{dt} + \overrightarrow{\mathrm{grad}}\,\phi_k\right) = -\frac{\partial \vec{\omega}}{\partial t} + \frac{1}{2}\overrightarrow{\mathrm{rot}}\,\vec{\phi} \tag{8.19}$$

By entering into this relationship a vector flow of charges $\vec{J}$, responsible for the temporal variation of the rotational plastic distortions, or simply a *flux of rotating charges*, defined with respect to the lattice and from the antisymmetric part of the tensor of flow of dislocation charges by the relation

$$\vec{J} = \frac{d\vec{\omega}^{pl}}{dt} = -\frac{1}{2}\sum_k \vec{e}_k \wedge \vec{J}_k \tag{8.20}$$

the geometro-kinetic equation for the vector of elastic and inelastic rotations $\vec{\omega}$ in the presence of rotational charges becomes

$$\vec{J} = -\frac{d\vec{\omega}}{dt} + \frac{1}{2}\overrightarrow{\mathrm{rot}}\,\vec{\phi} \tag{8.21}$$

This equation relating the vector $\vec{\omega}$ to the vector of flow of rotation charges $\vec{J}$ is of the same tensorial order and has a strong analogy with Maxwell's equation $\vec{j} = -\partial\vec{D}/\partial t + \overrightarrow{\mathrm{rot}}\,\vec{H}$ relating the electric displacement field $\vec{D}$ to the electric charge flow $\vec{j}$.

The equation of continuity for the charges of rotation

As for the equation of continuity of rotational charges, it is sufficient to combine the divergence of equation *(8.21)* with the material derivative *(8.17)* to get

$$\frac{d\lambda}{dt} = \left[\frac{d}{dt}(\mathrm{div}\,\vec{\omega}) - \mathrm{div}\left(\frac{d\vec{\omega}}{dt}\right)\right] - \mathrm{div}\,\vec{J} \tag{8.22}$$

wherein, as above, there is a source of rotational charge equal to the spatio-temporal commutator of $\vec{\omega}$.

We can verify that this continuity equation also represents half of the trace of the continuity equation *(8.14)* for the dislocation charges, i.e.

$$\frac{1}{2}\sum_k \vec{e}_k \frac{d\vec{\lambda}_k}{dt} = \frac{1}{2}\sum_k \vec{e}_k \left[\frac{d}{dt}\left(\overrightarrow{\mathrm{rot}}\,\vec{\beta}_k\right) - \overrightarrow{\mathrm{rot}}\left(\frac{d\vec{\beta}_k}{dt}\right)\right] - \frac{1}{2}\sum_k \vec{e}_k \overrightarrow{\mathrm{rot}}\,\vec{J}_k \tag{8.23}$$

The scalar source of lattice sites

The operation which consists in taking the trace of the equation of geometro-kinetic *(8.9)*

$$\sum_k \vec{e}_k \vec{J}_k = \sum_k \vec{e}_k \left(-\frac{d\vec{\beta}_k}{dt} + \overrightarrow{\mathrm{grad}}\,\phi_k\right) = -\frac{d\tau}{dt} + \mathrm{div}\,\vec{\phi} \tag{8.24}$$

shows a scalar corresponding to the trace of the *flow of dislocation charges* $\vec{J}_i$.

Using the relation *(5.64)*, it appears as the explicit *scalar volume source of sites of lattice* S_n linked to the trace of the tensorial flow of dislocation charges by the relationship

$$S_n = n\frac{d\tau^{pl}}{dt} = n\sum_k \vec{e}_k \vec{J}_k \tag{8.25}$$

Table 8.1 - Distortions and charges of dislocation in a solid

Tensors of elastic and anelastic distortions

$$\begin{cases} \vec{\beta}_i = \vec{\beta}_i^{(\delta)} + \vec{e}_i \wedge \vec{\omega}_O(t) = \vec{\beta}_i^{el} + \vec{\beta}_i^{an} + \vec{e}_i \wedge \vec{\omega}_O(t) & (1) \\ \vec{\omega} = -\dfrac{1}{2}\sum_k \vec{e}_k \wedge \vec{\beta}_k = \vec{\omega}^{(\delta)} + \vec{\omega}_0(t) = \vec{\omega}^{el} + \vec{\omega}^{an} + \vec{\omega}_0(t) & (2) \\ \tau = \sum_k \vec{\beta}_k \vec{e}_k = \sum_k \vec{\beta}_k^{(\delta)} \vec{e}_k = \tau^{el} \quad (\tau^{an} \equiv 0 \text{ by hypothesis}) & (3) \\ \vec{\varepsilon}_i = \vec{\beta}_i - \vec{e}_i \wedge \vec{\omega} = \vec{\beta}_i^{(\delta)} - \vec{e}_i \wedge \vec{\omega}^{(\delta)} = \vec{\varepsilon}_i^{el} + \vec{\varepsilon}_i^{an} & (4) \\ \vec{\alpha}_i = \vec{\varepsilon}_i - \dfrac{1}{3}\tau \vec{e}_i = \vec{\alpha}_i^{el} + \vec{\alpha}_i^{an} & (5) \end{cases}$$

Geometro-kinetic of distortions,

flow of dislocation charges, flow of rotational charges, and source of lattice sites

$$\begin{cases} \vec{J}_i = -\dfrac{d\vec{\beta}_i}{dt} + \overrightarrow{\mathrm{grad}}\,\phi_i & (6) \\ \vec{J} = -\dfrac{1}{2}\sum_k \vec{e}_k \wedge \vec{J}_k = -\dfrac{d\vec{\omega}}{dt} + \dfrac{1}{2}\overrightarrow{\mathrm{rot}}\,\vec{\phi} & (7) \\ \dfrac{S_n}{n} = \sum_k \vec{e}_k \vec{J}_k = -\dfrac{d\tau}{dt} + \mathrm{div}\,\vec{\phi} & (8) \end{cases}$$

Geometro-compatibility of distortions

and density of charges of dislocation, of flexion and of rotation

$$\begin{cases} \vec{\lambda}_i = \overrightarrow{\mathrm{rot}}\,\vec{\beta}_i & (9) \\ \vec{\lambda} = -\sum_k \vec{e}_k \wedge \vec{\lambda}_k = -\sum_k \vec{e}_k \wedge \overrightarrow{\mathrm{rot}}\vec{\beta}_k & (10) \\ \lambda = \dfrac{1}{2}\sum_k \vec{e}_k \vec{\lambda}_k = \mathrm{div}\,\vec{\omega} & (11) \end{cases}$$

Conservation and continuity equations of charges of dislocation and of rotation

$$\begin{cases} \mathrm{div}\,\vec{\lambda}_i = 0 & (12) \\ \dfrac{d\vec{\lambda}_i}{dt} = \left[\dfrac{d}{dt}\left(\overrightarrow{\mathrm{rot}}\,\vec{\beta}_i\right) - \overrightarrow{\mathrm{rot}}\left(\dfrac{d\vec{\beta}_i}{dt}\right)\right] - \overrightarrow{\mathrm{rot}}\,\vec{J}_i & (13) \\ \dfrac{d\lambda}{dt} = \left[\dfrac{d}{dt}(\mathrm{div}\,\vec{\omega}) - \mathrm{div}\left(\dfrac{d\vec{\omega}}{dt}\right)\right] - \mathrm{div}\,\vec{J} & (14) \end{cases}$$

Everything happens as if in fact the flow of dislocation $\vec{J}_i$ actually left a *"trace"* of its passage in the solid in the form of creation or annihilation of lattice sites. So we will write the geometro-kinetic equation for the scalar of elastic volume expansion $\tau = \tau^{el}$ (assuming $\tau^{an} \equiv 0$) in the form

$$\frac{S_n}{n} = -\frac{d\tau}{dt} + \operatorname{div}\vec{\phi} \tag{8.26}$$

In table 8.1, we wrote all the relations concerning distortions and dislocation densities and flow of charges (or charges of plastic distortion) in a solid. In figure 8.6, we show schematically all the operations which allow us, using the geometro-compatibility and geometro-kinetic of $\vec{\beta}_i$ to deduct this set of equations of compatibility, of continuity and of geometro-kinetic of the dislocation charges.

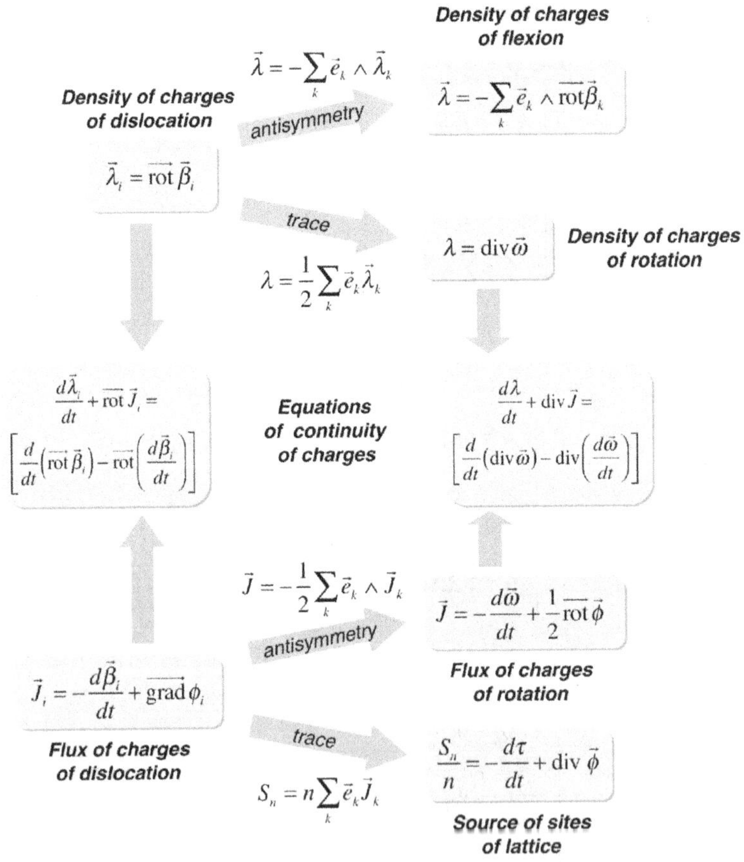

Figure 8.6 - *decompositions of the densities and flows of charges and equations of continuity*

8.2 - On the macroscopic concept of charges of plastic contortions

In an elastically and anelastically deformed media, with dislocation charges, we can also express the tensor of elastic and anelastic contortions $\vec{\chi}_i$.

The tensorial density of charges of plastic contortions of a solid

The relationships in table 3.1 show that the tensor expressions for $\vec{\chi}_i$ deriving from the deformation field $\vec{\varepsilon}_i$ or from the shear field $\vec{\alpha}_i$ and the volume expansion τ, are not changed, but that, instead, its expression from the field $\vec{\omega}$ of elastic and anelastic rotations shall make use of the charge density tensors $\vec{\lambda}_i$ and $\vec{\lambda}$. So in the presence of dislocation charges we have for $\vec{\chi}_i$

$$\vec{\chi}_i = \left[\overrightarrow{\mathrm{rot}}\,\vec{\varepsilon}_i\right]^{\mathrm{T}} = \left[\overrightarrow{\mathrm{rot}}\,\vec{\alpha}_i\right]^{\mathrm{T}} + \frac{1}{3}\vec{e}_i \wedge \overrightarrow{\mathrm{grad}}\,\tau = \overrightarrow{\mathrm{grad}}\,\omega_i + \left[\vec{\lambda}_i\right]^{\mathrm{T}} - \vec{e}_i\lambda \tag{8.27}$$

The *tensorial density of charges* $\left[\vec{\lambda}_i\right]^{\mathrm{T}} - \vec{e}_i\lambda$ which appears here is thus responsible for the plastic contortions of the solid. We will therefore call them *density of contortion charges*. They take the following shape

$$\left[\vec{\lambda}_i\right]^{\mathrm{T}} - \vec{e}_i\lambda = \vec{\lambda}_i + \vec{e}_i \wedge \left(\sum_k \vec{e}_k \wedge \vec{\lambda}_k\right) - \vec{e}_i\lambda = \vec{\lambda}_i - \vec{e}_i \wedge \vec{\lambda} - \vec{e}_i\lambda \tag{8.28}$$

The tensor of contortion in a charged solid can thus be written as

$$\vec{\chi}_i = \left[\overrightarrow{\mathrm{rot}}\,\vec{\varepsilon}_i\right]^{\mathrm{T}} = \left[\overrightarrow{\mathrm{rot}}\,\vec{\alpha}_i\right]^{\mathrm{T}} + \frac{1}{3}\vec{e}_i \wedge \overrightarrow{\mathrm{grad}}\,\tau = \overrightarrow{\mathrm{grad}}\,\omega_i + \left(\vec{\lambda}_i - \vec{e}_i \wedge \vec{\lambda} - \vec{e}_i\lambda\right) \tag{8.29}$$

The tensorial density of disclination charges

The compatibility equation in table 8.5 for elastic and anelastic contortions is also modified in the presence of plastic charges, and is written

$$\overrightarrow{\mathrm{rot}}\left(\vec{\lambda}_i - \vec{e}_i \wedge \vec{\lambda} - \vec{e}_i\lambda\right) = \overrightarrow{\mathrm{rot}}\,\vec{\chi}_i = \overrightarrow{\mathrm{rot}}\left[\overrightarrow{\mathrm{rot}}\,\vec{\varepsilon}_i\right]^{\mathrm{T}} = \overrightarrow{\mathrm{rot}}\left[\overrightarrow{\mathrm{rot}}\,\vec{\alpha}_i\right]^{\mathrm{T}} - \frac{1}{3}\overrightarrow{\mathrm{rot}}\,\overrightarrow{\mathrm{rot}}\left(\vec{e}_i\tau\right) \tag{8.30}$$

which is in a form similar to the compatibility of equation obtained for the elastic and anelastic distortions.

This relationship shows a new tensorial density of charges $\vec{\theta}_i$

$$\vec{\theta}_i = \overrightarrow{\mathrm{rot}}\left(\vec{\lambda}_i - \vec{e}_i \wedge \vec{\lambda} - \vec{e}_i\lambda\right) = \overrightarrow{\mathrm{rot}}\,\vec{\lambda}_i - \vec{e}_i\,\mathrm{div}\,\vec{\lambda} + \left(\vec{e}_i\vec{\nabla}\right)\vec{\lambda} + \vec{e}_i \wedge \overrightarrow{\mathrm{grad}}\,\lambda \tag{8.31}$$

Since this density of charges is equal to the rotational of $\vec{\lambda}_i - \vec{e}_i \wedge \vec{\lambda} - \vec{e}_i\lambda$, it satisfies the following relation

$$\mathrm{div}\,\vec{\theta}_i = \mathrm{div}\left[\overrightarrow{\mathrm{rot}}\left(\vec{\lambda}_i - \vec{e}_i \wedge \vec{\lambda} - \vec{e}_i\lambda\right)\right] \equiv 0 \tag{8.32}$$

which we will call the *conservation equation* of the tensorial charges.

Working again on relations *(3.38)* and *(3.40)* in the presence of these charges

$$\vec{\Omega} = \oint_C \delta\vec{\omega}^{\varepsilon} = \sum_k \vec{e}_k \iint_S \overrightarrow{\mathrm{rot}}\,\vec{\chi}_k\,d\vec{S} = \sum_k \vec{e}_k \iint_S \vec{\theta}_i\,d\vec{S} = \sum_k \vec{e}_k \iint_S \overrightarrow{\mathrm{rot}}\left(\vec{\lambda}_k - \vec{e}_k \wedge \vec{\lambda} - \vec{e}_k\lambda\right)d\vec{S} \neq 0 \tag{8.33}$$

there appears a vectorial closure defect $\vec{\Omega} = \vec{\omega}_B^{\varepsilon} - \vec{\omega}_A^{\varepsilon} \neq 0$ of the rotation field by deformation taken on the closed contour C within the medium (fig. 3.7). The differential $\delta\vec{\omega}^{\varepsilon}$ is no longer a total differential, but a *Pfaff differential form*, which only has meaning under local and differential form, where it is associated with an elementary rotation of elastic and anelastic deformations and rotations.

It is still possible to transform the integral over the surface S in an integral on the closed contour C in the form

$$\vec{\Omega} = \oint_C \delta\vec{\omega}^{\varepsilon} = \sum_k \vec{e}_k \iint_S \overrightarrow{\mathrm{rot}}\vec{\chi}_k \, d\vec{S} = \sum_k \vec{e}_k \oint_C \left(\vec{\lambda}_k - \vec{e}_k \wedge \vec{\lambda} - \vec{e}_k \lambda\right) d\vec{r} \neq 0 \qquad (8.34)$$

Thus, the integral of the *density* $\left(\vec{\lambda}_i - \vec{e}_i \wedge \vec{\lambda} - \vec{e}_i\lambda\right)$ *of charges of contortion* on a closed contour C results in the existence of a non-zero macroscopic angular quantity $\vec{\Omega}$, called a *macroscopic Frank vector*[3] defined on the contour C, and which corresponds to the macroscopic rotation necessary to accommodate the environment in the presence of the plastic charges. This discontinuity $\vec{\Omega}$ is called a macroscopic disclination of the solid in the sense of macroscopic defects of Volterra. From equation *(18.33)*, it is now possible to interpret the tensorial density of charges $\vec{\theta}_i = \overrightarrow{\mathrm{rot}}\left(\vec{\lambda}_i - \vec{e}_i \wedge \vec{\lambda} - \vec{e}_i\lambda\right)$. Indeed, as the integral over the surface S of the charge density provides the macroscopic Frank vector of a disclination, it will be called the density of disclination charges $\vec{\theta}_i$.

Creation of macroscopic disclinations

It is easy to imagine the realization of a disclination at the macroscopic scale in a solid continuum by locally cutting this solid and rotating a jaw of the cut relative to the other, before gluing them back together. This process is illustrated schematically in figure 8.7,

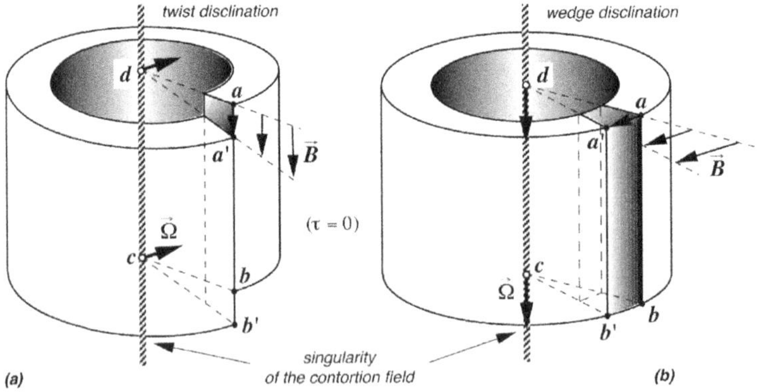

Figure 8.7 - *creation of the twist* **(a)** *and wedge* **(b)** *disclinations by cutting and welding back*

using a tube $abcd$ material being cut and glued together in two different ways:

- either by shear of the plane $a'b'cd$ without the addition or subtraction of material (fig. 19.3a),

[3] *R. DeWit, «Theory of disclinations II, III and IV», J. of the Nat. Bureau of Standards A, vol. 77A, p. 49-100, p. 359-368, and p. 607-658, 1973*

which leads to a one-dimensional topological singularity located on the axis cd, called *twist disclination*

- by rotation of one of the interfaces around the edge cd with the addition or subtraction of material (fig. 19.3b), which leads to a one-dimensional topological singularity located on the axis, called a *wedge disclination.*

Under the condition that the elastic volume expansion τ of the medium remained zero during the plastic deformation process, the vector $\vec{\Omega}$, obtained by the integral *(8.34)* on a contour C surrounding the singularity corresponds exactly to the macroscopic rotation underwent by the jaw $a'b'cd$. As the vector $\vec{\Omega}$ must remain constant if we vary the diameter of the contour of integration C or if we move this contour vertically, we deduce that disclination charges must be confined to the immediate vicinity of the axis cd of the tube, and that their tensorial density must be constant along the axis.

Topological singularities thus obtained are responsible for a distortion field within the solid. Therefore, they require a non-zero energy formation. They are stabilized in the solid by re-gluing of the two jaws of the cut, and thus by the bonds within the solid.

Comparing figures 8.2 and 8.7, there is a striking resemblance between screw dislocations and twist disclinations and between edge dislocations and wedge disclinations. This similarity is not accidental, since the operations used to generate these discontinuities are very similar. It is interesting to note in particular that the macroscopic disclinations also have a displacement vector $\vec{B}$ going from a to a' (fig. 8.7), as well as the macroscopic dislocation (fig. 8.2). However, this vector $\vec{B}$, in the case of disclinations, increases linearly with the diameter of the integrating loop C used to calculate it. This means that in the presence of a macroscopic disclination associated with a disclination charge distributed along the axis of the pipe cd, there must also be a dislocation charge density, as seen in the last term of equation *(8.34)*. But the latter, instead of being located on the axis of the tube as is the case for a macroscopic dislocation, will be evenly distributed on a surface in the cut plane $abcd$ (fig. 8.7) such that the Burgers vector $\vec{B}$ increases linearly with the diameter of the integration loop C.

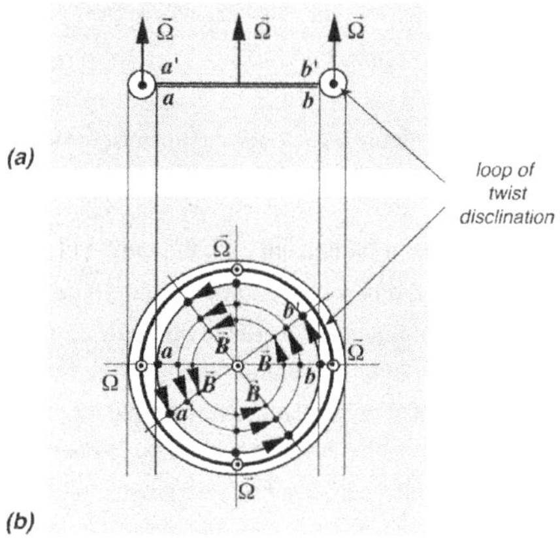

Figure 8.8 *- creation of a twist disclination loop by rotation of jaws*

One can also proceed as in figure 8.4a, and move the two jaws by rotating them relative to each other by a rotational angle $\vec{\Omega}$ in the cleavage plane, as shown in figure 8.8a. After re-bonding, the medium is deformed by the $\vec{\Omega}$ rotation and the torus then contains a *macroscopic twist disclination loop.*

Note here that the field of displacement of the media on each side of the cut plane is tangential to this plane and that the curvilinear displacement motion vector $\vec{B}$ on the cut plane increases from a zero value at the center to a maximum value on the edges of the torus. On the torus, the local displacement field $\vec{B}$ closely resembles the displacement field of a screw dislocation closed in on itself, but it is actually a *pseudo-dislocation* because the curvilinear Burgers vector, tangential to the dislocation line is not preserved [4] (fig. 8.8b).

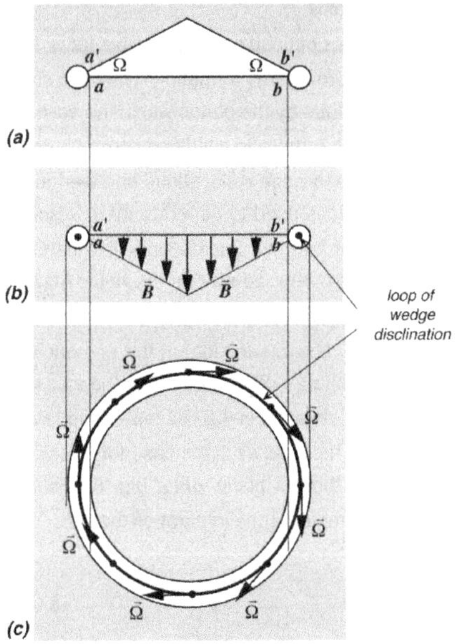

***Figure 8.9** - creation of a wedge disclination loop*
by adding or removing a part of the media in conical shape

One could also remove a piece of medium from the center of the torus, with a lenticular or conical shape with angle Ω to the base, as shown in figure 8.9a. In this case, the plane of re-bonding has a local displacement field $\vec{B}$ corresponding to perpendicular Burgers vectors whose lengths have a circular symmetry (fig. 8.9b). On the torus, the required deformation for re-bonding is a rotation $\vec{\Omega}$ tangential to the torus, which would therefore correspond to a *macroscopic wedge disclination loop*, but which is actually a *pseudo-disclination* because the Frank vector, still tangential to the disclination line is not retained along the line (figure 8.9c).

---

[4] *W. Huang, T. Mura, Elastic J. of Applied Physics, vo. 41(13), p. 5175, 1970; A. Unzicker, arXiv:gr-qc/0011064, 2000; A. Unzicker, arXiv:gr-qc/9612061v2, 2010*

The vectorial density of flexion charges

As for the flexion vector $\vec{\chi}$ due to elastic and anelastic bending, it is not modified when expressed as a function of the deformation field $\vec{\varepsilon}_i$ or a function of shear fields $\vec{\alpha}_i$ and volume expansion τ, while its expression from the vector of elastic and anelastic rotations $\vec{\omega}$ depends on a new vector $\vec{\lambda}$ corresponding to the anti-symmetric part of $\vec{\lambda}_i$, the density tensor of dislocation charges

$$\begin{aligned} \vec{\chi} &= \sum_k \vec{e}_k \wedge \vec{\chi}_k = -\sum_k \vec{e}_k \wedge \overrightarrow{\mathrm{rot}}\, \vec{\varepsilon}_k = -\sum_k \vec{e}_k \wedge \overrightarrow{\mathrm{rot}}\, \vec{\alpha}_k - \frac{2}{3}\overrightarrow{\mathrm{grad}}\, \tau \\ &= \sum_k \vec{e}_k \operatorname{div} \vec{\alpha}_k - \frac{2}{3}\overrightarrow{\mathrm{grad}}\, \tau = -\overrightarrow{\mathrm{rot}}\, \vec{\omega} - \sum_k \vec{e}_k \wedge \vec{\lambda}_k = -\overrightarrow{\mathrm{rot}}\, \vec{\omega} + \vec{\lambda} \end{aligned} \tag{8.35}$$

We deduce the interpretation of this $\vec{\lambda}$ vector which is nothing other than the density of flexion charges, given from the density $\vec{\lambda}_i$ of dislocation charges by the relation

$$\vec{\lambda} = -\sum_k \vec{e}_k \wedge \vec{\lambda}_k = -\sum_k \vec{e}_k \wedge \overrightarrow{\mathrm{rot}}\vec{\beta}_k \tag{8.36}$$

From equations *(8.29)* and *(8.35)*, we deduce that the contortions $\vec{\chi}_i$ and flexions $\vec{\chi}$ associated to elastic and anelastic deformation $\vec{\varepsilon}_i$ of the lattice, cannot be deduced directly from the elastic and anelastic rotations $\vec{\omega}$, since it adds terms dependent on the $\vec{\lambda}_i$, $\vec{\lambda}$ and λ density charges.

The scalar density of curvature charges

The compatibility equation in table 3.1 for elastic and anelastic curvature is also modified in the presence of plastic charges, and is written

$$\operatorname{div} \vec{\chi} = -\sum_k \vec{e}_k \overrightarrow{\mathrm{rot}}\, \vec{\chi}_k = \sum_k \vec{e}_k \overrightarrow{\mathrm{rot}}\, \overrightarrow{\mathrm{rot}}\, \vec{\varepsilon}_k = \sum_k \vec{e}_k \overrightarrow{\mathrm{rot}}\, \overrightarrow{\mathrm{rot}}\, \vec{\alpha}_k - \frac{2}{3}\Delta\tau = \sum_k \vec{e}_k \overrightarrow{\mathrm{rot}}\, \vec{\lambda}_k \tag{8.37}$$

This relationship shows that it is possible to introduce a new *scalar density of charges* θ, responsible for the plastic contortions of the solid, defined from the divergence of the vector density $\vec{\lambda}$ of flexion charges

$$\theta = \operatorname{div} \vec{\lambda} = -\operatorname{div}\left(\sum_k \vec{e}_k \wedge \vec{\lambda}_k \right) = \sum_k \vec{e}_k \overrightarrow{\mathrm{rot}}\, \vec{\lambda}_k = -\sum_k \vec{e}_k \vec{\theta}_k \tag{8.38}$$

Note that this charge density θ can also be connected directly to the trace of the disclination density charge $\vec{\theta}_i = \overrightarrow{\mathrm{rot}}\left(\vec{\lambda}_i - \vec{e}_i \wedge \vec{\lambda} - \vec{e}_i \lambda \right)$, which is easily verified from equation *(8.31)*.

The geometro-compatibility equation *(8.37)* for the flexion vector $\vec{\chi}$ is similar in shape to that obtained for the vector rotation $\vec{\omega}$, and will play later an important role. Now we can write

$$\theta = -\sum_k \vec{e}_k \vec{\theta}_k = \operatorname{div} \vec{\lambda} = \operatorname{div} \vec{\chi} \tag{8.39}$$

According to figure 3.8, this scalar density θ of plastic charges allows us to write

$$\iiint_V \theta\, dV = \iiint_V \operatorname{div} \vec{\lambda}\, dV = \iiint_V \operatorname{div} \vec{\chi}\, dV = \oiint_S \vec{\chi}\, d\vec{S} = \oiint_S \chi_\perp\, dS \neq 0 \tag{8.40}$$

Thus, the localized presence of a scalar density of plastic charges $\theta = \operatorname{div} \vec{\lambda}$ gives rise to a *divergence of the flexion field* surrounding this charge and therefore *a divergent "curvature"*

around the charge. This is why we call θ the *scalar of curvature charge.*
From equation *(8.35)*, we can also deduce the rotational field of flexion $\vec{\chi}$

$$\overrightarrow{\mathrm{rot}}\vec{\chi} = -\sum_k \overrightarrow{\mathrm{rot}}\left(\vec{e}_k \wedge \overrightarrow{\mathrm{rot}}\,\vec{\varepsilon}_k\right) = -\sum_k \overrightarrow{\mathrm{rot}}\left(\vec{e}_k \wedge \overrightarrow{\mathrm{rot}}\,\vec{\alpha}_k\right) = -\overrightarrow{\mathrm{rot}}\,\overrightarrow{\mathrm{rot}}\,\vec{\omega} + \overrightarrow{\mathrm{rot}}\,\vec{\lambda} \tag{8.41}$$

and it is found to depend on the vectorial density charges $\overrightarrow{\mathrm{rot}}\,\vec{\lambda}$, which represents the *density of rotational flexion charges*, which is put on equal footing with the density of divergent flexion charges, namely the density of charges of curvature. However, this equation *(8.41)* differs from equation *(8.38)* by the fact that it is not a geometro-compatible equation since the rotational field $\vec{\omega}$ appears in it.

8.3 - Complete topological description of charged solids

In figure 8.10, we show all the operations that give all the distortion and contortion tensors governing the topology of the charged solids as well as the compatibility equations that these tensors must satisfy. It shows a beautiful symmetry in the arrangement of relations to derive the tensor of distortion and contortion, as well as in the geometro-compatible equations. It also shows the *torsion tensor* $\left[\vec{\chi}_i\right]^S$, and it is easily shown, using *(18.29)*, that this tensor is expressed from the rotational field $\vec{\omega}$, and a *density of torsion charges* $\vec{\lambda}_i - \left(\vec{e}_i \wedge \vec{\lambda}\right)/2 - \vec{e}_i\lambda$.
The different tensor of density of dislocation and disclination charges are not independent of each other, since all are derived from the initial tensorial density dislocation charge $\vec{\lambda}_i$, as illustrated by figure 8.11.
In fact, *only the two scalar densities* $\lambda = \mathrm{div}\,\vec{\omega}$ and $\theta = \mathrm{div}\,\vec{\lambda}$ are completely independent of one another, because the density of rotational charges λ is deducted from the trace of $\vec{\lambda}_i$, trace that is not involved in the charge curvature θ as it is derived from the anti-symmetric part of θ via the divergence. This independence of the scalar charges λ and θ will play an important role in the future.

Symmetry of the tensor of density of disclination charges

It should also be noted that the *tensorial density* $\vec{\theta}_i$ *of disclination charges is symmetrical.* Indeed, equation *(8.31)* gives the tensor $\vec{\theta}_i$, and using the conservation equation of dislocation charges $\mathrm{div}\,\vec{\lambda}_k \equiv 0$, we show that the anti-symmetric part of $\vec{\theta}_i$ is identically zero

$$\begin{aligned}
&\sum_k \vec{e}_k \wedge \left[\overrightarrow{\mathrm{rot}}\,\vec{\lambda}_k - \vec{e}_k\,\mathrm{div}\,\vec{\lambda} + \left(\vec{e}_k\vec{\nabla}\right)\vec{\lambda} + \vec{e}_k \wedge \overrightarrow{\mathrm{grad}}\,\lambda\right] = \\
&\sum_{k,i}\left\{\vec{e}_k \wedge \left(\vec{\nabla}\wedge\vec{\lambda}_k\right) - \vec{e}_k \wedge \left[\vec{e}_i \wedge \left(\vec{e}_k\vec{\nabla}\right)\vec{\lambda}_i\right] + \frac{1}{2}\vec{e}_k \wedge \left[\vec{e}_k \wedge \vec{\nabla}\left(\vec{e}_i\vec{\lambda}_i\right)\right]\right\} = \\
&\sum_k \left\{\vec{\nabla}\left(\vec{e}_k\vec{\lambda}_k\right) - \left(\vec{e}_k\vec{\nabla}\right)\vec{\lambda}_k - \left(\mathrm{div}\,\vec{\lambda}_k\right)\vec{e}_k + \left(\vec{e}_k\vec{\nabla}\right)\vec{\lambda}_k + \frac{1}{2}\vec{\nabla}\left(\vec{e}_k\vec{\lambda}_k\right) - \frac{3}{2}\vec{\nabla}\left(\vec{e}_k\vec{\lambda}_k\right)\right\} \equiv 0
\end{aligned} \tag{8.42}$$

To finish this chapter, all the relations concerning contortions and the charges of contortions are reported in table 8.2.

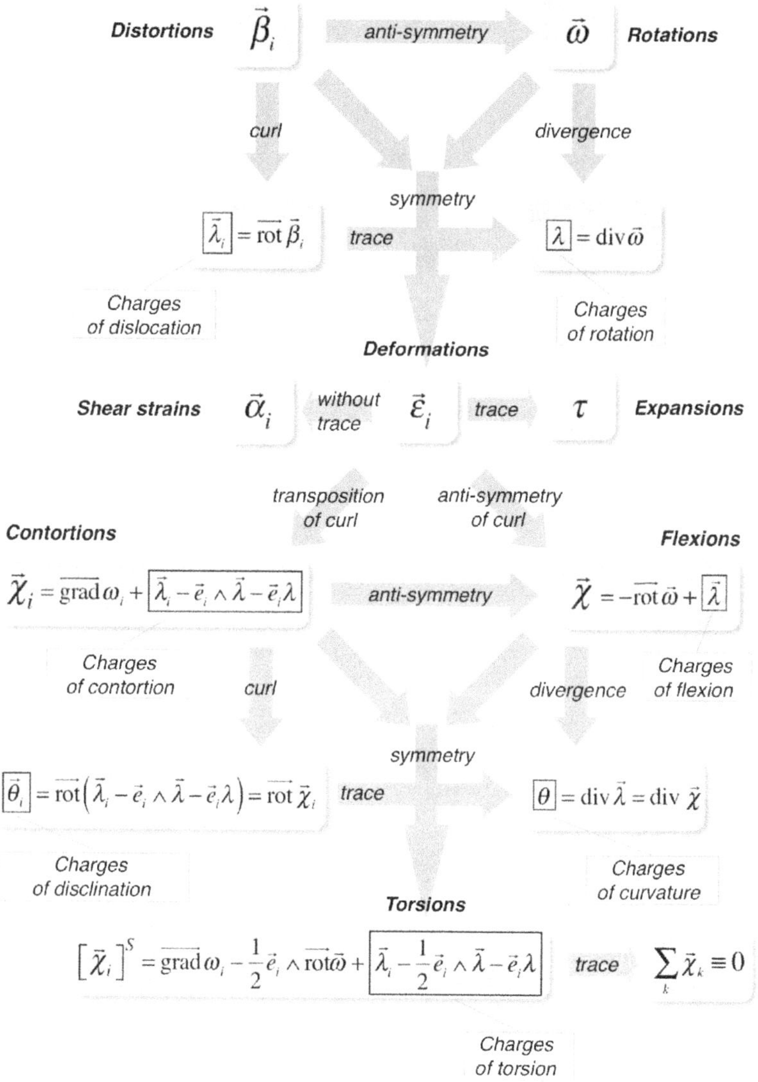

Figure 8.10 - *complete decomposition of the tensors of distortion and contortion.*

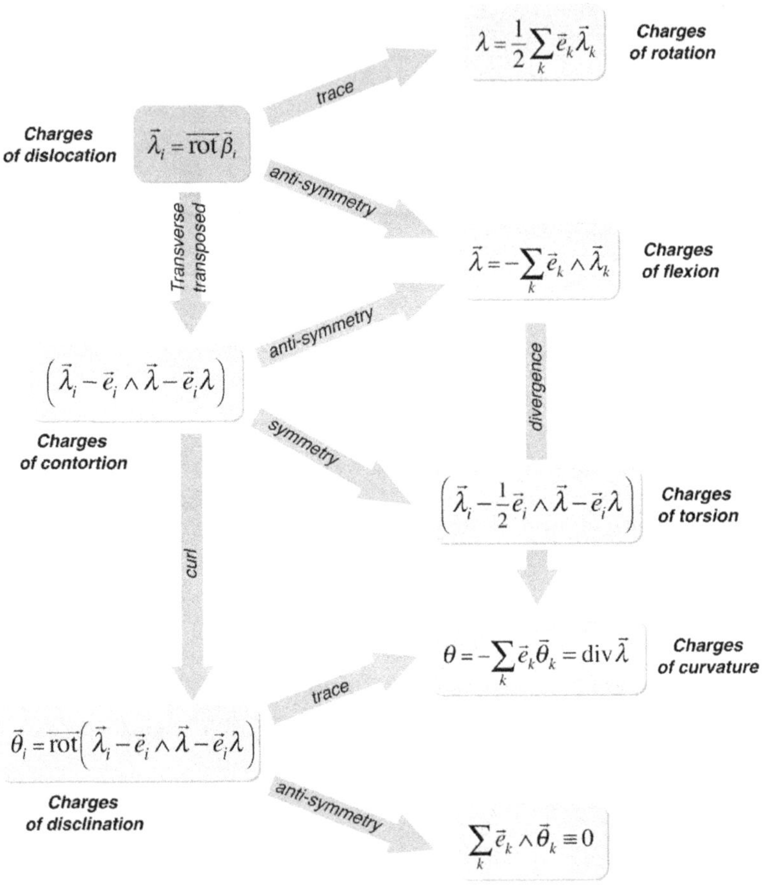

Figure 8.11 *- Complete decomposition of the tensorial charges*

Table 8.2 - Contortions and charges of disclination in a solid

Transverse tensor of contortion and density charges of contortion

$$\begin{aligned}\vec{\chi}_i &= \overrightarrow{\text{grad}}\,\omega_i + \left(\vec{\lambda}_i - \vec{e}_i \wedge \vec{\lambda} - \vec{e}_i \lambda\right) \\ &= \left[\overrightarrow{\text{rot}}\,\vec{\varepsilon}_i\right]^{\text{T}} = \left[\overrightarrow{\text{rot}}\,\vec{\alpha}_i\right]^{\text{T}} + \frac{1}{3}\vec{e}_i \wedge \overrightarrow{\text{grad}}\,\tau\end{aligned} \quad (1)$$

Flexion vector and density of charges of flexion

$$\begin{aligned}\vec{\chi} &= \sum_k \vec{e}_k \wedge \vec{\chi}_k = -\sum_k \vec{e}_k \wedge \overrightarrow{\text{rot}}\,\vec{\varepsilon}_k = -\sum_k \vec{e}_k \wedge \overrightarrow{\text{rot}}\,\vec{\alpha}_k - \frac{2}{3}\overrightarrow{\text{grad}}\,\tau \\ &= \sum_k \vec{e}_k \,\text{div}\,\vec{\alpha}_k - \frac{2}{3}\overrightarrow{\text{grad}}\,\tau = -\overrightarrow{\text{rot}}\,\vec{\omega} + \vec{\lambda}\end{aligned} \quad (2)$$

Transversality of the tensor of contortions

$$\sum_k \vec{e}_k \vec{\chi}_k = \sum_k \vec{e}_k \left[\overrightarrow{\text{rot}}\,\vec{\varepsilon}_k\right]^{\text{T}} = \sum_k \vec{e}_k \overrightarrow{\text{rot}}\,\vec{\varepsilon}_k \equiv 0 \quad (3)$$

Transverse symmetric tensor of torsion and density of torsion charges

$$\begin{aligned}\left[\vec{\chi}_i\right]^{\text{S}} &= \vec{\chi}_i + \frac{1}{2}\vec{e}_i \wedge \vec{\chi} = \overrightarrow{\text{grad}}\,\omega_i - \frac{1}{2}\vec{e}_i \wedge \overrightarrow{\text{rot}}\,\vec{\omega} + \left(\vec{\lambda}_i - \frac{1}{2}\vec{e}_i \wedge \vec{\lambda} - \vec{e}_i \lambda\right) \\ &= \left[\overrightarrow{\text{rot}}\,\vec{\varepsilon}_i\right]^{\text{S}} = \left[\overrightarrow{\text{rot}}\,\vec{\alpha}_i\right]^{\text{S}}\end{aligned} \quad (4)$$

Geometro-compatibility of contortions and density of disclination charges

$$\begin{aligned}\vec{\theta}_i &= \overrightarrow{\text{rot}}\left(\vec{\lambda}_i - \vec{e}_i \wedge \vec{\lambda} - \vec{e}_i \lambda\right) = \overrightarrow{\text{rot}}\,\vec{\chi}_i \\ &= \overrightarrow{\text{rot}}\left[\overrightarrow{\text{rot}}\,\vec{\varepsilon}_i\right]^{\text{T}} = \overrightarrow{\text{rot}}\left[\overrightarrow{\text{rot}}\,\vec{\alpha}_i\right]^{\text{T}} - \frac{1}{3}\overrightarrow{\text{rot}}\,\overrightarrow{\text{rot}}\left(\vec{e}_i \tau\right)\end{aligned} \quad (4)$$

Symmetry of the density tensor of disclination charges

$$\sum_k \vec{e}_k \wedge \vec{\theta}_k = \sum_k \vec{e}_k \wedge \overrightarrow{\text{rot}}\left(\vec{\lambda}_k - \vec{e}_k \wedge \vec{\lambda} - \vec{e}_k \lambda\right) = \sum_k \vec{e}_k \wedge \overrightarrow{\text{rot}}\,\vec{\chi}_k \equiv 0 \quad (5)$$

Geometro-compatibility of flexions and density of charges of curvature

$$\begin{aligned}\theta &= -\sum_k \vec{e}_k \vec{\theta}_k = \text{div}\,\vec{\lambda} = \text{div}\,\vec{\chi} \\ &= \sum_k \vec{e}_k \overrightarrow{\text{rot}}\,\overrightarrow{\text{rot}}\,\vec{\varepsilon}_k = \sum_k \vec{e}_k \overrightarrow{\text{rot}}\,\overrightarrow{\text{rot}}\,\vec{\alpha}_k - \frac{2}{3}\Delta\tau\end{aligned} \quad (6)$$

Equation of conservation of charges of disclination

$$\text{div}\,\vec{\theta}_i = \text{div}\left[\overrightarrow{\text{rot}}\left(\vec{\lambda}_i - \vec{e}_i \wedge \vec{\lambda} - \vec{e}_i \lambda\right)\right] = 0 \quad (7)$$

Chapter 9

Topological singularities associated with the charges

Since the tensorial density $\vec{\lambda}_i$ of distortion charges must satisfy a conservation equation $\operatorname{div}\vec{\lambda}_i \equiv 0$, it is impossible for it to appear in a punctual way. It must always occupy a field of *nonzero volume* in the solid medium. We therefore treat mathematically the different possible topologies of charged domains, namely *strings, membranes, torus loops* and *clusters*, by illustrating them with quantified topological singularities such as *dislocations, disclinations* and *loops* that appear within a structured lattice.

On the other hand, writing of the tensor density of disclination charges $\vec{\theta}_i = \overrightarrow{\operatorname{rot}}\left(\vec{\lambda}_i - \vec{e}_i \wedge \vec{\lambda} - \vec{e}_i \lambda\right)$ implies that the appearance of disclinations in a place of a solid medium is a direct consequence of the presence in this place of dislocations, so that there cannot be isolated disclinations within a lattice.

Finally, we show that the vector density $\vec{\lambda}$ of flexion charges, the scalar density λ of rotation charges and the scalar density $\theta = \operatorname{div}\vec{\lambda}$ of curvature charges are the three fundamental quantities necessary and sufficient to describe the long-range effects of topological singularities localized within a lattice.

9.1 – Strings and dislocation lines

As the density of charges of distortion $\vec{\lambda}_i$ is a tensorial quantity, the non-zero density areas of charges must be extended in space, and their spatial topology is constrained in order to satisfy the conservation equation $\operatorname{div}\vec{\lambda}_i \equiv 0$. In this section, we will analyze a particular type of field topology of charges, namely tubular areas that can be called *strings*.

The dislocation strings and their charge density $\vec{\lambda}_i$

Suppose a region of a solid lattice, which is approximately in homogeneous volume expansion τ, that is to say wherein the spatial variation of the volume expansion τ is very small. This hypothesis is one that almost always applies to the case of real solids such as metals. But suppose also that significant rotations are possible at great distances in this solid, which is perfectly feasible in the case of real solids. If there exists within this solid a localized area in *string* form, a tubular shape, containing a non-zero density of dislocation charges, shown in figure 9.1 by a shaded area, it is possible to define in the frame $O\xi_1\xi_2\xi_3$ of ***GO*** a volume V partially containing this domain. The integral over the volume V of the conservation equation of tensorial charge density can be transformed into an integral over the surface surrounding the integration volume

$$\iiint_V \operatorname{div}\vec{\lambda}_i \, dV = \oiint_S \vec{\lambda}_i \, d\vec{S} \equiv 0 \qquad (9.1)$$

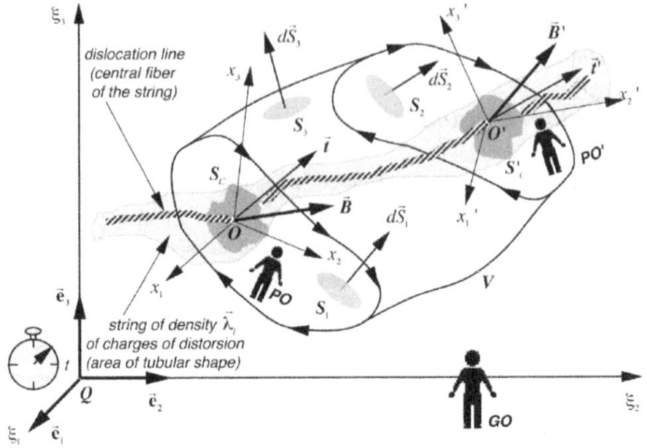

***Figure 9.1** - string (domain of tubular shape) of density $\vec{\lambda}_i$ of charges of distortion and the dislocation line (central fiber of the string)*

With the geometry described in figure 9.1, the surface S may be divided into a side surface S_3, on which the charge density is zero, and therefore does not contribute to the surface integral, and two cutting surfaces S_1 and S_2, areas of non-zero density of charges, so that the curved surface integral can be decomposed by the difference of surface integrals S_1 and S_2

$$\oiint_S \vec{\lambda}_i \, d\vec{S} = \iint_{S_2} \vec{\lambda}_i \, d\vec{S}_2 - \iint_{S_1} \vec{\lambda}_i \, d\vec{S}_1 \equiv 0 \tag{9.2}$$

The integrals on S_1 and S_2 can be made in the local reference frames $Ox_1x_2x_3$ and $Ox_1'x_2'x_3'$ of observers ***PO*** and ***PO'*** respectively. According to the integral relationship of compatibility *(8.11)*, these integrals then represent the projections of the Burgers vector on the axes of the local frames $Ox_1x_2x_3$ and $Ox_1'x_2'x_3'$, so we have the following consequence of the equation of conservation of the tensorial density $\vec{\lambda}_i$ of charges of dislocation

$$B_i = -\iint_{S_1} \vec{\lambda}_i \, d\vec{S}_1 = -\iint_{S_2} \vec{\lambda}_i \, d\vec{S}_2 = B_i' \tag{9.3}$$

This relationship has a very important topological result concerning the geometrical forms acceptable for a non-zero density field of charges $\vec{\lambda}_i$. Indeed, as the projections B_i of the Burgers vector $\vec{B}$ on the axes of local frames, projections defined in these local repositories by a transversal surface to the nonzero density of field $\vec{\lambda}_i$, these are invariant scalars throughout the area where density is not null, and thus this area must necessarily be continuous (connected). It cannot end abruptly in the solid as the scalar B_i must be kept on any plane intersecting the domain. Therefore, this area has the form of a *tubular string*, which must necessarily *cross the solid through-and-through*, or be a *loop-shaped ring*.

As for the Burgers vector $\vec{B}$, defined by the relationship $\vec{B} = \sum B_i \vec{e}_i$, *it is not necessarily an invariant vector in the absolute frame* along the dislocation since only its scalar components B_i

are invariant along the string, while the associated base vectors $\vec{e}_i$ associated with local frame may change along the string in the case where the medium undergoes strong rotations from one place to another. That is, if the volume expansion τ remains homogeneous, the vector $\vec{B}$ is of invariant norm in the absolute frame, but its direction can change along the string in exactly the same way as the local frames do along the string in the solid.

On the other hand, the Burgers vector $\vec{B}$ is a real invariant within any local frame $Ox_1x_2x_3$ defined along the dislocation since within the scope of a local frame, τ and $\vec{\omega}$ cannot present significant variations. In other words, *the Burgers vector $\vec{B}$ can be considered as an invariant in local frames*, and this regardless of the direction $\vec{t}$ that the dislocation can take in the local frame , where $\vec{t}$ is the unit vector tangent to the central fiber (figure 9.1).

In relation *(8.11)* giving the Burgers vector $\vec{B}$ of a dislocation from the tensorial density $\vec{\lambda}_i$, it is always possible to integrate on a surface S perpendicular to the direction $\vec{t}$ of the central fiber by writing $d\vec{S} = \vec{t}dS$ (figure 9.1)

$$\vec{B} = -\sum_i \vec{e}_i \left(\vec{t} \iint_S \vec{\lambda}_i \, dS \right) \tag{9.4}$$

Dislocation lines, or dislocations, and their linear charge $\vec{\Lambda}_i$

The field of tubular string of charges $\vec{\lambda}_i$ can be modeled in the form of a *dislocation line*, commonly called *dislocation*, represented by a one-dimensional core fiber in the center of the non-zero density $\vec{\lambda}_i$ of the string of charges (fiber is represented by line hatched in the drawing of figure 9.1). This *dislocation line* must necessarily either pass through the solid from end to end, or form a dislocation loop closed on itself. In the local frame, the vector $\vec{B}$ must remain constant, if ever the diameter C of the contour of integration is varied or this contour along the string is displaced, with the *sine qua non* condition that this contour is always located outside the central area containing the charge density $\vec{\lambda}_i$, and that the volume expansion is uniform in all the local frame. We deduce that if the dislocation string is sufficiently thin (of sufficiently small cross-section), the charge density $\vec{\lambda}_i$ can be represented by a magnitude confined to the immediate vicinity of the central fiber of the string, which will be called *dislocation line*, by introducing the concept of *linear tensorial charge of dislocation* $\vec{\Lambda}_i$, namely a set of three vectors defined on the central fiber by the following relationship

$$\vec{\Lambda}_i = \iint_{S'_C} \vec{\lambda}_i \, dS \tag{9.5}$$

where the surface S'_C is the area of the core of the string. These relationships are then used to deduce the Burgers vector $\vec{B}$ from $\vec{\Lambda}_i$, as well as its components B_i (which are invariant along the dislocation line)

$$\vec{B} = \sum_i B_i \vec{e}_i = -\sum_i \left(\vec{t} \vec{\Lambda}_i \right) \vec{e}_i \quad \Rightarrow \quad B_i = -\vec{t} \vec{\Lambda}_i \tag{9.6}$$

To satisfy the relations *(9.6)*, it is necessary and sufficient to state that the vectors $\vec{\Lambda}_i$ representing the linear tensorial charge of the dislocation line have *conservative tangential components* $\vec{\Lambda}_i$ that satisfy the following relationships

$$\vec{\Lambda}_i = -\vec{t} \, B_i \tag{9.7}$$

Thus, regardless of the direction $\vec{t}$ of the central fiber of a dislocation, the components $\vec{\Lambda}_i$, tangential to the dislocation line are directly deducted from the knowledge of the components B_i of the Burgers vector $\vec{B}$ of the string.

From the *trace* of the linear tensorial charge $\vec{\Lambda}_i$ we deduce linear scalar charge of rotation of the dislocation, using the relations *(8.17)* and *(9.7)*, and note that this scalar charge depends on the scalar dot product of the Burgers vector $\vec{B}$ by the natural vector $\vec{t}$ tangent to the dislocation line

$$\Lambda = \frac{1}{2}\sum_i \vec{e}_i \vec{\Lambda}_i = \frac{1}{2}\sum_i \vec{e}_i\left(-\vec{t}\,B_i\right) = -\frac{1}{2}\vec{B}\vec{t} \tag{9.8}$$

As for the *antisymmetric part* of the linear tensorial charge $\vec{\Lambda}_i$, it can be represented by a *linear vector charge of flexion* of the dislocation, which depends on the cross product of the vector $\vec{B}$ by the natural vector tangent to the dislocation line $\vec{t}$

$$\vec{\Lambda} = -\sum_i \vec{e}_i \wedge \vec{\Lambda}_i = -\sum_i \vec{e}_i \wedge\left(-\vec{t}\,B_i\right) = \vec{B}\wedge\vec{t} \tag{9.9}$$

Thus, the relative orientation of the Burgers vector $\vec{B}$ of the dislocation and of the natural tangent vector $\vec{t}$ to its central fiber plays an important role in the expression of different linear tensorial charges, and actually determines the nature of the dislocation line.

Decomposing the Burgers vector into its parallel part $\vec{B}_{||}$ and perpendicular part $\vec{B}_\perp$ to the tangent vector $\vec{t}$, we have the following relationship

$$\vec{B} = \underbrace{\left(\vec{B}\vec{t}\right)\vec{t}}_{\vec{B}_{||}} + \underbrace{\vec{t}\wedge\left(\vec{B}\wedge\vec{t}\right)}_{\vec{B}_\perp} = -2\Lambda\vec{t} + \vec{t}\wedge\vec{\Lambda} \tag{9.10}$$

which shows that a dislocation can be perfectly and completely described by the data of the *linear charge of rotation* Λ, which is linked to the parallel portion $\vec{B}_{||}$ of the Burgers vector, and its *linear flexion charge* $\vec{\Lambda}$, which is related to the perpendicular part $\vec{B}_\perp$ of the Burgers vector.

On the quantification and taxonomy of dislocations in a lattice

The connected areas of non-zero dislocation density charges can be modeled in the simplest way as thin strings. In this section, we will show that the straight strings appearing in a solid lattice are *quantified* on a microscopic scale (fig. 9.2 and 9.3), and that these strings then represent *elementary plastic distortion singularity fields*, ie "*elementary particles*" of the plastic deformation of the lattice.

If we consider the case of an ordered array of particles on the microscopic scale, one can introduce dislocations by cutting links in a plane of the lattice, parallel movement of the jaws and reconstruction of bonds, as shown in figures 9.9 and 9.10 in the case of a simple cubic lattice.

The Burgers vector $\vec{B}$ of singularities thus obtained is deduced by considering a closed circuit C on the solid, which surrounds the singularity, and seeking the closure $\vec{B}$ of the corresponding open circuit in the undistorted virtual lattice.

With figures 9.2 and 9.3 , we see that the microscopic singularities of the lattice have an essential characteristic: *their Burgers vector is quantized*, that is to say, its components can only be integer multiples of the unit cell length a of the lattice.

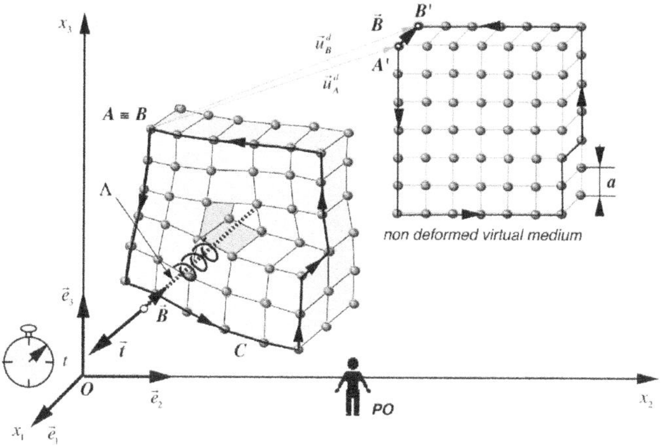

Figure 9.2 *- screw dislocation is quantified in a cubic lattice*

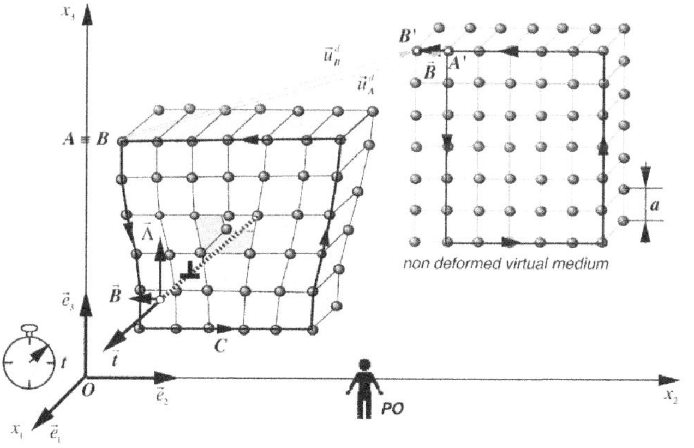

Figure 9.3 *- edge dislocation is quantified in a cubic lattice*

The nature of the microscopic plastic singularity can change according to the respective directions being taken in the local coordinate system, the Burgers vector $\vec{B}$ and the unit vector $\vec{t}$ tangent to the line:

- when $\vec{B}$ is parallel to $\vec{t}$ (fig. 9.2) , the *linear dislocation charge* $\vec{\Lambda}_i$ has a non-zero trace $(\Lambda \neq 0)$, so has rotational charge, and zero antisymmetric part $(\vec{\Lambda} = 0)$. One speaks in this case of *screw dislocations*[1], and of linear charge of rotation Λ of the screw dislocation. We

[1] *J. M. Burgers, Proc. Kon. Ned. Akad. Weten, vol.42, p. 293, 378, 1939*

symbolically represent it by a screw on the dislocation line. As $\Lambda = -\vec{B}\vec{t}\,/\,2$ when the screw dislocation has a *right rotation*, identical to the rotational direction of a normal screw or corkscrew, Λ is positive and vectors $\vec{t}$ and $\vec{B}$ are oriented in opposite directions. On the other hand, if the screw dislocation has a *left rotation*, identical to the rotational direction opposite to that of a normal screw or a corkscrew, Λ is negative and vectors $\vec{t}$ and $\vec{B}$ are oriented in the same direction (fig. 9.4). Note that the choice of a given direction $\vec{t}$ is quite arbitrary as only the sign of Λ is fixed.

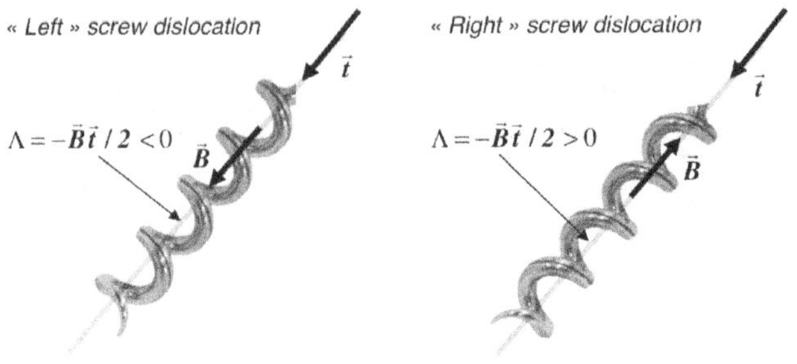

Figure 9.4 *- screw dislocation, left and right respectively*

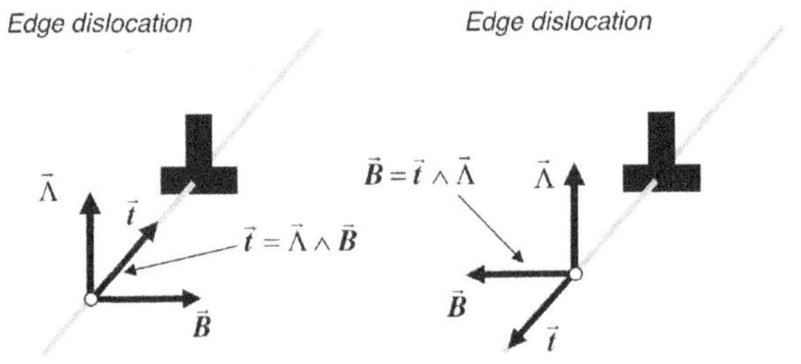

Figure 9.5 *- edge dislocation*

- when $\vec{B}$ is perpendicular to $\vec{t}$ (fig. 9.3), the linear dislocation charge Λ_i has zero trace ($\Lambda = 0$)), so no charge of rotation, but a non-zero antisymmetric part ($\vec{\Lambda} \neq 0$). One speaks in this case of *edge dislocations*[2], and of *linear flexion charge* $\vec{\Lambda}$ *of the edge dislocation*, and symbolically we represent the latter by a sign ⊥ on the dislocation line, oriented so as to represent the additional particle plane. The vector $\vec{\Lambda}$ always has the direction of the additional plane ⊥ of the edge dislocation. This is the only fixed size, so that $\vec{B} = \vec{t} \wedge \vec{\Lambda}$ if the direction $\vec{t}$ is given, and $\vec{t} = \vec{\Lambda} \wedge \vec{B}$ if the Burgers vector $\vec{B}$ is given (fig. 9.5).
- when $\vec{B}$ is neither parallel nor perpendicular to $\vec{t}$ the *linear charge of dislocation* Λ_i has a

[2] *E. Orowan, Z. Phys., vol. 89,p.605,614 et 634, 1934; M. Polanyi, Z. Phys., vol.89, p. 660, 1934; G. I. Taylor, Proc. Roy. Soc. London, vol. A145, p. 362, 1934*

non-zero trace $(\Lambda \neq 0)$, but also a non-zero anti-symmetric part $(\vec{\Lambda} \neq 0)$, such that it behaves both as a source of rotations and elastic and anelastic flexion. We speak in this case of *mixed dislocations*.

In a discrete lattice, a dislocation may well change direction. In other words, along the dislocation line, the tangent vector $\vec{t}$ is not necessarily preserved. In this case, as the Burgers vector $\vec{B}$ is conserved in the local frame, this means that the dislocation should change type. For example, figure 9.6 shows a simple cubic lattice model in which a screw dislocation enters the front of left, turns inside the lattice by becoming mixed, and emerges as an edge dislocation on the adjacent face to the right.

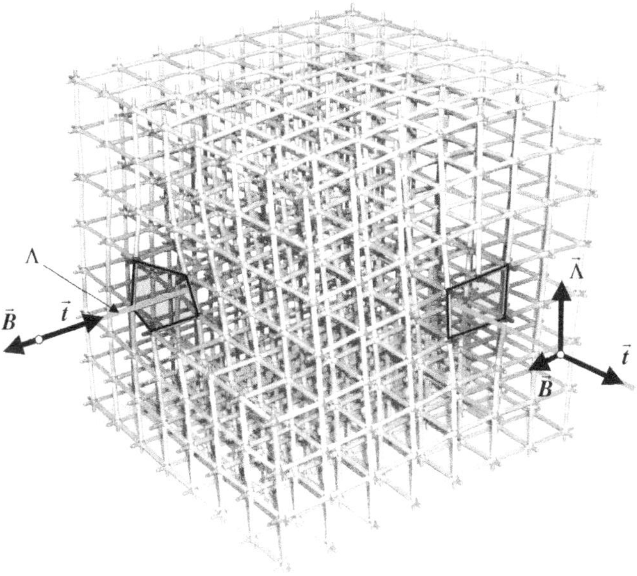

Figure 9.6 - *dislocation changing from a screw type to an edge type in a cubic lattice*

The lattice dislocations, the «elementary particles» of the plastic deformation

Quantized dislocations are the most basic vectors of the plastic deformation of a lattice. In this sense, they might be called the *"elementary particles"* of plastic deformation. Moreover, any dislocation string has its *"anti-string"*. Indeed, it is easy to see that two parallel dislocations in the same direction $\vec{t}$ and Burgers vectors $\vec{B}$ and $-\vec{B}$ respectively, annihilate completely if they come to meet within the lattice.

Using an integration of the compatibility relationship *(8.17)* on a volume V surrounding a screw dislocation on a length L (fig. 9.7) we have

$$\Lambda = \frac{1}{L}\iiint_V \lambda dV = \frac{1}{L}\iiint_{V_C} \lambda dV = \frac{1}{L}\iiint_V \operatorname{div}\vec{\omega} dV = \frac{1}{L}\oiint_S \vec{\omega} d\vec{S} = \frac{1}{L}\iint_{S_3} \vec{\omega} d\vec{S}_3 \tag{9.11}$$

We notice that screw dislocations, carrying a linear scalar charge $\Lambda \neq 0$, are sources of a *field of divergent local rotations*, which is, as we have seen, the analog of the electric field. Thus, at a distance R of the string, the norm of the rotation field is simply $|\vec{\omega}| = |\Lambda| / 2\pi R$.

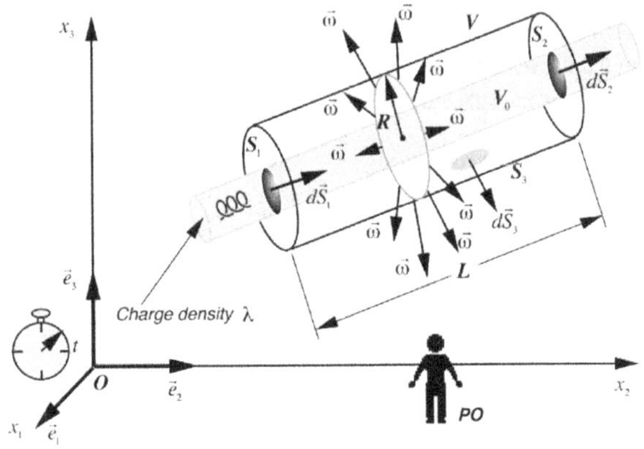

Figure 9.7 *- the radial field of rotation around a screw dislocation*

As to the edge dislocations, which are carriers of a flexion charge $\vec{\Lambda} \neq 0$, they are a source of a *lattice flexion*, and therefore of a local curvature of the lattice in their vicinity as the flexion vector $\vec{\chi}$ satisfies the relationship $\vec{\chi} = -\overrightarrow{\mathrm{rot}}\,\vec{\omega} + \vec{\lambda}$.

On the importance of the microscopic structure of the lattice

The *microscopic structure* of a solid plays a significant role on the nature of dislocations that may appear in this solid, but it is beyond the scope of this book to describe in detail these microscopic aspects. In the excellent book[3] by Hirth and Lothe, the reader will find a detailed description of most of these aspects of lattice dislocations. One can also find an excellent history of the discovery of dislocations in the twentieth century in the article by Hirth[4] (1985).

In fact, it is essentially the structure of the core of the quantized dislocation which will depend on the microscopic structure of the network. A first important effect of the periodic structure of the lattice is that the energy of a line dislocation depends strongly on the direction in which the line points in the lattice. For example, consider an edge dislocation like the one shown in figure 9.3. It seems quite clear that the movement of its position along the axis Ox_2 must change the energy of this dislocation periodically with a spatial frequency which is that of the lattice. The minimum and maximum energy positions are respectively called Peierls valleys and hills, named Peierls[5] who was the first in 1940 to have considered this problem. For example, to move a

[3] *J. P. Hirth and J. Lothe, theory of dislocations, second edition, John Wiley & Sons, New-York, 1982*

[4] *J.-P. Hirth, A Brief History of Dislocation Theory, Metallurgical Transactions A, vol. 16A, p.2085, 1985*

[5] *R. E. Peierls, Proc. Phys. Soc., vol. 52, p. 23, 1940*

dislocation from a Peierls valley to another, we must then apply a force called the *Peierls-Nabarro force*[6].

It is also conceivable that a dislocation is oblique to the Peierls valleys, and that it must therefore overcome these valleys. In this case, it is logical to think that the oblique line dislocation does not necessarily have the lowest energy configuration, and the dislocation could be interested in taking a polygonal shape by the formation of *kinks*[7] that minimize the portions of the dislocation lying on the hills of Peierls, and therefore help to minimize the overall energy (fig.9.8).

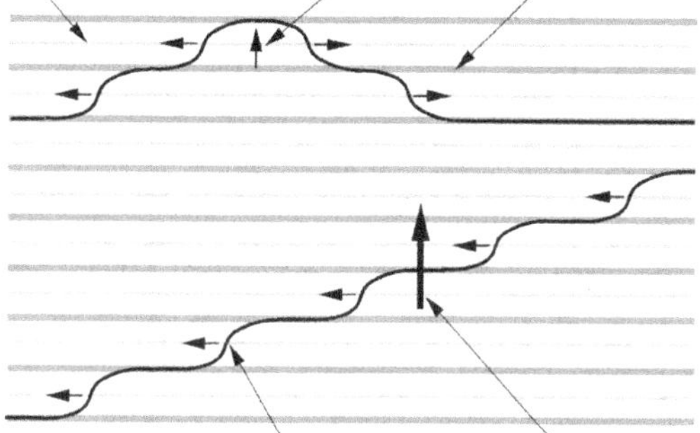

Figure 9.8 *- The kinks and their mobility within the Peierls hills*

The existence of these kinks also has the advantage of allowing the mobility of the dislocation perpendicularly to the valleys Peierls simply by the movement of the kinks along the dislocation line (fig. 9.8). In conventional materials such as metals of different structures we can change the mobility of dislocations through a process of creation and annihilation of pairs of kinks, a process that is usually thermally activated, and is responsible for some of the macroscopic anelastic and plastic properties of these materials. These processes have been extensively discussed in the literature [3].

On the existence of partial dislocation and the 'strong force' that bind them

Dislocations appearing in a more complex structures than the simple cubic lattice, such as face-centered cubic lattices, cubic or hexagonal centered lattices, generally have much more complicated structures within them. Due to energetic reasons and according to the crystal system considered, there will appear a dissociation of the core of the dislocation into two or more *partial dislocations*[8] whose Burgers vectors are individual fractions of the lattice translation vectors.

[6] *F. R. N. Nabarro, Proc. Phys. Soc. London, vol. 59, p. 256, 1947*

[7] *J. Weertman, Phys. Rev., vol. 101, p. 1429, 1956; A. Seeger, Phil. Mag., vol.1, p. 651, 1956*

[8] *J. Frenkel,T. Kontorava, Fiz. Zh., vol.1 , p. 137, 1939; R. D. Heidenreich, W. Shockley, Report of Conf. on Strength of Solids, Phys. Soc., London, p. 57, 1948; F. C. Frank, Proc. Phys. Soc. London, vol. 62A, p. 202, 1949; N. Thompson,Proc. Phys. Soc. London, vol. 66B, P. 481, 1953*

For example, in the face-centered cubic metals (FCCs), the stacking of atoms is characterized by sequences *abc abc abc* ... (fig. 9.9). The Burgers vector $\vec{B}_p$ of a perfect dislocation should in principle be connecting two lattice nodes. But for energetic reasons, the most favorable Burgers vectors are those who have a minimum length because the distortion energy stored in the lattice by a dislocation is proportional to the square of the Burgers vector as we will discuss later. Thus in the case of figure 9.9, the dislocations dissociate on their sliding plane into two partial of Burgers vectors $\vec{B}_1$ and $\vec{B}_2$, so that $\vec{B}_p = \vec{B}_1 + \vec{B}_2$. In the case of this dissociation $\vec{B}_1\vec{B}_2 > 0$ so we have $\vec{B}_p^2 = \vec{B}_1^2 + \vec{B}_2^2 + 2\vec{B}_1\vec{B}_1 > \vec{B}_1^2 + \vec{B}_2^2$.

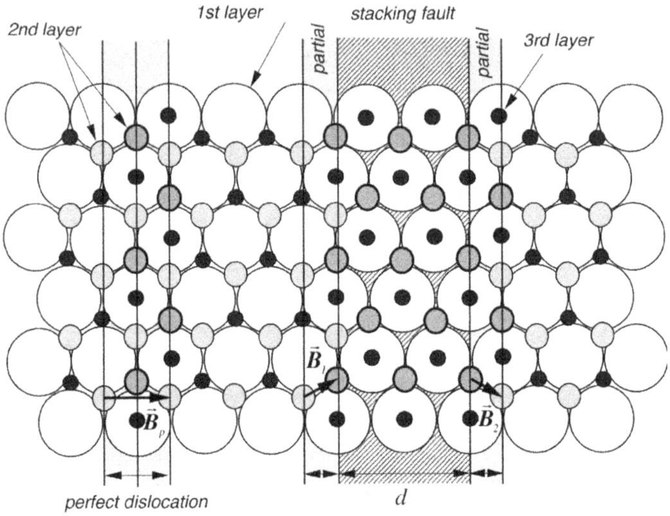

Figure 9.9 - *partial dislocations and ribbon of stacking faults in a FCC*

The two partials obtained in figure 9.9 by the dissociation are called of *Shockley type* [8]. The distance between the two partials is then controlled by a competition between the energy decrease associated with the increased distance between the partials that repel and the energy increase due to the formation of an energetic ribbon of stacking faults (abc abc abc ac ...) between the two partial dislocations, as illustrated by figure 9.10.

As the ribbon has a stacking fault energy γ per unit area, the total energy $E_t(d)$ per unit length of dislocation for a dissociated dislocation over a distance d is written $E_t(d) = \gamma d + E_d(d)$, where $E_d(d)$ is the energy of the two partials depending on the distance that separates them, which is a monotonically decreasing function, decreasing from E_0 for $d = 0$ to $2E_0/3$ pour $d \to \infty$ in the case of the Shockley partials illustrated in figure 9.9.

The energy $E_t(d)$, therefore, presents a minimum for the distance $d = d_0$ (fig. 9.10), which is the equilibrium distance between the two partials, controlled by the competition between the energy decrease associated with the increase in distance between the partials and the energy increase due to the formation of a ribbon of stacking faults between the two partials. This behavior of the energy $E_t(d)$ induces an interaction force between the two partials that could be called a "*strong force*" in the sense that the energy of the pair of partials has a minimum which fixes the position of equilibrium d_0, but increases if one tries to increase the separation distance beyond it. The "strong force" qualifier proposed here as the behavior of the interaction force between the partials presents an interesting analogy with the strong force acting between quarks in the Standard Model of elementary particles as we shall see later.

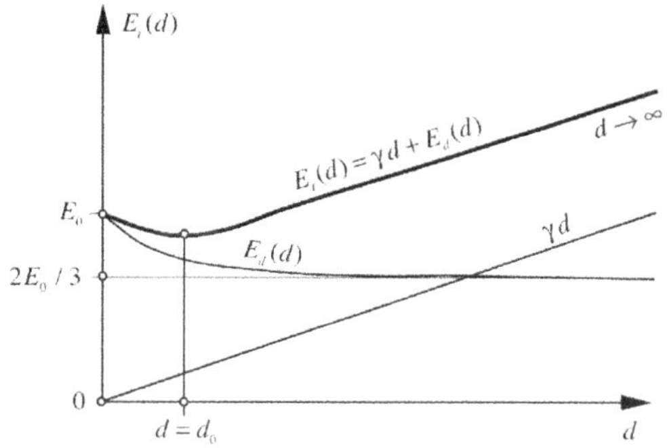

Figure 9.10 *- The energy of two partial dislocations as a function of the distance that separates them*

As an example, consider the illustration (fig. 9.11) of a model of a mixed dislocation (with edge and screw at once) separated into two partial in a face-centered cubic structure. It clearly shows the existence of a stacking fault between the two partials. And since this is a mixed dislocation, both partials present a series of steps (kinks), such as we have described previously. Moreover, one can even observe network flexion induced by the edge portion of the dissociated dislocation.

Figure 9.11 *- Model of a mixed dislocation in a FCC,*
showing a dissociation between two partials, as well the kinks on the partials

The set of all the consequences related to the network structure are obviously too specific to the type of crystal structure to be treated here in detail. But they can be addressed in any book dealing with dislocations in the crystal structures.

The effects of a spatial variation of volume expansion

In the case of imaginary medias in which the volume expansion is no longer uniform, say with large variations of τ, it is not possible to represent a dislocation of the lattice by a linear density of charge $\vec{\Lambda}_i$ focused on a line on the center of the dislocation. Indeed, in this case, the integral over a surface S surrounding the center of the dislocation

$$\vec{B} = -\sum_i \vec{e}_i \left(\vec{t} \iint_S \vec{\lambda}_i \, dS \right) \tag{9.12}$$

will not give the same result at different locations along the dislocation line, and does not give the same result according to the size of the selected area S, which implies that the charge density $\vec{\lambda}_i$ is not localized in the center of dislocation, but is now disseminated in a tube more or less bulky around the center, so that the integral (9.12) gives the actual Burgers vector associated with S, the selected surface in the local referential $Ox_1x_2x_3$ we used to measure it (fig. 9.12).

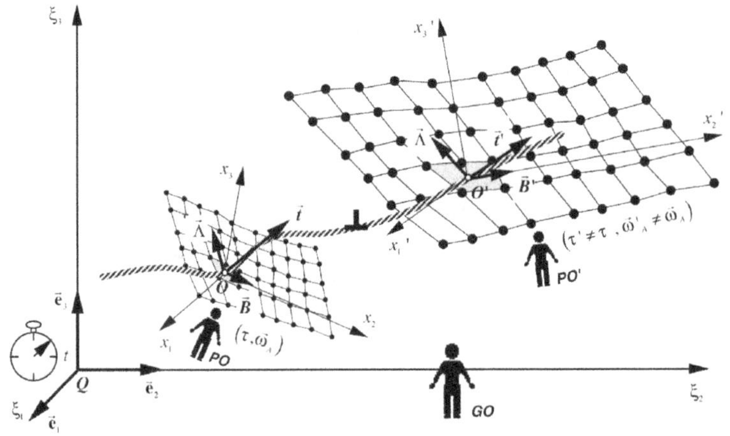

***Figure 9.12** - effects of a spatial variation of the volume expansion on a lattice dislocation*

But it is clear that throughout this tube, the condition (9.2) remains valid if the surfaces S_1 and S_2 are chosen broadly enough to encompass the charge density $\vec{\lambda}_i$ surrounding the core of the lattice dislocation. It goes without saying that this kind of situations where the volume expansion and local rotations are not homogeneous within a large portion of the solid with dislocations, is much more difficult to treat than the case where there is homogeneity of τ and $\vec{\omega}$ over great distances or when the lattice dislocations are confined to small enough regions so that they can be considered homogeneous.

However, it will always be possible to describe locally any dislocation as a localized string in a coordinate system $Ox_1x_2x_3$ as shown in figure 9.12, by using locally relations *(9.8)* to *(9.10)* to calculate the linear densities $\vec{\Lambda}_i$, $\vec{\Lambda}$ and Λ, and by assuming that the volume expansion τ

may be regarded as homogeneous in the referential $Ox_1x_2x_3$. In this case, the Burgers vector length is given locally by the relation

$$\left|\vec{B}\right|_{\tau\neq 0}=\left|\vec{B}_0\right|_{\tau=0}\mathrm{e}^{\tau/3} \tag{9.13}$$

so that the linear density in a region where the volume expansion τ is not zero will be connected to the same linear density of the dislocation at $\tau=0$ by the relations

$$\left|\vec{\Lambda}_i\right|_{\tau\neq 0}=\left|\vec{\Lambda}_{0i}\right|_{\tau=0}\mathrm{e}^{\tau/3} \quad ; \quad \left|\vec{\Lambda}\right|_{\tau\neq 0}=\left|\vec{\Lambda}_0\right|_{\tau=0}\mathrm{e}^{\tau/3} \quad ; \quad \left|\Lambda\right|_{\tau\neq 0}=\left|\Lambda_0\right|_{\tau=0}\mathrm{e}^{\tau/3} \tag{9.14}$$

Since linear density $\vec{\Lambda}_i$ is derived as the product of the volume density $\vec{\lambda}_i$ by a surface, it is deduced that the volume density $\vec{\lambda}_i$ in a region compressed or expanded to a non-zero value of τ is connected to the initial volume density of the same region with $\tau=0$ by the following relations

$$\left|\vec{\lambda}_i\right|_{\tau\neq 0}=\frac{\left|\vec{\Lambda}_i\right|_{\tau\neq 0}}{\left|S\right|_{\tau\neq 0}}=\frac{\left|\vec{\Lambda}_{0i}\right|_{\tau=0}\mathrm{e}^{\tau/3}}{\left|S_0\right|_{\tau=0}\mathrm{e}^{2\tau/3}}=\left|\vec{\lambda}_{0i}\right|_{\tau=0}\mathrm{e}^{-\tau/3} \tag{9.15}$$

We also deduce that

$$\left|\vec{\lambda}\right|_{\tau\neq 0}=\left|\vec{\lambda}_0\right|_{\tau=0}\mathrm{e}^{-\tau/3} \quad ; \quad \left|\lambda\right|_{\tau\neq 0}=\left|\lambda_0\right|_{\tau=0}\mathrm{e}^{-\tau/3} \tag{9.16}$$

On the impossibility of finding isolated disclination lines

It can be shown that there cannot exist *isolated disclination lines* in a solid, i.e. charged tubes whose overall Frank vector $\vec{\Omega}$ is not zero. Indeed, *as the density of dislocation charges $\vec{\lambda}_i$ and disclination charges $\vec{\theta}_i$ are confined in the same tube*, it is possible to replace $\vec{\theta}_i$ by its expression $\vec{\theta}_i=\overrightarrow{\mathrm{rot}}\left[\vec{\lambda}_i-\vec{e}_i\wedge\vec{\lambda}-\vec{e}_i\lambda\right]$ in equation (8.34). It is then found that the Frank vector $\vec{\Omega}$ may be derived by the integral over the contour C_1 (figure 9.1) of the density $\vec{\lambda}_i-\vec{e}_i\wedge\vec{\lambda}-\vec{e}_i\lambda$, and that this density is zero on this contour on the outside of the tube, so that $\vec{\Omega}=0$. It therefore follows that there cannot be isolated charged tube whose overall Frank vector $\vec{\Omega}$ is not zero.

9.2 – Membranes of dislocation and torsion, flexion and accommodation boundaries

We call *charged membrane* a thin interface containing charges, which separates two media containing no charges (fig. 9.13). It is clear that these membranes can be any surfaces in space (infinite surfaces, closed spheroidal surfaces or torus, ribbons or hollow tubes, thin plates, etc.), with the only topological condition that on any point of the membrane the dislocation charge conservation equation is satisfied $\mathrm{div}\,\vec{\lambda}_i\equiv 0$ and with the charges of disclination derived from the dislocation charges via the equation $\vec{\theta}_i=\overrightarrow{\mathrm{rot}}\left[\vec{\lambda}_i-\vec{e}_i\wedge\vec{\lambda}-\vec{e}_i\lambda\right]$. *Passthrough conditions* for topological tensor through such a charged membrane are derived from the geometro-compatibility equations and are easy to establish. It is sufficient to write the full compatibility of the equation taking on a thin volume, with the movable membrane, of which each face is located on either side of said membrane (Fig. 9.13).

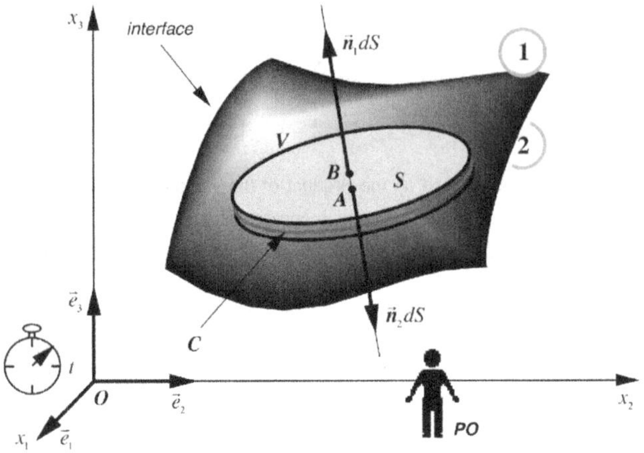

***Figure 9.13** - bi-dimensional charged membrane*

The surface charge $\vec{\Pi}_i$ of dislocation of a joint in the presence of gradients of $\vec{B}$

To establish the passthrough conditions for the distortion tensor, we must integrate its compatibility equation in the volume V of the membrane, which contains the dislocation density $\vec{\lambda}_i$

$$\iiint_V \overrightarrow{\mathrm{rot}}\,\vec{\beta}_i\, dV = \iiint_V \vec{\lambda}_i\, dV = \oiint_S \left(d\vec{S} \wedge \vec{\beta}_i\right) \qquad (9.17)$$

Provided that the membrane is sufficiently thin, the surface integral can be written as the sum of the surface integrals on both sides of S and the volume integral can be decomposed as follows

$$\oiint_S \left(d\vec{S} \wedge \vec{\beta}_i\right) = \iint_{S_1} \left(\vec{n}_1 \wedge \vec{\beta}_i^{(1)}\right) dS + \iint_{S_2} \left(\vec{n}_2 \wedge \vec{\beta}_i^{(2)}\right) dS = \iint_S dS \int_A^B \vec{\lambda}_i\, dl \qquad (9.18)$$

wherein $\vec{n}_1$ and $\vec{n}_2$ are the unit vectors normal to the surface from the membrane on each side of the membrane.

Thus, if the charged membrane is very thin, it is possible to introduce the concept of *surface charge of dislocation* $\vec{\Pi}_i$ by writing

$$\vec{\Pi}_i = \int_A^B \vec{\lambda}_i\, dl \qquad (9.19)$$

which allows us to deduce that

$$\vec{n}_1 \wedge \vec{\beta}_i^{(1)} + \vec{n}_2 \wedge \vec{\beta}_i^{(2)} = \vec{\Pi}_i \qquad (9.20)$$

The topological interpretation of this passthrough condition is simple: the existence of a surface charge of dislocation $\vec{\Pi}_i$ in the membrane leads to a discontinuity of the tangential components of the distortion vector $\vec{\beta}_i$ on both sides of it. But on both sides of the membrane, outside the volume, the medium contains no plastic charges, so that the tensor $\vec{\beta}_i$ must derive from the gradient of the components of the displacement field $\vec{u}$, i.e. $\vec{\beta}_i = \overrightarrow{\text{grad}}\, u_i$. It is therefore possible to write

$$\vec{n}_1 \wedge \overrightarrow{\text{grad}}\, u_i^{(1)} + \vec{n}_2 \wedge \overrightarrow{\text{grad}}\, u_i^{(2)} = \vec{\Pi}_i \tag{9.21}$$

In the local framework, this expression implies that the presence of the charged membrane locally induces a dislocation of the displacement field $\vec{u}$ associated with the elastic and anelastic distortions whose Burgers vector $\vec{B}$ is then given by

$$\vec{B} = \int_A^B \delta\vec{u} = \vec{u}^{(1)} - \vec{u}^{(2)} \tag{9.22}$$

so that

$$\vec{\Pi}_i = \vec{n}_1 \wedge \overrightarrow{\text{grad}}\, u_i^{(1)} - \vec{n}_1 \wedge \overrightarrow{\text{grad}}\, u_i^{(2)} = \vec{n}_1 \wedge \overrightarrow{\text{grad}}\, \boldsymbol{B}_i \tag{9.23}$$

According to this relationship, the existence of a nonzero surface charge $\vec{\Pi}_i$ in the membrane is therefore subject to the condition that *there is a gradient of the components of the Burgers vector on the surface of the membrane.* This means among others that if the Burgers vector is constant over the whole membrane, meaning if the membrane is a simple translation of the two surfaces S_1 and S_2 relative to each other, the surface density $\vec{\Pi}_i$ is zero.
Finally, it is noted that the three vectors $\vec{\Pi}_i$ making up the surface tensor charge *must be tangent vectors to the surface of the membrane,* which is in fact a direct result of the dislocation charge conservation equation $\text{div}\,\vec{\lambda}_i \equiv 0$. Indeed, considering the integral over a volume V in figure 9.13 we have

$$\iiint_V \text{div}\,\vec{\lambda}_i\, dV = \oiint_S \vec{\lambda}_i\, d\vec{S} = \iint_{S_1} \vec{\lambda}_i\, \vec{n}_1\, dS + \iint_{S_2} \vec{\lambda}_i\, \vec{n}_2\, dS \equiv 0 \tag{9.24}$$

which can be satisfied for any volume V which closely surrounds the charged interface only if the vectors $\vec{\lambda}_i$ are perpendicular to $\vec{n}_1$ and $\vec{n}_2$ on either side of the interface, in other words if they are tangent to the surface of the membrane.
As in the case of one-dimensional dislocation lines, one can extract the trace and the anti-symmetric part of the tensor $\vec{\Pi}_i$ of dislocation surface charge, by writing

$$\begin{cases} \vec{\Pi} = -\sum_i \vec{e}_i \wedge \vec{\Pi}_i = -\sum_i \vec{e}_i \wedge \left(\vec{n}_1 \wedge \overrightarrow{\text{grad}}\, \boldsymbol{B}_i\right) \\ \Pi = \dfrac{1}{2}\sum_i \vec{e}_i \vec{\Pi}_i = \dfrac{1}{2}\sum_i \vec{e}_i \left(\vec{n}_1 \wedge \overrightarrow{\text{grad}}\, \boldsymbol{B}_i\right) \end{cases} \tag{9.25}$$

The two-dimensional modeling of a thin membrane obtained by calculating the surface charges is usually called a *joint* or a *boundary*[9]. The boundary is then fully characterized by the given surface tensor $\vec{\Pi}_i$ of charges of dislocation, whose vectors are tangential to the surface. But it can also be given by the anti-symmetric part $\vec{\Pi}$ (the surface charge of a flexion boundary) and the trace Π (the surface charge of a rotation boundary) of the tensor of charges $\vec{\Pi}_i$, as in the case of the one-dimensional lines of dislocation. This point is well illustrated in Figure 9.14, wherein we show three thin membranes whose Burgers vectors linearly increase along the axis

[9] *C. Somigliana, Atti. Accad. naz. Lincei Rc., vol. 23, p. 463, 1912, et vol. 24, p.655, 1915; W. T. Read, W. Shockley, Phys. Rev. , vol. 78, p. 275, 1950; F. C. Frank, in Report of the Conf. on Defects in Crystalline Solids, Phys. Soc. London, p. 159, 1955*

Ox_1, and which are respectively oriented along the axes Ox_3 and Ox_1. As these thin membranes of dislocation actually allow us to disorient or to accomodate the solid grains located on either side of the membrane, they are generally called *grain boundaries.*

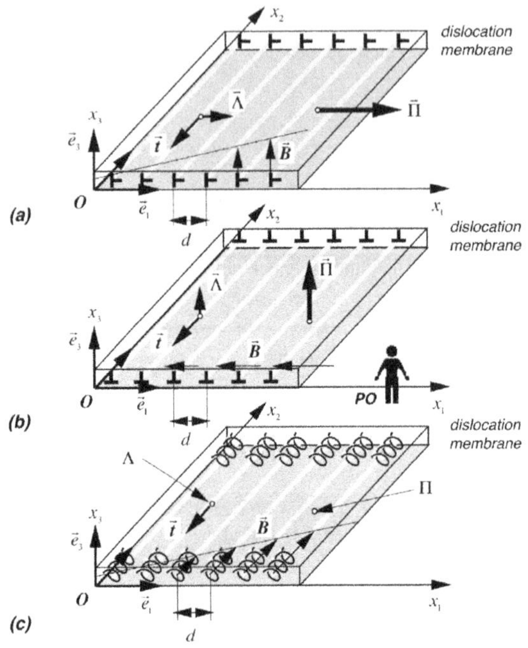

Figure 9.14 *- dislocation membranes of edge type **(a** and **b)** or screw type **(c)***

One can for example consider that these membranes are actually charged with edge and screw dislocations oriented parallel to the axis Ox_2. If the Burgers vectors of each individual dislocation have a length B and the distance between these individual dislocations is d, we have, $\left|\overrightarrow{\operatorname{grad}} B_i\right| = B / d = \alpha$, which allows us the calculation of expressions *(9.25).*

But we can also simply represent each individual dislocation by a line charge vector $\vec{\Lambda}$ if we are dealing with an edge dislocation or by a line charge scalar Λ if dealing with a screw dislocation. It then verifies that:

- the edge-type thin membrane with a Burgers vector perpendicular to the surface and increasing along the axis Ox_1 (fig. 9.14a) can be completely characterized by a *vector* $\vec{\Pi}$ *of surface charges of flexion* $\vec{\Pi}$, which is a vector tangential to the plane of the membrane, directed along Ox_1, and worth

$$\vec{\Pi} = \vec{\Lambda} / d \quad ; \quad \vec{\Pi} \,||\, Ox_1 \tag{9.26}$$

Since this type of edge membrane allows us to disorient the solid grains located on either side of the membrane, it is called a disorientation boundary, and in this particular case, as disorientation corresponds to a flexion of the solid, one speaks of a *tilt boundary*.

- the edge-type thin membrane with a Burgers vector parallel to the surface and increasing along the axis Ox_1 (fig. 9.14b) can be fully characterized by a vectorial surface charge of flexion $\vec{\Pi}$, whose vector is perpendicular to the membrane, and which satisfies

$$\vec{\Pi} = \vec{\Lambda} / d \quad ; \quad \vec{\Pi} \parallel Ox_3 \tag{9.27}$$

As this type of edge membrane area actually allows us to change, in direction Ox_1, the density of crystal planes of the solid grains located on either side of the membrane, it can be described as a *misfit boundary*.

- the screw-type thin membrane with a Burgers vector parallel to the membrane and increasing along the axis Ox_1 (fig. 9.14c) is fully characterized by the scalar surface charge of rotation Π, satisfying

$$\Pi = \Lambda / d \tag{9.28}$$

This type of screw membrane also corresponds to a disorientation boundary between the solid grains located on either side of the membrane. In this particular case, as disorientation corresponds to a rotation of the grains with respect to each other, it is called a *twist boundary*.

On the quantification of membranes of dislocation on a lattice

Dislocation membranes also quantify on a microscopic lattice, as shown in figure 9.15 representing a screw dislocation ribbon consisting of three aligned screw dislocations, corresponding to the case of figure 9.14c.

It is of course also possible to imagine ribbons of similar quantified dislocations, but made of edge dislocations, corresponding to the cases of figures 9.14a and 9.14b.

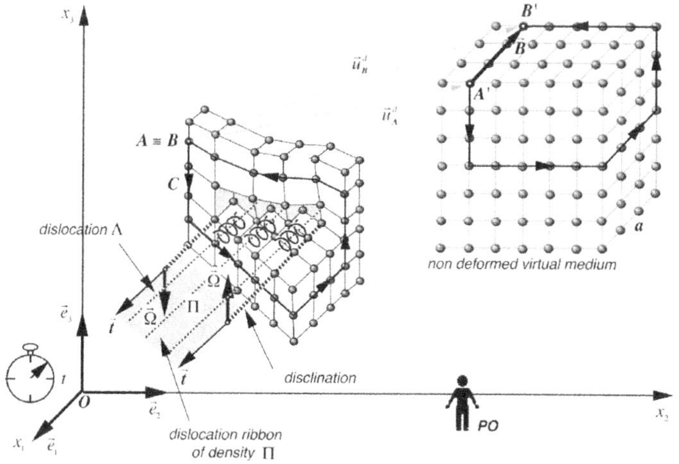

Figure 9.15 - *quantified bi-dimensional dislocation ribbon made up of three dislocations of screw type*

The effects of a volume expansion

In case a medium expands or contracts locally, it is useful to know how the charges of dislocation change. Knowing that the surface charges of flexion $\vec{\Pi}$ and rotation Π of the joint is given by the relationships $\vec{\Lambda}\,/\,d$ and $\Lambda\,/\,d$ by using equations *(9.16)* and the fact that $d = d_0\,\mathrm{e}^{\tau/3}$, it is found that

$$\left|\vec{\Pi}\right|_{\tau\neq 0} = \frac{\left|\vec{\Lambda}\right|_{\tau\neq 0}}{d\big|_{\tau\neq 0}} = \frac{\left|\vec{\Lambda}_0\right|_{\tau=0}\mathrm{e}^{\tau/3}}{d_0\big|_{\tau=0}\mathrm{e}^{\tau/3}} = \left|\vec{\Pi}_0\right|_{\tau=0} \tag{9.29}$$

It follows that the surface densities of dislocation charges are conserved when the medium expands or contracts, which leads to the relations

$$\vec{\Pi}_i\Big|_{\tau\neq 0} = \vec{\Pi}_{0i}\Big|_{\tau=0} \quad ; \quad \vec{\Pi}\Big|_{\tau\neq 0} = \vec{\Pi}_0\Big|_{\tau=0} \quad ; \quad \Pi\Big|_{\tau\neq 0} = \Pi_0\Big|_{\tau=0} \tag{9.30}$$

9.3 – Strings and lines of disclination at the frontiers of membranes of dislocation

While isolated strings of disclinations cannot exist *(19.19)*, it will appear disclinations *in the presence of an extended domain of dislocation charges, such as a dislocation membrane.* One can consider for example the case of a flat membrane of dislocation charges having two regions with different surface densities of charges separated by a boundary (fig. 9.16).

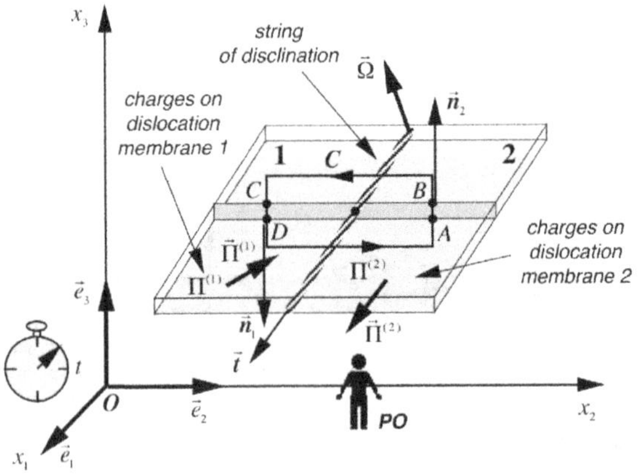

***Figure 9.16** - string of disclination at the frontier of two membranes of dislocations*

Assuming that the surface charges of dislocation $\vec{\Pi}_i^{(1)}$ and $\vec{\Pi}_i^{(2)}$ of the two regions are constant but with different values in each region, it is possible to calculate the Frank vector $\vec{\boldsymbol{\Omega}}$ on a surface S intersecting the membrane surrounding the boundary between the two zones and perpendicular to this border, using the relationship

$$\vec{\Omega} = \sum_i \vec{e}_i \iint_S \vec{\theta}_i \, d\vec{S} = \sum_i \vec{e}_i \iint_S \overrightarrow{\mathrm{rot}}\left[\vec{\lambda}_i - \vec{e}_i \wedge \vec{\lambda} - \vec{e}_i \lambda\right] d\vec{S} = \sum_i \vec{e}_i \oint_C \left(\vec{\lambda}_i - \vec{e}_i \wedge \vec{\lambda} - \vec{e}_i \lambda\right) d\vec{r} \quad (9.31)$$

This relation transforms in the following way

$$\vec{\Omega} = \sum_i \vec{e}_i \int_A^B \left(\underbrace{\vec{\lambda}_i \vec{n}_2}_{=0} - \vec{e}_i \left(\vec{\lambda} \wedge \vec{n}_2\right) - \vec{e}_i \vec{n}_2 \lambda \right) dl + \sum_i \vec{e}_i \int_C^D \left(\underbrace{\vec{\lambda}_i \vec{n}_1}_{=0} - \vec{e}_i \left(\vec{\lambda} \wedge \vec{n}_1\right) - \vec{e}_i \vec{n}_1 \lambda \right) dl \quad (9.32)$$

In this relationship, the terms $\vec{\lambda}_i \vec{n}_1$ and $\vec{\lambda}_i \vec{n}_2$ are zero since the charge densities $\vec{\lambda}_i$ in a membrane can only be vectors parallel to the membrane *(9.25)*. We have for $\vec{\Omega}$

$$\vec{\Omega} = \int_A^B \left(-\left(\vec{\lambda} \wedge \vec{n}_2\right) - \vec{n}_2 \lambda\right) dl + \int_C^D \left(-\left(\vec{\lambda} \wedge \vec{n}_1\right) - \vec{n}_1 \lambda\right) dl = -\vec{\Pi}^{(1)} \wedge \vec{n}_1 - \Pi^{(1)} \vec{n}_1 - \vec{\Pi}^{(2)} \wedge \vec{n}_2 - \Pi^{(2)} \vec{n}_2 \quad (9.33)$$

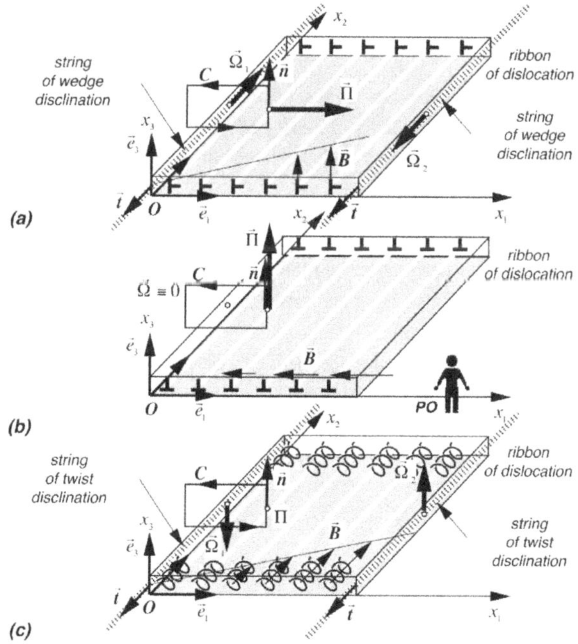

Figure 9.17 *- ribbons of dislocation bordered by disclinations of wedge type* ***(a)*** *and of twist type* ***(c)****. The ribbon of dislocation* ***(b)*** *is not bordered by lines of disclination*

It is found that the value for $\vec{\Omega}$ is independent of the contour chosen as long as this contour surrounds the border between the two regions of the membrane. Thus, for a contour closely surrounding the border, one obtains a line having a constant Frank vector along the line, which can therefore be called *disclination string* or *line of disclination* or simply *disclination*[10]. In figure 9.17, the disclination lines are plotted, and they border the dislocation ribbon similar to those presented in figure 9.14. In this figure, we see the following:

- the ribbon of edge dislocations with a Burgers vector perpendicular to the membrane and increasing along the axis Ox_1 (fig 9.17a) corresponds to a *localized tilt boundary*. It is bordered by two wedge disclinations since $\vec{\Omega}_1 = -\vec{\Omega}_2 = -\vec{\Pi} \wedge \vec{n} = -|\vec{\Pi}|\vec{t}$,

- the ribbon of edge dislocations with Burgers vector parallel to the membrane and increasing along the axis Ox_1 (fig 9.17b) is a localized accommodation boundary. It is not bordered by disclination since $\vec{\Omega}_1 = \vec{\Omega}_2 \equiv 0$,

- the ribbon of screw dislocations with Burgers vector parallel to the membrane and increasing along the axis Ox_1 (fig 9.17c) corresponds to a *localized twist boundary*. It is bordered by *two twist disclinations* since $\vec{\Omega}_1 = -\vec{\Omega}_2 = -\Pi\vec{n}$.

Note that the disclination of figure 9.17a is the macroscopic disclination shown in figure 8.7b, while the disclination of figure 9.17c corresponds to the macroscopic disclination of figure 8.7a. Furthermore, the quantization of a ribbon of dislocations similar to that of figure 9.17c has already been illustrated in figure 9.15, in which are drawn the two disclinations bordering the quantized dislocation ribbon.

Finally, the relation *(9.33)* justifies *a fortiori* that the surface densities are retained in case of expansion or contraction of the volume, as expressed by the relationship *(9.30)*, since the rotation angles $\vec{\Omega}$ must be an invariant under this kind of transformation.

The scalar charge of curvature of a wedge disclination line

In the previous chapter, we have introduced the notion of scalar density of curvature charges defined by the relation $\theta = \operatorname{div}\vec{\lambda}$. Then one can legitimately ask whether the disinclination obtained at the border of a dislocation membrane as seen in figure 9.16, can be represented by a scalar linear charge Θ, in the same way that a screw dislocation can be fully represented by its linear scalar charge Γ . To answer this question, we need to resume with relationship *(9.31)*, and consider only the terms in $\vec{\lambda}$ and λ since the term in $\vec{\lambda}_i$ is not involved in the case of a disclination on the border of a membrane dislocation, as shown in equation *(9.32)*. We then have

$$\begin{aligned}\vec{\Omega} &= -\sum_i \vec{e}_i \iint_S \overrightarrow{\operatorname{rot}}\left[\vec{e}_i \wedge \vec{\lambda} + \vec{e}_i \lambda\right] d\vec{S} \\ &= -\iint_S \operatorname{div}\vec{\lambda}\, d\vec{S} + \sum_k \iint_S \left(\overrightarrow{\operatorname{grad}}\,\lambda_k\right) dS_k + \iint_S \overrightarrow{\operatorname{grad}}\,\lambda \wedge d\vec{S}\end{aligned} \qquad (9.34)$$

In the first term, one can replace $\operatorname{div}\vec{\lambda}$ by θ . The second term is zero by the fact that, in the configuration shown in figure 9.16, there cannot be a gradient of $\vec{\lambda}$ in the direction of the disclination, and the third term can be obtained on the contour surrounding the surface integration, so that

---

[10] *R. DeWit, «Theory of disclinations II, III and IV», J. of the Nat. Bureau of Standards A, vol. 77A, p. 49-100, p. 359-368, and p. 607-658, 1973; E. Kröner, K.-H. Anthony, «Dislocations and Disclinations in Material Structures: The Basic Topological Concepts», Annu. Rev. Mater. Sci., vol. 5, p. 43, 1975*

$$\vec{\Omega} = -\iint_S \theta \, d\vec{S} - \oint_C \lambda d\vec{r} = \vec{\Omega}\Big|_{\| \vec{t}} + \vec{\Omega}\Big|_{\perp \vec{t}} \tag{9.35}$$

The first term corresponds to the case of a wedge disclination (figure 9.17a) and the second term corresponds to the case of a twist disclination (figure 9.17c). By then introducing the linear scalar curvature charge Θ, we can write for a wedge disclination, in the case of figure 9.16

$$\vec{\Omega}\Big|_{\| \vec{t}} = \vec{\Omega}_{wedge} = -\iint_S \theta \, d\vec{S} = -\Theta \vec{t} = -\vec{\Pi}^{(1)} \wedge \vec{n}_1 - \vec{\Pi}^{(2)} \wedge \vec{n}_2 \tag{9.36}$$

from which we deduce that it is possible to calculate the *linear charge of curvature* Θ via the expression

$$\Theta = -\vec{\Omega}_{wedge} \vec{t} = \left(\vec{\Pi}^{(1)} \wedge \vec{n}_1\right)\vec{t} + \left(\vec{\Pi}^{(2)} \wedge \vec{n}_2\right)\vec{t} = \vec{\Pi}^{(1)}\left(\vec{n}_1 \wedge \vec{t}\right) + \vec{\Pi}^{(2)}\left(\vec{n}_2 \wedge \vec{t}\right) \tag{9.37}$$

Introducing here the unit vector $\vec{m}$ perpendicular to the disclination line and radial with respect to its center, the following expression is obtained for Θ

$$\Theta = -\vec{\Omega}_{wedge} \vec{t} = \vec{\Pi}^{(1)} \vec{m} + \vec{\Pi}^{(2)} \vec{m} \tag{9.38}$$

The effects of a variation of the volume expansion

In the case where the medium expands or contracts locally, knowing that the surface densities of dislocation of charges $\vec{\Pi}$ on a membrane does not depend on τ *(9.30)*, and using the relation *(9.38)*, we find

$$\Theta\Big|_{\tau \neq 0} = \Theta_0\Big|_{\tau = 0} \tag{9.39}$$

The quantification of disclinations of wedge type in a lattice

While isolated disclinations cannot exist, it is possible to imagine a structured solid media which contain line disclinations quantified on the lattice in the case of wedge disclinations, as illustrated by figure 9.18. In this figure, two wedge disclinations are shown with $\vec{\Omega} = \mp 90°$ in a simple cubic lattice, and there we show the curvature vector $\vec{\chi}$ due to the charge Θ.

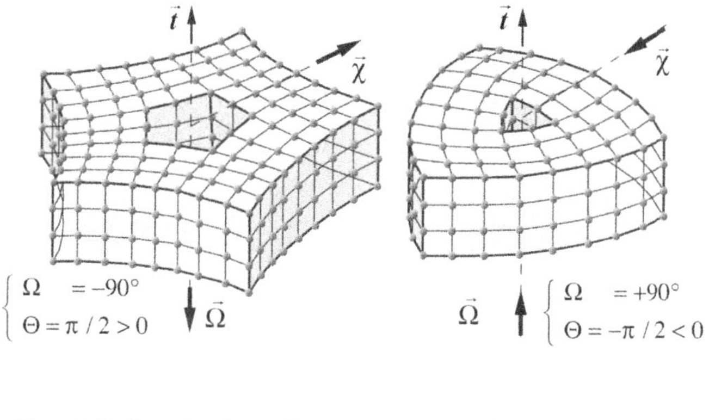

Figure 9.18 - *Examples of quantified wedge disclinations* $\Omega = \mp 90°$ *on a cubic lattice*

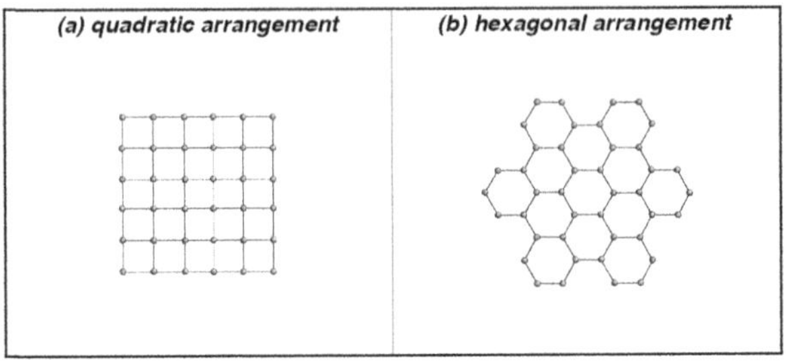

Figure 9.19 - *quadratic or hexagonal arrangements of particles in a plane*

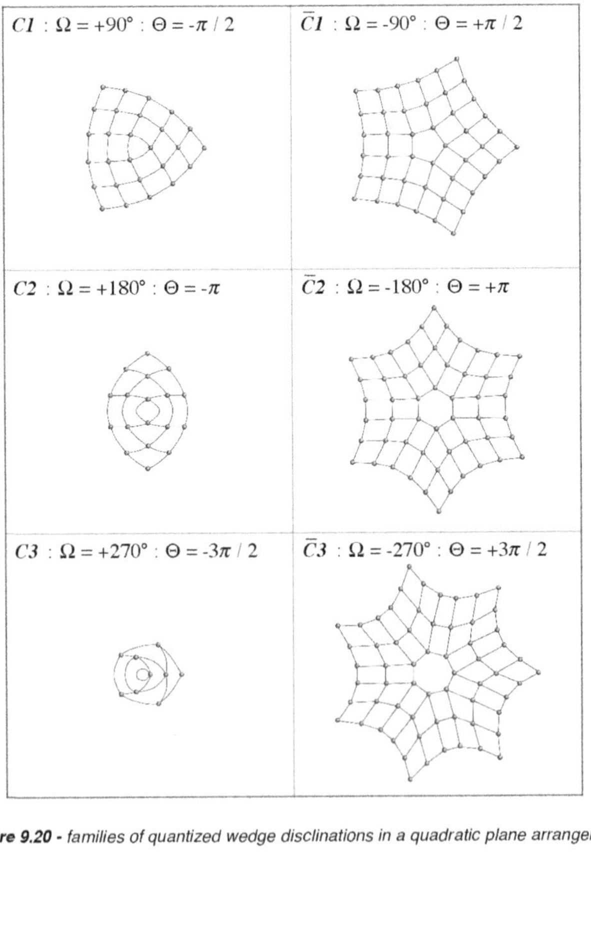

Figure 9.20 - *families of quantized wedge disclinations in a quadratic plane arrangement*

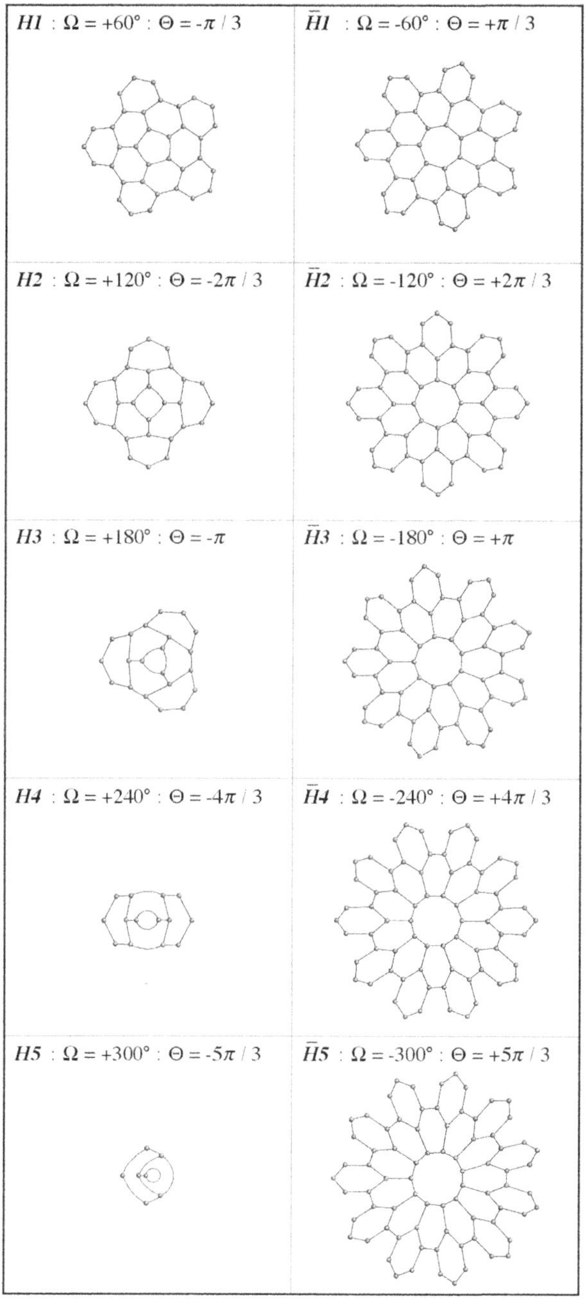

***Figure 9.21** - families of quantized wedge disclinations in a hexagonal plane arrangement*

One can then imagine that there may be various families of quantized wedge disclinations by considering a solid media with different arrangements of the particles in a secant plane to the disclination line[11]. For example, we will consider here simple arrangements such as quadratic or hexagonal arrangement of the particles in a plane (figure 9.19). But one could obviously be considering more complex arrangements such as three-dimensional centered cubic, hexagonal or cubic face centered structures.

In the case of quadratic arrangement, there may be at most 3 different quantized wedge disclinations , which we will call $C1$, $C2$ and $C3$, with rotation angles Ω of + 90 °, + 180 ° and + 270 ° to which correspond three quantized wedge anti-disclinations $\overline{C1}$, $\overline{C2}$ and $\overline{C3}$, with rotation angles Ω of -90 °, -180 ° and -270 ° (fig. 9.20).

In the case of hexagonal arrangement, there are at most 5 quantized wedge disclinations $H1$, $H2$, $H3$, $H4$ and $H5$, with rotation angles Ω of + 60 °, + 120 °, + 180 °, + 240 ° and + 300 °, to which correspond 5 quantified wedge anti-disclinations $\overline{H1}$, $\overline{H2}$, $\overline{H3}$, $\overline{H4}$ and $\overline{H5}$, with rotation angles Ω of -60 °, -120 °, -180 °, -240 ° and -300 ° (fig. 9.21).

In both cases of figures 9.20 and 9.21, we show the disclinations with a size calculated so that the volume expansion τ is the same in all figures. Note also that the disclinations of + 270 ° in the quadratic arrangement and of +300 ° in the hexagonal arrangement may exist or not exist according to the imaginary medium considered, because their existence is linked to the ability to connect between them two bonds of the same "particle" in the solid medium we consider.

The multiplets of quantized wedge disclinations

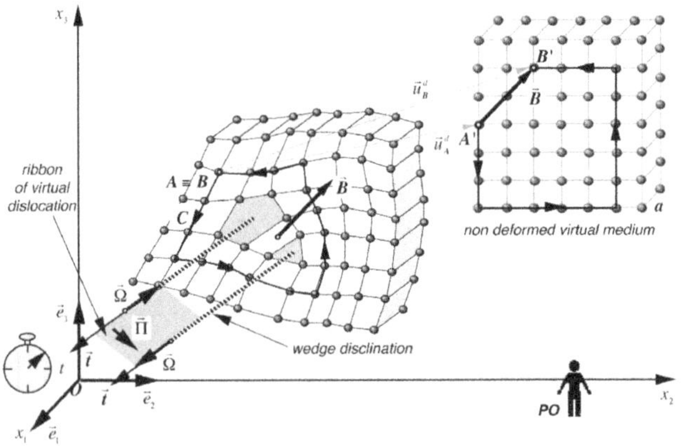

***Figure 9.22** - doublet of quantized wedge disclinations with a ribbon of virtual dislocation.*

It has been shown that there cannot be isolated disclinations with nonzero Frank vector. Therefore, it is necessary to combine several disclinations near each other so that the Frank vector

[11] *The disclinations have been observed essentially in the lattices of flux lines in superconductors of type II, in certain polymers and in the nematic liquid crystals. See: E. Kröner, K.-H. Anthony, Annu. Rev. Mater. Sci., vol. 5, p. 43, 1975; M. Kleman, J. Friedel, Reviews of Modern Physics, vol. 80, p.61-115, 2008*

obtained on a contour surrounding these disclinations is zero.
The example shown in figure 9.22 illustrates this fact: by coupling two quantized wedge disclinations of type $C1$ and $\overline{C1}$ in a simple cubic structure, the Frank vector becomes zero, and there appears a virtual ribbon of edge dislocation between the two disclinations with nonzero overall Burgers vector $\vec{B}$. The ribbon of edge dislocation is similar to the one shown in figure 9.17a and therefore contains a vectorial surface charge of flexion $\vec{\Pi}$. But this flexion charge $\vec{\Pi}$ is not associated with a real quantized dislocation network, but to a virtual ribbon of edge dislocations.
We can find the multiplets of disclinations with global null Frank vector that can be built on the basis of the quantized wedge disclinations that we have described in the case of simple cubic lattice in figure 9.19a. The basic multiplets with zero Frank vector, that is to say, those that cannot be further broken down into two or more multiplets of non null Frank vector, are reported in table 9.1. It is found that, in a simple cubic lattice, there may be three doublets, 4 triplets and 2 quadruplets.

3 doublets
+90° / -90°
+180° / -180°
+270° / -270°

4 triplets	
+90° / +90° / -180°	*-90° / -90° / +180°*
+90° / +180° / -270°	-90° / -180° / +270°

2 quadruplets	
+90° / +90° / +90° / -270°	-90° / -90° / -90° / +270°

***Table 9.1** - the multiplets of quantified wedge disclinations in a simple cubic plane structure*

In the case of the lattice of figure 9.19b, with plans having a hexagonal structure of the particles, there may be 5 doublets, 12 triplets, 8 quadruplets, 4 quintuplets and 2 sextuplets as shown in table 9.2.
In both tables, the multiplets made with disclinations $C3$ (+ 270 °) or $H5$ (+ 300 °) are grayed out because they could not exist, for example in structured media that do not allow a "self connection" of two bonds of a same particle.

5 doublets
+60° / -60°
+120° / -120°
+180° / -180°
+240° / -240°
+300° / -300°

12 triplets	
+60° / +60° / -120°	-60° / -60° / +120°
+60° / +120° / -180°	-60° / -120° / +180°
+60° / +180° / -240°	-60° / -180° / +240°
+120° / +120° / -240°	-120° / -120° / +240°
+60° / +240° / -300°	-60° / -240° / +300°
+120° / +180° / -300°	-120° / -180° / +300°

8 quadruplets	
+60° / +60° / +60° / -180°	-60° / -60° / -60° / +180°
+60° / +60° / +120° / -240°	-60° / -60° / -120° / +240°
+60° / +120° / +120° / -300°	-60° / -120° / -120° / +300°
+60° / +60° / +180° / -300°	-60° / -60° / -180° / +300°

4 quintuplets	
+60°/+60°/+60°/+60°/-240°	-60°/-60°/-60°/-60°/+240°
+60°/+60°/+60°/+120°/-300°	-60°/-60°/-60°/-120°/+300°

2 sextuplets	
60°/+60°/+60°/+60°/+60°/-300°	-60°/-60°/-60°/-60°/-60°/300°

Tableau 9.2 - *the multiplets of quantized wedge disclinations in a hexagonal plane structure.*

In both tables, the multiplets made with disclinations $C3$ (+ 270 °) or $H5$ (+ 300 °) are grayed out because they could not exist, for example in structured media that do not allow a "self connection" of two bonds of a same particle.

Examples of quantized multiplets of disclinations in the cubic structure

In this section, as examples, we study some imaginary cases of doublets and triplets of disclinations in the simple cubic structure (fig. 9.23 to 9.28).

For each case shown, we indicate in an inset the number of edge dislocations on each side of the quadrangle **ABCD**. This number of edge dislocations is composed of a fixed number depending on the number of disclinations in play and their respective arrangement, plus a number of edge dislocations due to the inclusion of additional n_i "particle" planes placed between the disclinations to vary the distance between them (these additional plans are shown in red in the figures).

In the figures 9.23 and 9.24, we show the doublets $C1-\bar{C}1$ and $C2-\bar{C}2$, by varying in each case the orientation of the pair relative to the crystal structure (with an angle of 45 ° and 0 ° respectively with respect to the structure). One immediately notices that in the case of doublets, there is always a fixed number of edge dislocations emerging from the quadrangular figure ***ABCD***, and that the addition of insert planes systematically increases this number.

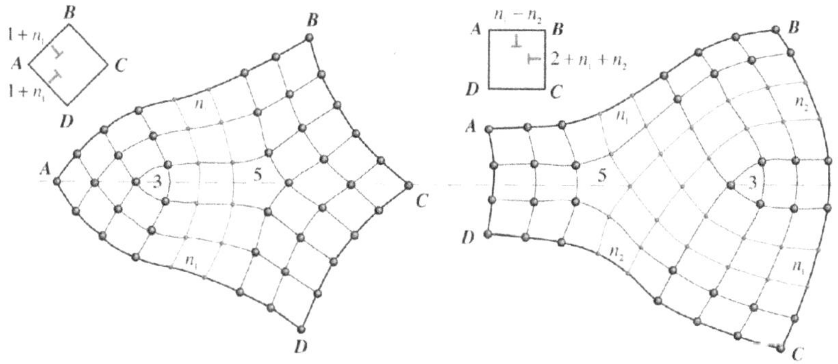

Figure 9.23 *- doublets $C1-\bar{C}1$ of wedge disclinations in a cubic lattice*

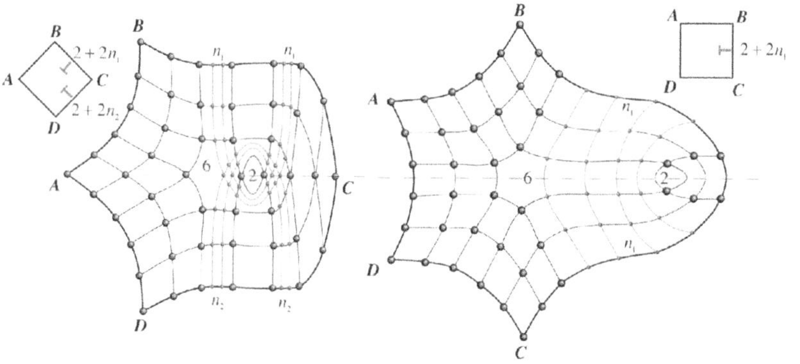

Figure 9.24 *- doublets $C2-\bar{C}2$ of wedge disclinations in a cubic lattice*

In figure 9.25, we plotted triplets $C1-\bar{C}2-C1$, varying the direction of alignment of these triplets with the structure (0 ° and 45 ° respectively), while in figure 9.26, the triplet $C1-\bar{C}2-C1$ and its "antiparticle "$\bar{C}1-C2-\bar{C}1$ are plotted.

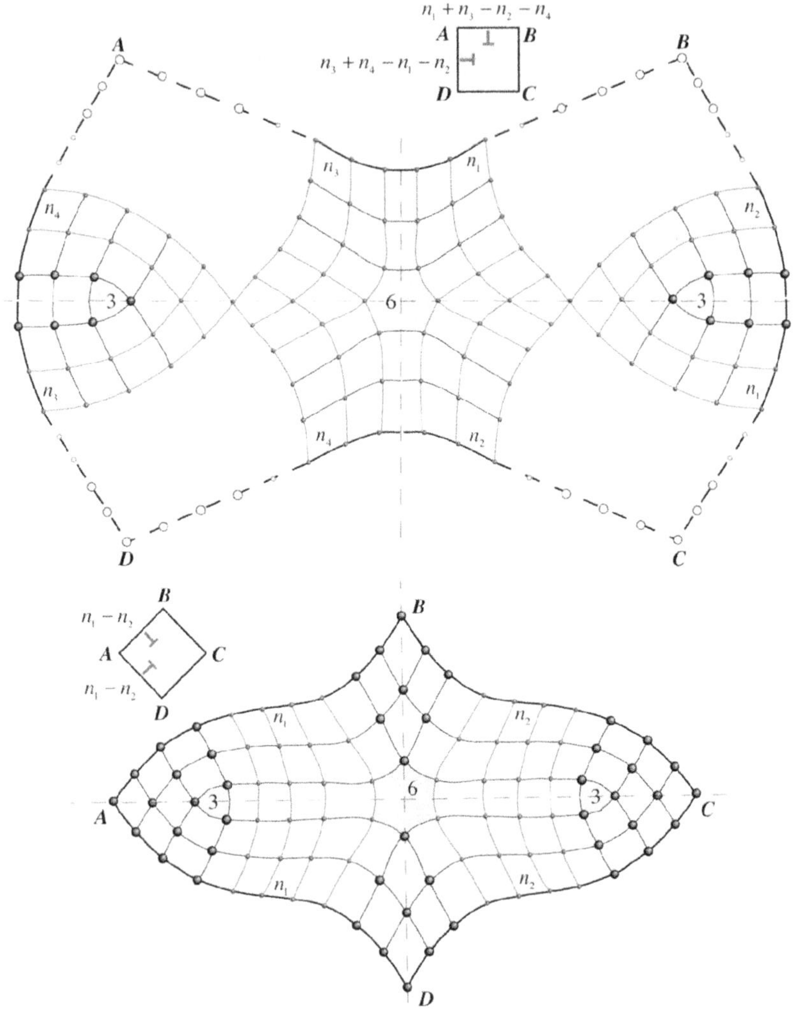

Figure 9.25 - *triplets $C1-\bar{C}2-C1$ of wedge disclinations in a cubic lattice*

In all these cases of triplets, for which disclinations components are aligned on an axis, we see that if the additional insert planes are placed symmetrically, these triplets remain perfectly symmetrical and there is no edge dislocation that emerges in quadrangle ***ABCD***.

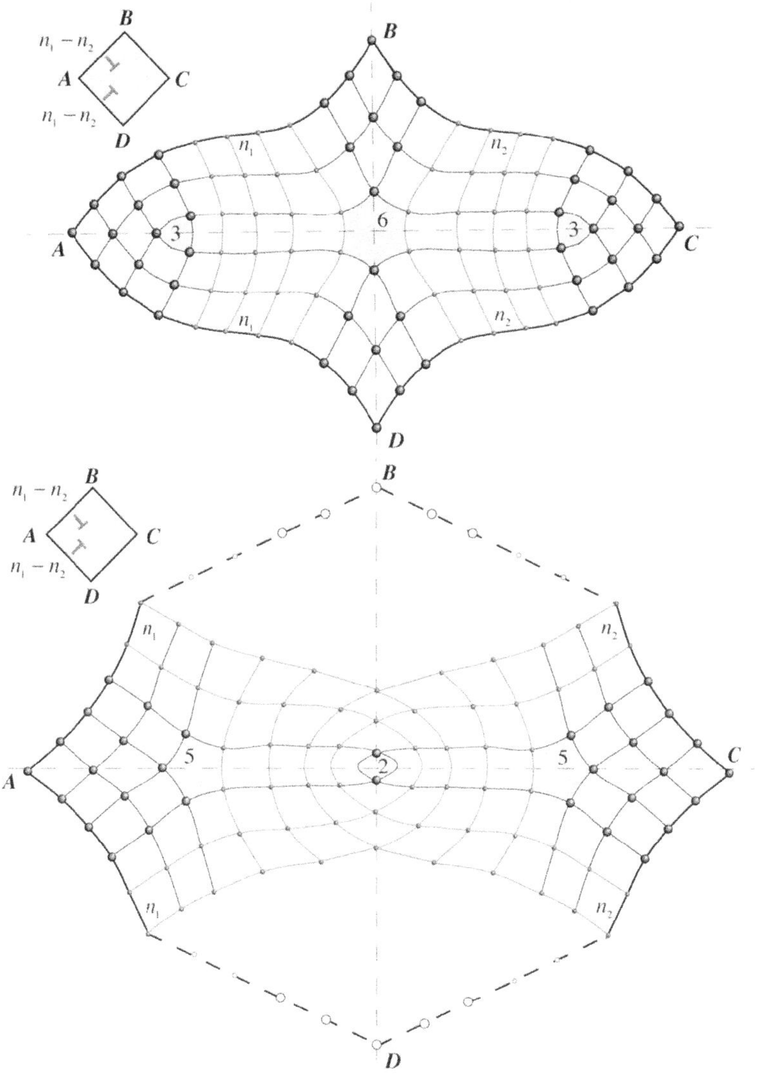

Figure 9.26 *- triplets* $C1 - \bar{C}2 - C1$ *and* $\bar{C}1 - C2 - \bar{C}1$ *of wedge disclinations in a cubic lattice*

However, even if no edge dislocation emerges, it is clear that there is strong local deformation of the lattice, which must correspond to a strain energy which is nothing but the energy of the triplet formation. In addition, the introduction of additional infill symmetrical planes has the effect of increasing the distance between the disclinations triplet, and thus increases the energy of the triplet.

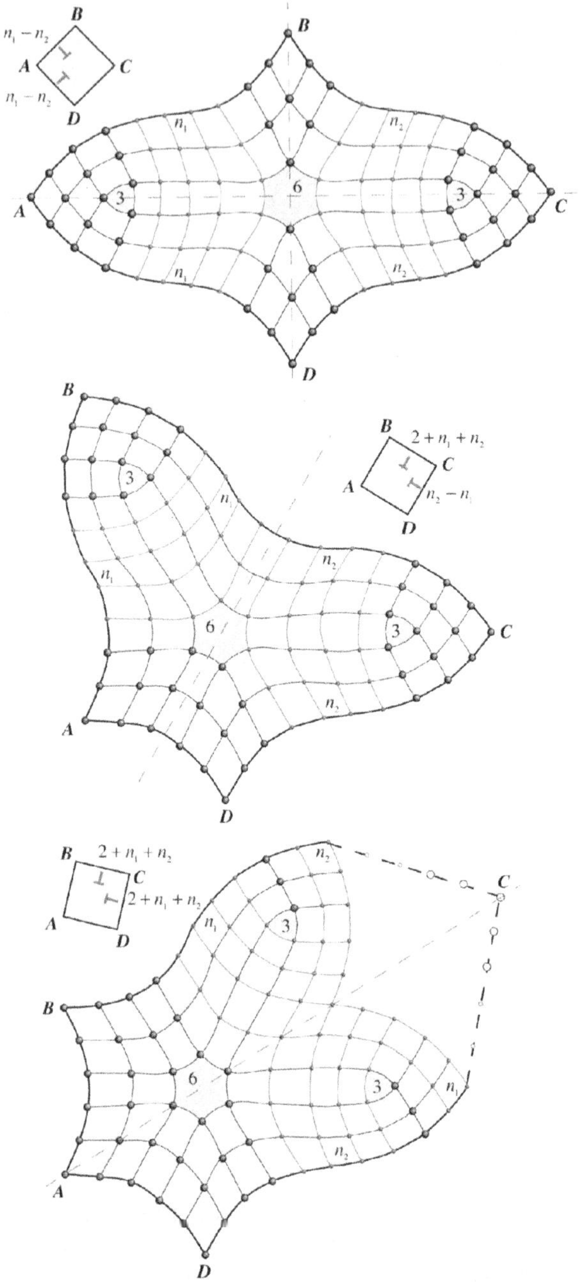

Figure 9.27 *- the different spatial positions of triplets* $C1 - \overline{C}2 - C1$

In figure 9.27, we show triplets $C1-\bar{C}2-C1$, but this time by varying the relative spatial arrangement of disclinations with each other. The angles formed by the straight lines connecting the two disclinations $C1$ and $\bar{C}2$ are 180 °, 120 ° and 60 ° respectively. It is found that the triplet with an asymmetry, in cases 120 ° and 60 °, shows a fixed number of edge dislocations emerging from the quadrangular figure ***ABCD***, and this number increases with the dissymmetry, since it goes from 2 to 4 when the angle goes from 120 ° to 60 °. It is easy to imagine on the basis of these figures that, in all cases, an increase in the distance between the disclinations of the triplet by adding intermediate planes may only increase the energy of the triplet!

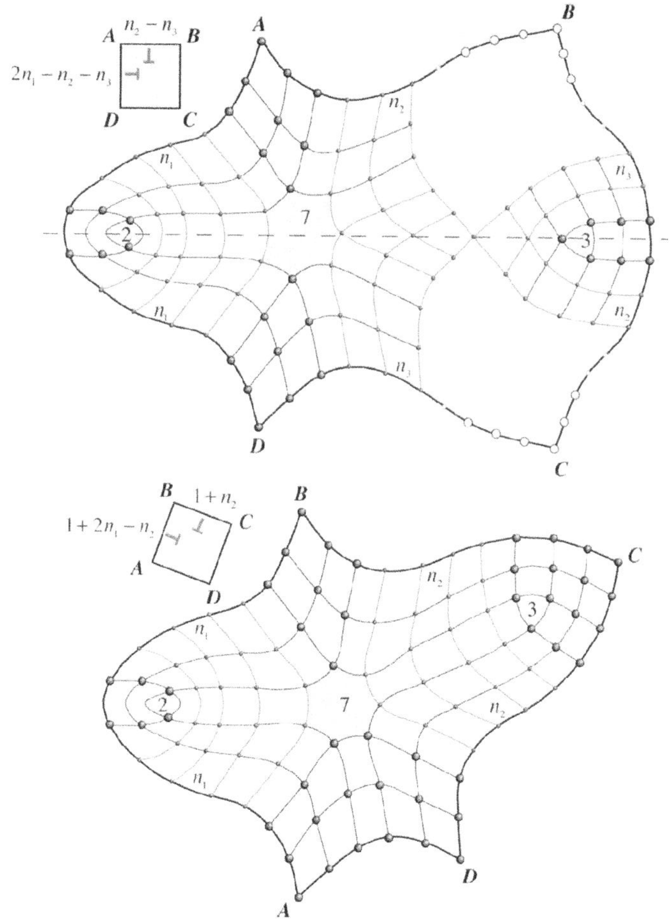

Figure 9.28 *- two spatial configurations of triplet $C2-\bar{C}3-C1$*

Finally, in figure 9.28, we show examples of more "exotic" triplets, by drawing two different spatial configurations of the triplet $C2-\bar{C}3-C1$.

The ribbons of virtual dislocations connecting the disclinations in a multiplet

All the examples we have shown in the previous section show that each multiplet of disclinations can have many different configurations, depending on the number of disclinations comprising the multiplet, the spatial arrangement of disclinations and the number and disposition of infill planes. Moreover, one can well imagine that, for the multiplets obtained in a hexagonal structure, there is a greater multitude of different configurations since the number of singlets, that is to say, quantified disclinations, is greater and the number of possible angles between the arms of the multiplets is also higher. It would therefore be very convenient to find a mathematical way to characterize the various configurations which may occur for a given multiplet in a given structure. We will treat the case of a doublet and a triplet wedge disclinations, the other multiplets can easily be treated similarly.

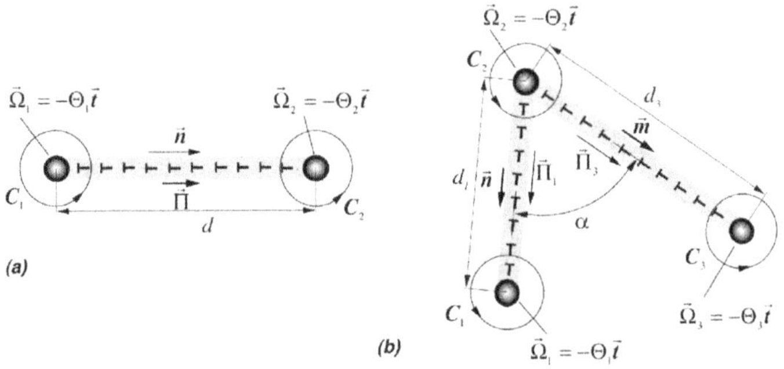

***Figure 9.29** - ribbons of virtual dislocations of a doublet **(a)** and a triplet **(b)** of wedge disclinations*

In the case of the doublet shown in figure 9.29a, both disclinations are separated by a distance d, and the direction of the dipole is the one given by the unit vector $\vec{n}$. The disclinations of the doublet are characterized by their Frank vectors $\vec{\Omega}_1 = -\Theta_1\vec{t}$ and $\vec{\Omega}_2 = -\Theta_2\vec{t}$. As we have seen previously, the doublet of disclinations must be linked by the existence of virtual dislocations, which we will represent by a dislocation membrane of length d connecting the disclinations and containing a vectorial surface density of dislocation charges $\vec{\Pi}$, with $\vec{\Pi} \parallel \vec{n}$. By applying equation *(9.38)* on the edges C_1 and C_2, the following relations are obtained

$$\begin{cases} \Theta_1 = -\vec{\Omega}_1\vec{t} = \vec{\Pi}\vec{n} \\ \Theta_2 = -\vec{\Omega}_2\vec{t} = -\vec{\Pi}\vec{n} \end{cases} \tag{9.40}$$

from which it is deduced that $\Theta_1 = -\Theta_2$.

At large distances, the doublet of disclinations may be considered as an edge dislocation with linear density charge $\vec{\Lambda}$ given by the relation

$$\vec{\Lambda} = d\vec{\Pi} = d\Theta_1\vec{n} \tag{9.41}$$

This shows that the dislocation equivalent to the doublet of disclinations has a Burgers vector $\vec{B}$ proportional to the scalar charge density Θ_1 and to the distance d between the disclinations

$$\vec{B} = \vec{t} \wedge \vec{\Lambda} = d\,\Theta_1 \left(\vec{t} \wedge \vec{n} \right) \tag{9.42}$$

As the Burgers vector $\vec{B}$ associated with the charge $\vec{\Lambda}$ must be quantized in a lattice, the minimum value $d_{\min}$ of the distance d obtained when there is no intermediate planes between the disclinations of the doublet, satisfied

$$d_{\min} = \left| \vec{B} \right|_{\min} / \Theta_1 \tag{9.43}$$

Consider then the case of doublets $C1 - \bar{C}1$, $C2 - \bar{C}2$ and $C3 - \bar{C}3$, in a simple cubic lattice, assuming that there are no intermediate planes like those placed between the disclinations of the doublets in figures 9.23 and 9.24. As $\Theta_1 = i\pi / 2$ in the cubic lattice with $i = 1, 2, 3$ depending on the doublet $Ci - \bar{C}i$, the value $d_{\min}$ satisfies

$$d_{\min} = \left| \vec{B} \right|_{\min} / \Theta_1 = 2 \left| \vec{B} \right|_{\min} / i\pi \tag{9.44}$$

in which $\left| \vec{B} \right|_{\min}$ must be a multiple of the lattice step a, or the diagonal $a\sqrt{2}$ of the base square. Thus we have:

- in the case of figure 19.30a, $\left| \vec{B} \right|_{\min} = a\sqrt{2}$ and $i = 1$, giving $d_{\min} = 2a\sqrt{2} / \pi = 0,9a$,
- in the case of figure 19.30b, $\left| \vec{B} \right|_{\min} = 2a$ and $i = 1$, giving $d_{\min} = 4a / \pi = 1,28a$,
- in the case of figure 19.31a, $\left| \vec{B} \right|_{\min} = 2a\sqrt{2}$ and $i = 2$, giving $d_{\min} = 4a\sqrt{2} / \pi = 1,8a$,
- in the case of figure 19.30a, $\left| \vec{B} \right|_{\min} = 2a$ and $i = 2$, giving $d_{\min} = 2a / \pi = 0,64a$.

In the case of a perfect hexagonal lattice[12], the Burgers vector must be quantified as a multiple of the distance b between two hexagons in the perfect lattice. For the doublet $H1 - \bar{H}1$ shown in figure 9.30, $\Theta_1 = \pi / 3$ and it was found that $\left| \vec{B} \right|_{\min} = b$, so we have the relationship $d_{\min} = \left| \vec{B} \right|_{\min} / \Theta_1 = 3b / \pi \cong 0,955b$.

In the case of the triplet shown in figure 9.29b, the three disclinations are separated by distances d_1 and d_3 respectively, and the directions of the two branches of the triplet are those given by the unit vector $\vec{n}$ and $\vec{m}$, in such a way that the angle α formed by the two arms is given by the relationship $\cos\alpha = \vec{n}\vec{m}$. The disclinations of the triplet are characterized by Frank vectors $\vec{\Omega}_1 = -\Theta_1 \vec{t}$, $\vec{\Omega}_2 = -\Theta_2 \vec{t}$ and $\vec{\Omega}_3 = -\Theta_3 \vec{t}$. They must be linked by the existence of virtual dislocations which are represented by two membranes of dislocations with length d_1 and d_3, linking the disclinations and containing a vector surface density of dislocation charges $\vec{\Pi}_1$ and $\vec{\Pi}_3$, with $\vec{\Pi}_1 \parallel \vec{n}$ and $\vec{\Pi}_3 \parallel \vec{m}$. By applying equation *(9.38)* on the contours C_1, C_2 and C_3, we have relationships

$$\begin{cases} \Theta_1 = -\vec{\Omega}_1 \vec{t} = -\vec{\Pi}_1 \vec{n} \\ \Theta_2 = -\vec{\Omega}_2 \vec{t} = \vec{\Pi}_1 \vec{n} + \vec{\Pi}_3 \vec{m} \\ \Theta_3 = -\vec{\Omega}_3 \vec{t} = -\vec{\Pi}_3 \vec{m} \end{cases} \tag{9.45}$$

from which we can see that $\sum \Theta_i = 0$.

---

[12] *Multiplets of wedge disclinations leading to edge dislocations can be clearly observed in the hexagonal lattice of the graphen. See: J. C. Meyer, C. Kisielowski, R. Erni, M. D. Rossell, M. F. Crommie, A. Zettl, Nano Lett., vol. 8(11), p. 3582, 2008; M. P. Ariza, M. Ortiz, Journal of the Mechanics and Physics of Solids, vol. 58, p. 710, 2010*

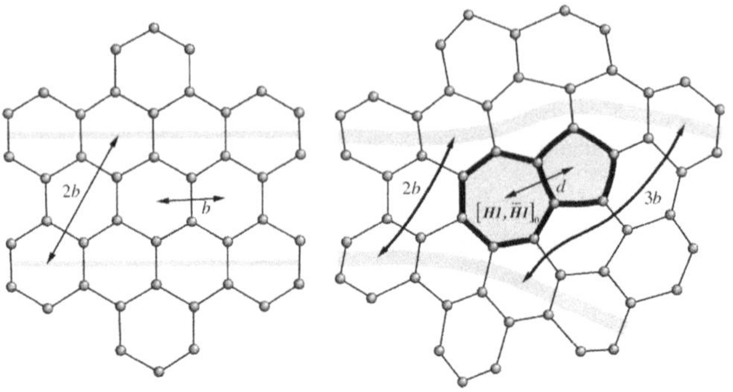

Figure 9.30 - *illustration of doublet* $H1 - \overline{H}1$ *and distances* b *and* d

At large distances, the triplet of disclinations may be considered as two parallel strings of edge dislocations with linear charge densities $\vec{\Lambda}_1$ and $\vec{\Lambda}_3$, given by the relations

$$\begin{cases} \vec{\Lambda}_1 = d_1 \vec{\Pi}_{(1)} = -d_1 \Theta_1 \vec{n} \\ \vec{\Lambda}_3 = d_3 \vec{\Pi}_{(3)} = -d_3 \Theta_3 \vec{m} \end{cases} \tag{9.46}$$

These two parallel dislocations can be replaced by a single dislocation, equivalent to the disclination triplet, which has a Burgers vector $\vec{B}$ given by the relation

$$\vec{B} = \vec{t} \wedge \vec{\Lambda} = -d_1 \Theta_1 (\vec{t} \wedge \vec{n}) - d_3 \Theta_3 (\vec{t} \wedge \vec{m}) \tag{9.47}$$

In the case of the symmetric triplet with alignement of three disclinations on a single line (figures 9.29 and 9.30), we have $\vec{m} = -\vec{n}$, and the total Burgers vector is written

$$\vec{B} = \vec{t} \wedge \vec{\Lambda} = (d_3 \Theta_3 - d_1 \Theta_1)(\vec{t} \wedge \vec{n}) \tag{9.48}$$

In this case, if $d_3 \Theta_3 = d_1 \Theta_1$, the triplet has no overall Burgers vector at a distance, but only two local Burgers vector which cancel each other.

According to equations *(9.41)* and *(9.46)*, the energy of a multiplet, associated with the local deformations induced in the lattice, may only increase if the distances d_i separating the disclinations of the multiplet increase. This implies that there is a force of attraction between the disclinations which increases the more one separates these disclinations. By analogy with the naming in the Standard Model of elementary particles, one could speak here of a force of strong nature which links the disclinations of the multiplet, and one could assimilate the virtual dislocation ribbons $\vec{\Pi}_i$ *to kinds of gluons,* since it is these virtual charges that bind together the disclinations, and prohibit the isolation of one of them.

9.4 – Strings of dispiration and solid lattices with axial symmetries

We will call dispiration strings, or lines of dispiration or simply dispirations[13], mixed strings resul-

[13] *W. F. Harris, Philos. Mag. 22, p. 949, 1970; W. F. Harris, Sci. Am. 237(12), p. 130, 1977*

ting from the combination of strings or lines of disclination with strings or lines of dislocation.

On the combination of wedge disclinations and edge dislocations

Figure 9.31 shows two examples of dispirations composed of a wedge disclination and an edge dislocation, represented in the dense plane of a face-centered cubic structure or a hexagonal structure. From the dense plane of a perfect crystal (a), we construct a wedge disclination with $\Omega = +60°$ ***(b)***. By then cutting the crystal along the line ***A-B***, and by adding a line ***A'-B'*** of particles, we introduce a linear density $\vec{\Lambda}$ of edge dislocation in the center of the disclination ***(c)***. Similarly, by removing the particles along the line ***C-D*** in figure 9.31 ***(b)***, and by gluing the jaw with the particles of the line ***A-B***, we introduce again a linear density $\vec{\Lambda}$ of edge dislocation in the center of the disclination, as shown in ***(d)*** of figure 9.31.
The absorption and emission of edge dislocations with a quantified wedge disclination is the basic process that explains the movement of a disclination in a lattice[14]. Here we leave the readers to try to illustrate this type of process!

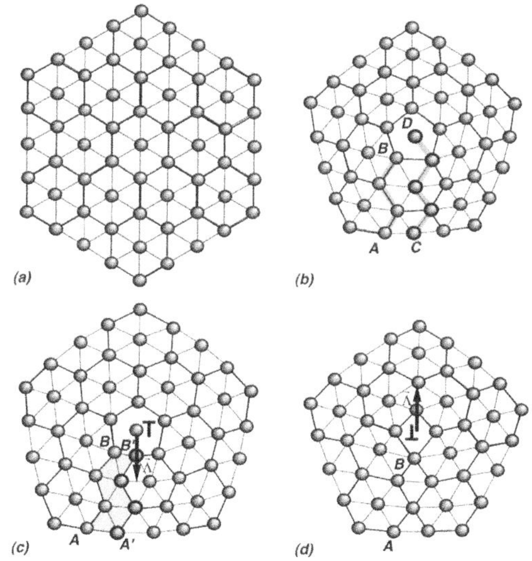

Figure 9.31 - *dispirations **(c)** and **(d)** by combination of a wedge disclination **(b)** with an edge dislocation in the dense plane of a FCC or a hexagonal lattice*

Lattices with axial symmetries

One can imagine lattices which have an axial symmetry in the particles, such as cubic lattices ***(a)*** and ***(b)*** and hexagonal lattices ***(c)*** and ***(d)*** shown in figure 9.32. This axial symmetry of the particles can simply present a preferred orientation of the particles in the plans of the structure,

[14] *W. F. Harris, L. E. Scriven, J. Appl. Phys. 42 , p 3309, 1971; R. deWit, J. Appl. Phys. 42, p. 3304, 1971*

as in the cases ***(a)*** and ***(c)***, or may have an orientation and a preferential direction as in cases ***(b)*** and ***(d)***. This type of lattice thus has an alternating structure of layers ***a, b, c, d,*** ...
In addition, the rotation axes of the particles along the vertical axis produces a directed medium, that is referred to as *dextrorotatory* (right-handed) in the case ***(b)*** and *levorotatory* (left-handed) in the cases ***(c)*** and ***(d)***.

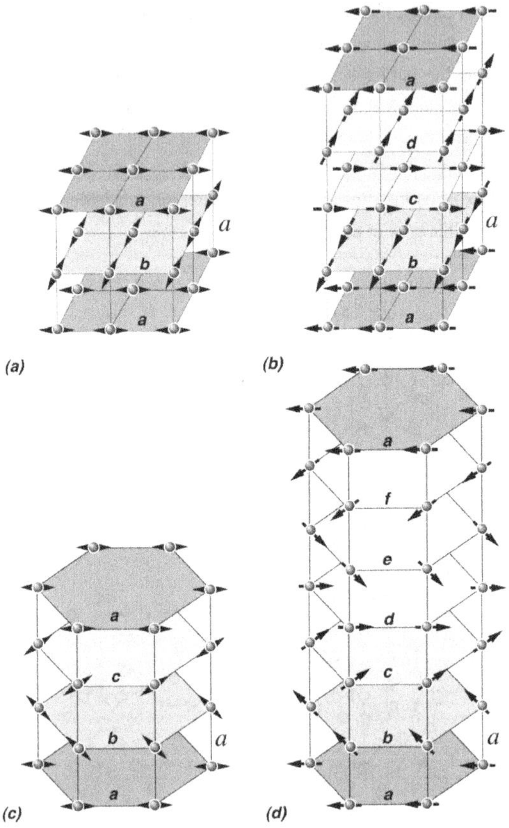

Figure 9.32 *- cubic and hexagonal lattices presenting an axial symmetry of the particles of dextrorotatory type (right-handed) in* ***(b)*** *and of levorotatory type (left-handed) in* ***(c)*** *and* ***(d)***

If it is forbidden to break the axial orientation of the particles in a plane, it is not possible to introduce a vertically oriented screw dislocation with any Burgers vector. Indeed, if the distance between the horizontal planes is a , in order to ensure the continuity of the particle orientation, and also their direction in cases ***(b)*** and ***(d)***, the length of the Burgers vector of the screw dislocation $\vec{B}_{vis}$ has to be equal to $\pm 2a$ in case ***(a)***, $\pm 4a$ in case ***(b)***, $\pm 3a$ in case ***(c)*** and $\pm 6a$ in case ***(d)***.

The dissociation of screw dislocations and the "strong force" linking the partials.

In the axially symmetric environments such as those reported in figure 9.32, screw dislocations must have Burgers vectors $\vec{B}_{vis}$ whose lengths are multiples of the unit length of the lattice a. In this case, the screw dislocations have an incentive to split into partial with Burgers vectors of length $\vec{B}_1$, $\vec{B}_2$ and $\vec{B}_3$, forming respectively 2, 4, 3 or 6 partial in cases ***(a)*** to ***(d)*** respectively. Between the partial dislocations form *ribbons of connection faults* between axial planes **ab, bc, cd,** etc. The separation distance between partials then depends on the energy γ per unit area of the connection fault.

For example, figure 9.33 illustrates the dissociation in the case of axial lattice ***(c)***. In this case, the screw dislocations have an incentive to dissociate into three partial Burgers vectors, $\vec{B}_1$, $\vec{B}_2$ and $\vec{B}_3$, all of length a, so that $\vec{B}_{vis} = \vec{B}_1 + \vec{B}_2 + \vec{B}_3$. In the case of this separation, since the length $\vec{B}_{vis}$ is $3a$, we have $\vec{B}_{vis}^2 = 3\left(\vec{B}_1^2 + \vec{B}_2^2 + \vec{B}_2^2\right)$ and since the energy due to the dislocation is proportional to the square of the Burgers vector, the energy of the three isolated partial is 3 times lower than the energy of the original screw dislocation. If the connection fault ribbon has an energy γ per unit area, the total energy $E_t(d)$ per unit length of dislocation for a screw dislocation dissociated over distances d is written $E_t(d) = 2\gamma d + E_d(d)$ where $E_d(d)$ is the energy of the three partials depending on the distance d separating them. This function is monotonically decreasing from E_0 for $d = 0$ to $E_0 / 3$ for $d \to \infty$ in the case of the partials shown in figure 9.33.

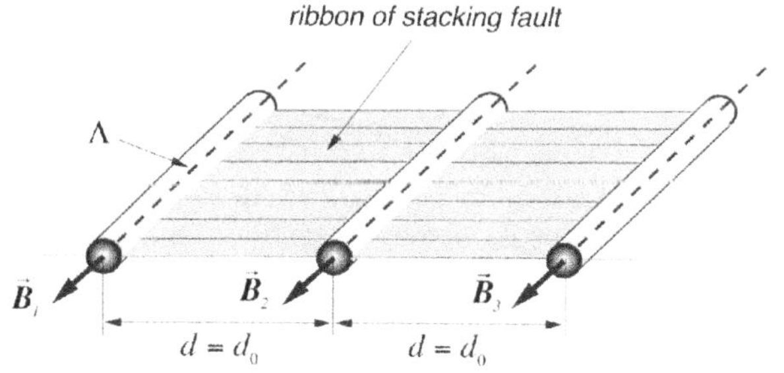

Figure 9.33 *- dissociation of a screw dislocation in three partials in the case of an axial lattice* ***(c)***

The energy $E_t(d)$ thus has a minimum similar to that reported in figure 9.10 for the distance $d = d_0$, which is the equilibrium distance between the three partial, controlled by the competition between energy decrease associated with the increased distance between the partials and increased energy due to the formation of an energy ribbon of connection faults between the three partials.

This behavior of the energy $E_t(d)$ induces and interaction strength between the partials that we qualify as *strong force* in the sense that the energy of the triplet partials will increase if we try to increase the distance separation beyond d_0. This therefore presents a kind of strong force in its behavior and an interesting analogy with the strong force acting between quarks in the Standard Model of elementary particles.

Structural dispirations in a lattice with axial symmetry.

Then we immediately imagine that there must also be connection conditions ensuring the continuity of the axial symmetry if we want to introduce a disclination in such a lattice. In fact, to ensure this continuity, it will be necessary to associate a screw dislocation with the Burgers vector $\vec{B}$ to the disclination. There then appears here a structural need to introduce dispirations in such a media[15].

Figure 9.34 illustrates this perfectly. In order to introduce a disclination $\vec{\Omega} = +90°$ in the medium shown in figure 9.32 **(b)**, it is necessary to add a screw dislocation correctly oriented, with Burgers vector $\vec{B}$ and length a, which ensures the continuity of the axial orientation of particles on the middle of the planes. Note that a screw dislocation with an inverted Burgers vector direction and length $3a$ could also have ensured the continuity of the axial orientation of the particles, so there are two different dispirations with rotation $\vec{\Omega}$ =+90°, both with a linear charge of curvature $\Theta = -\pi / 2$, but differing in their Burgers vector $\vec{B}$ associated with a rotation charge per unit length Λ equal to $\Lambda = a$ or $\Lambda = -3a$.

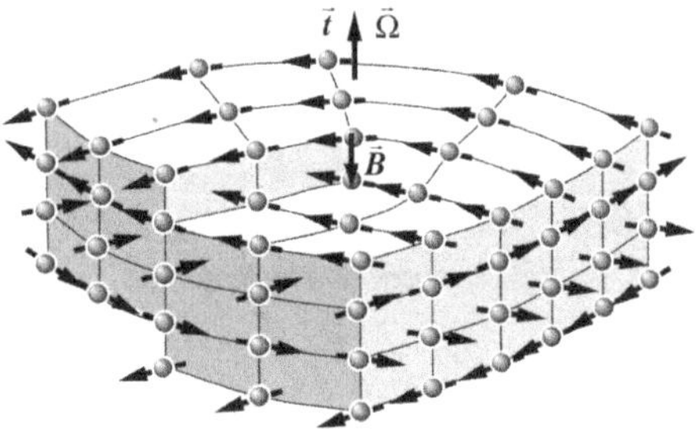

***Figure 9.34** - dispiration of +90° introduced in the axial medium of figure 9.32 **(b)** showing the need to add a translation vector $\vec{B}$ to insure axial continuity*

It is not too difficult to find what linear charges of torsion Λ must be associated with the different wedge dispirations with charge of curvature Θ that can be introduced in the cubic and hexagonal structures shown in figure 9.32.

In table 9.3, we plotted these charges for the cubic structures of figures 9.32 ***(a)*** and ***(b)***. In case ***(a)***, the structure shows no difference between dextrorotatory and levorotatory direction orientation of the lattice particles. On the other hand, there is a difference between these two orientations in case ***(b)***, which implies a change of the sign of the charge Λ between dextrorotatory and levorotatory medias.

In table 9.4, we show the charges Θ and Λ for the hexagonal structures of figures 9.32 ***(c)*** and ***(d)***. Since there is in both cases a difference between the dextrorotatory and levorotatory orientations of the medium, it appears in both cases a sign change of the charge Λ between these two axial directions.

[15] *W. F. Harris, «Geometry of Desinclinations in Crystals», in Surface and Defect Properties of Solids, volume 3, ed. by M. W. Roberts, J. M. Thomas, The Chemical Society, London, 1974*

	Θ	$\Lambda_{(a)}$	$\Lambda_{(b)}^{(dextrorotatory)}$	$\Lambda_{(b)}^{(levorotatory)}$
vis	0	$\pm 2a$	$\pm 4a$	$\pm 4a$
$C1$	$-\pi / 2$	$\pm a$	$+a\ ,\ -3a$	$-a\ ,\ +3a$
$\bar{C}1$	$+\pi / 2$	$\pm a$	$-a\ ,\ +3a$	$+a\ ,\ -3a$
$C2$	$-\pi$	$0\ ,\ \pm 2a$	$+2a\ ,\ -2a$	$+2a\ ,\ -2a$
$\bar{C}2$	$+\pi$	$0\ ,\ \pm 2a$	$+2a\ ,\ -2a$	$+2a\ ,\ -2a$
$C3$	$-3\pi / 2$	$\pm a$	$-a\ ,\ +3a$	$+a\ ,\ -3a$
$\bar{C}3$	$+3\pi / 2$	$\pm a$	$+a\ ,\ -3a$	$-a\ ,\ +3a$

Table 9.3 *- linear charges of curvature Θ and of rotation Λ of the dispirations in the cubic structures of figures 9.39* ***(a)*** *and* ***(b)***

	Θ	$\Lambda_{(c)}^{(dextrorotatory)}$	$\Lambda_{(c)}^{(levorotatory)}$	$\Lambda_{(d)}^{(dextrorotatory)}$	$\Lambda_{(d)}^{(levorotatory)}$
vis	0	$\pm 3a$	$\pm 3a$	$\pm 6a$	$\pm 6a$
$H1$	$-\pi / 3$	$+a\ ,\ -2a$	$-a\ ,\ +2a$	$+a\ ,\ -5a$	$-a\ ,\ +5a$
$\bar{H}1$	$+\pi / 3$	$-a\ ,\ +2a$	$+a\ ,\ -2a$	$-a\ ,\ +5a$	$+a\ ,\ -5a$
$H2$	$-2\pi / 3$	$-a\ ,\ +2a$	$+a\ ,\ -2a$	$+2a\ ,\ -4a$	$-2a\ ,\ +4a$
$\bar{H}2$	$+2\pi / 3$	$+a\ ,\ -2a$	$-a\ ,\ +2a$	$-2a\ ,\ +4a$	$+2a\ ,\ -4a$
$H3$	$-\pi$	$0\ ,\ \pm 3a$	$0\ ,\ \pm 3a$	$\pm 3a$	$\pm 3a$
$\bar{H}3$	$+\pi$	$0\ ,\ \pm 3a$	$0\ ,\ \pm 3a$	$\pm 3a$	$\pm 3a$
$H4$	$-4\pi / 3$	$+a\ ,\ -2a$	$-a\ ,\ +2a$	$-2a\ ,\ +4a$	$+2a\ ,\ -4a$
$\bar{H}4$	$+4\pi / 3$	$-a\ ,\ +2a$	$+a\ ,\ -2a$	$+2a\ ,\ -4a$	$-2a\ ,\ +4a$
$H5$	$-5\pi / 3$	$-a\ ,\ +2a$	$+a\ ,\ -2a$	$-a\ ,\ +5a$	$+a\ ,\ -5a$
$\bar{H}5$	$+5\pi / 3$	$+a\ ,\ -2a$	$-a\ ,\ +2a$	$+a\ ,\ -5a$	$-a\ ,\ +5a$

Tableau 9.4 *- linear charges of curvature Θ and rotation Λ of the dispirations in the hexagonal structures of figures 9.35* ***(c)*** *and* ***(d)***

With the dispirations of tables 9.3 and 9.4, it would be possible to build, in each of these four medias, doublets and triplets of dispiration strings having zero overall line charge Θ and some overall line charge Λ which may be zero or not, on the basis of the models shown in figure 9.29.

9.5 – Loops of dislocation and disclination

To satisfy the conservation equation $\operatorname{div}\vec{\lambda}_i = 0$, a line of dislocation or disclination cannot suddenly stop in the middle. However a line closing on itself to form a localized loop still meets the conservation equation. In this section, we will therefore present such loops and their properties in a solid lattice. To simplify the mathematical treatment of loops, it is wise to start by developing the appropriate mathematical tools.

On the vectorial geometry of loops of charged strings

To develop a mathematical tool to simply describe the vectorial geometry of straight or curved strings, we consider a radius of circular string in three-dimensional space (fig. 9.35).
This string is "charged" with vectorial volume densities $\vec{a}$ and $\vec{b}$ and with a scalar density c. We want to write the vector operators acting on the charge densities contained in the string, ie, $c = \operatorname{div}\vec{b}$ et $b = \overrightarrow{\operatorname{rot}}\,\vec{a}$ as operators acting on the linear charges $\vec{A}$, $\vec{B}$ and C, and defined on the central fiber of the string

$$\vec{A} = \iint_{S_1} \vec{a}\, d\vec{S} \quad ; \quad \vec{B} = \iint_{S_1} \vec{b}\, d\vec{S} \quad ; \quad C = \iint_{S_1} c\, dS \tag{9.49}$$

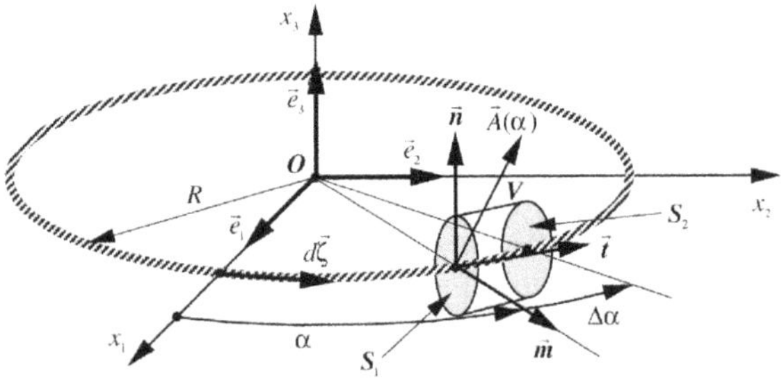

Figure 9.35 *- vectorial geometry of a loop of string*

It is useful to define in advance the linear charges $\vec{A}$, $\vec{B}$ and C, as functions of the angle α along the loop, and to express them in the natural local frame tangent to the string, defined by the unit vectors $\vec{m}$, $\vec{t}$ and $\vec{n}$, by writing

$$\begin{cases} \vec{A} = A_m(\alpha)\vec{m} + A_t(\alpha)\vec{t} + A_n(\alpha)\vec{n} \\ \vec{B} = B_m(\alpha)\vec{m} + B_t(\alpha)\vec{t} + B_n(\alpha)\vec{n} \\ C = C(\alpha) \end{cases} \tag{9.50}$$

In the case of the divergence operator, it is possible to calculate the integral of $c = \operatorname{div}\vec{b}$ on a volume element V contained in the angular region $\left[\alpha, \alpha + \Delta\alpha\right]$ (fig. 9.35). With the divergence theorem we have

$$\iiint_V c\, dV = \iiint_V \operatorname{div}\vec{b}\, dV = \oiint_S \vec{b}\, d\vec{S} = \vec{t}\,\vec{B}\Big|_{S_2} - \vec{t}\,\vec{B}\Big|_{S_1} \tag{9.51}$$

which we can write by expressing the integral on the volume as

$$R\Delta\alpha \iint_{S_1} c\, dS = R\Delta\alpha C = \left[\vec{B}(\alpha + \Delta\alpha) - \vec{B}(\alpha)\right]\vec{t} \tag{9.52}$$

so that C is given by the relationship

$$C = \lim_{\Delta\alpha \to 0} \frac{1}{R\Delta\alpha}\left[\vec{B}(\alpha + \Delta\alpha) - \vec{B}(\alpha)\right]\vec{t} \tag{9.53}$$

Taking the limit $\Delta\alpha \to 0$, we have the expression for the divergence operator applied to the linear density $\vec{B}$ which represents the central fibre in the local frame of the string

$$c = \operatorname{div}\vec{b} \quad \Rightarrow \quad C = \frac{1}{R}\frac{\partial \vec{B}}{\partial \alpha}\vec{t} \tag{9.54}$$

In the case of the operator $b = \overrightarrow{\operatorname{rot}}\,\vec{a}$, a similar procedure gives us

$$R\Delta\alpha \iint_{S_1} \vec{b}\, dS = \iiint_V \overrightarrow{\operatorname{rot}}\,\vec{a}\, dV = \oiint_S d\vec{S} \wedge \vec{a} = \vec{t} \wedge \vec{A}\Big|_{S_2} - \vec{t} \wedge \vec{A}\Big|_{S_1} \tag{9.55}$$

which, when we use the fact that $\vec{t} \wedge \vec{A} = (\vec{t} \wedge \vec{m})A_m + (\vec{t} \wedge \vec{n})A_n = A_n\vec{m} - A_m\vec{n}$, can be written

$$\vec{B} = \lim_{\Delta\alpha \to 0} \frac{1}{R\Delta\alpha}\left\{\left[A_n(\alpha + \Delta\alpha) - A_n(\alpha)\right]\vec{m} - \left[A_m(\alpha + \Delta\alpha) - A_m(\alpha)\right]\vec{n}\right\} \tag{9.56}$$

Taking the limit $\Delta\alpha \to 0$, we have the rotational operator expression applied to the linear density $\vec{A}$ of the line representing the central fibre in the local frame of the string

$$b = \overrightarrow{\operatorname{rot}}\,\vec{a} \quad \Rightarrow \quad \vec{B} = \frac{1}{R}\left(\frac{\partial A_n}{\partial \alpha}\vec{m} - \frac{\partial A_m}{\partial \alpha}\vec{n}\right) \tag{9.57}$$

Operators acting on the linear charges of a loop *(9.54)* and *(9.57)* still deserve a few comments:

- choosing the orientation of the loop as shown in figure 9.42, i.e. $\vec{n}$ parallel to $\vec{e}_3$, the natural unit vectors of the natural frame tangent to the string can be expressed in the local frame of the observer ***PO***, and vice versa

$$\begin{cases} \vec{t} = \cos\alpha\,\vec{e}_2 - \sin\alpha\,\vec{e}_1 \\ \vec{m} = \cos\alpha\,\vec{e}_1 + \sin\alpha\,\vec{e}_2 \\ \vec{n} = \vec{e}_3 \end{cases} \quad \text{and} \quad \begin{cases} \vec{e}_1 = -\vec{t}\sin\alpha + \vec{m}\cos\alpha \\ \vec{e}_2 = \vec{t}\cos\alpha + \vec{m}\sin\alpha \\ \vec{e}_3 = \vec{n} \end{cases} \tag{9.58}$$

- the derivatives of the unit vectors $\vec{m}$, $\vec{t}$ and $\vec{n}$ with respect to the angle α can be written

$$\frac{\partial \vec{t}}{\partial \alpha} = -\vec{m} \quad ; \quad \frac{\partial \vec{m}}{\partial \alpha} = \vec{t} \quad ; \quad \frac{\partial \vec{n}}{\partial \alpha} = 0 \tag{9.59}$$

- in the case of a rectilinear string, the operators *(9.54)* and *(9.57)* can be expressed by replacing the derivative in α by a derivative in ζ, *the curvilinear index along the central fibre of the string*, by using the fact that

$$\lim_{R\to 0} \frac{1}{R}\frac{\partial}{\partial \alpha} = \frac{\partial}{\partial \zeta} \tag{9.60}$$

Glide loops and prismatic loops of dislocations

Now if we want to study the effect of the curvature of a dislocation in the local framework, the easiest method is to consider a circular dislocation loop, and describe it using mathematical tools developed in the previous section. For a circular dislocation loop of radius R, the tensorial linear charge $\vec{\Lambda}_i$ can be connected to the Burgers vector using equation *(9.7)*

$$\vec{\Lambda}_i = -\vec{t}\, \boldsymbol{B}_i \tag{9.61}$$

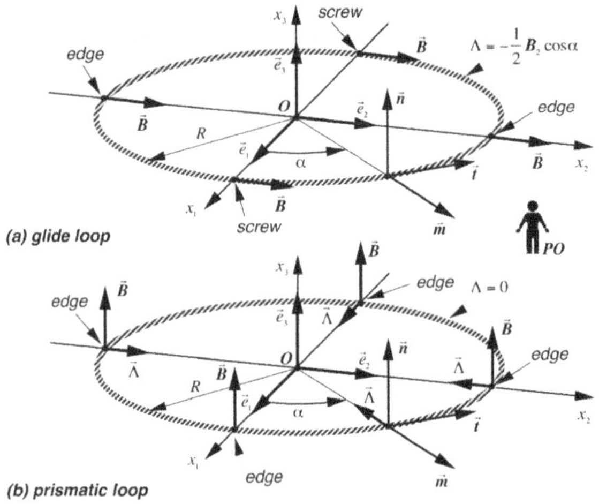

Figure 9.36 *- glide loops and prismatic loops of dislocation*

It appears then three types of dislocation loops according to the orientation of the Burgers vector with respect to the normal $\vec{n}$ to the surface of the loop, as shown clearly in figure 9.36:

- the *glide loops* when $\vec{B} \perp \vec{n}$, which have edge portions (where $\vec{B} \parallel \vec{m}$)), screw portions (where $\vec{B} \parallel \vec{t}$) and mixed portions, which have a kind of "vector" nature as their Burgers vector can take any orientation in the plane perpendicular to $\vec{n}$,
- the *prismatic loops* when $\vec{B} \parallel \vec{n}$, which have a kind of "scalar" nature, since their Burgers vector have a given direction,
- the *mixed loops* when $\vec{B}$ has a component in the direction $\vec{n}$ and a component in the plane of the loop.

We can verify that these dislocation loops satisfy the conservation equation $\operatorname{div} \vec{\lambda}_i = 0$ for the charges of distortion. Indeed using relation *(9.54)*, we have

$$\operatorname{div} \vec{\lambda}_i = 0 \;\Rightarrow\; \frac{1}{R}\frac{\partial \vec{\Lambda}_i}{\partial \alpha}\vec{t} = -\frac{1}{R}\frac{\partial\left(\boldsymbol{B}_i \vec{t}\right)}{\partial \alpha}\vec{t} = -\frac{1}{R}\boldsymbol{B}_i\frac{\partial \vec{t}}{\partial \alpha}\vec{t} - \frac{1}{R}\frac{\partial \boldsymbol{B}_i}{\partial \alpha} = -\frac{1}{R}\frac{\partial \boldsymbol{B}_i}{\partial \alpha} = 0 \tag{9.62}$$

This relationship implies that $\partial \boldsymbol{B}_i / \partial \alpha \equiv 0$, and therefore that $\boldsymbol{B}_i$ does not depend on the angle α, which corresponds to the expression of the invariant component of the Burgers vector of a dislocation line. We can deduce the linear scalar charge Λ of rotation of the dislocation loops by using relation *(9.8)*

$$\Lambda = -\frac{1}{2}\vec{\boldsymbol{B}}\vec{\boldsymbol{t}} = -\frac{1}{2}\vec{\boldsymbol{B}}\left(\cos\alpha\,\vec{e}_2 - \sin\alpha\,\vec{e}_1\right) = -\frac{1}{2}\left(\boldsymbol{B}_2\cos\alpha - \boldsymbol{B}_1\sin\alpha\right) \qquad (9.63)$$

As $\vec{\boldsymbol{B}}$ does not depend on α, it is deduced that the slip dislocation loops have a *dipole moment of rotation charges*. Integrating the linear density Λ of torsional charge on the contour of the loop, it is deduced that the dislocation loops show *no global scalar charge rotation* q_λ ($q_\lambda = 0$), whichever their nature

$$q_\lambda = \oint_C \Lambda R\, d\alpha = -\frac{1}{2}R\oint_C\left(\boldsymbol{B}_2\cos\alpha - \boldsymbol{B}_1\sin\alpha\right)d\alpha \equiv 0 \qquad (9.64)$$

We can still deduce the vectorial linear charge $\vec{\Lambda}$ of flexion of the dislocation loops using the relations *(9.9)* and *(9.58)*. It follows

$$\vec{\Lambda} = \vec{\boldsymbol{B}} \wedge \vec{\boldsymbol{t}} = \vec{\boldsymbol{B}} \wedge \left(\cos\alpha\,\vec{e}_2 - \sin\alpha\,\vec{e}_1\right) \qquad (9.65)$$

We then deduce the linear scalar curvature charge Θ of the dislocation through the relationship $\theta = \operatorname{div}\vec{\lambda}$ and the expressions *(9.54)* and *(9.65)*

$$\theta = \operatorname{div}\vec{\lambda} \;\Rightarrow\; \Theta = \frac{1}{R}\frac{\partial\left(\vec{\boldsymbol{B}} \wedge \vec{\boldsymbol{t}}\right)}{\partial\alpha}\vec{\boldsymbol{t}} = \frac{1}{R}\left(\vec{\boldsymbol{B}} \wedge \frac{\partial\vec{\boldsymbol{t}}}{\partial\alpha}\right)\vec{\boldsymbol{t}} = -\frac{1}{R}\left(\vec{\boldsymbol{B}} \wedge \vec{\boldsymbol{m}}\right)\vec{\boldsymbol{t}} \qquad (9.66)$$

so that

$$\Theta = -\frac{1}{R}\vec{\boldsymbol{n}}\vec{\boldsymbol{B}} = -\frac{1}{R}\boldsymbol{B}_3 \qquad (9.67)$$

As $\boldsymbol{B}_3$ may only be a constant independent of α, this relationship implies that the *prismatic dislocation loops* have a total scalar curvature charge q_θ given by the integral of the linear density Θ on the contour of the loop

$$q_\theta = \oint_C \Theta R\, d\alpha = -\oint_C \vec{\boldsymbol{n}}\vec{\boldsymbol{B}}\, d\alpha = -2\pi\vec{\boldsymbol{n}}\vec{\boldsymbol{B}} = -2\pi\boldsymbol{B}_3 \qquad (9.68)$$

As $\vec{\boldsymbol{B}} = \vec{\boldsymbol{t}} \wedge \vec{\Lambda}$, we also have

$$q_\theta = -2\pi\vec{\boldsymbol{n}}\left(\vec{\boldsymbol{t}} \wedge \vec{\Lambda}\right) = 2\pi\vec{\Lambda}\vec{\boldsymbol{m}} \qquad (9.69)$$

The effect of a uniform variation of volume expansion on the curvature charge q_θ

As the scalar charge of curvature q_θ of a prismatic loop does not depend on the radius R of the dislocation loop, but only on the component $\boldsymbol{B}_3$ of the Burgers vector, the dependency of the curvature charge q_θ on the local volume expansion τ of the medium is easily deduced

$$q_\theta = -2\pi\,\boldsymbol{B}_3 = -2\pi\,\boldsymbol{B}_{03}\,\mathrm{e}^{\tau/3} = q_{\theta 0}\,\mathrm{e}^{\tau/3} \qquad (9.70)$$

Quantification of the dislocation loops on a lattice

At the microscopic scale of a solid lattice, the Burgers vectors of dislocation loops are quantized, as shown schematically in figure 9.44 for prismatic loops and glide loops in a cubic lattice. In figure 9.37, there is also clear that:

- the prismatic dislocation loops, which are of a "scalar" nature are obtained by the addition or the deletion of a plane of particles within the loop (translation perpendicular to the plane of the loop), so that the system presents extra material in the loop plane. Note that the overall scalar charge q_θ of diverging curvature of the prismatic loop is directly related to the existence of this "extra material".
- the glide dislocation loops, which are of "vectorial" nature are obtained by slip (translation parallel to the plane of the loop) in the direction of the Burgers vector, so that the lattice does not have "extra-material" in this case. However, the presence of a screw component in regions where $\vec{B} \parallel \vec{t}$ induces a dipolar field of rotation in the vicinity of the glide loop.

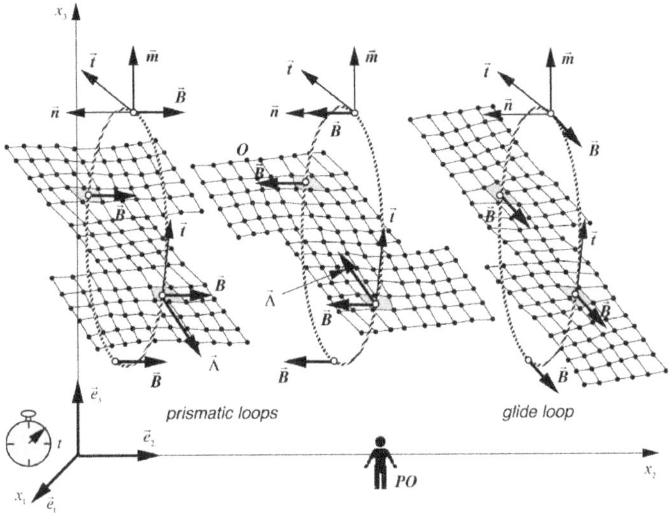

***Figure 9.37** - quantification of the dislocation loops on a lattice*

The loops of wedge disclinations

In figure 8.9, we have shown an embodiment of a macroscopic loop of wedge disclination. Now let's try to express the mathematics of it. For this, consider a loop consisting of a doublet of wedge disclinations bound by a virtual ribbon of dislocation, as shown in figure 9.38. According to the relationship *(9.40)*, the linear density of scalar charge of curvature Θ_1 and Θ_2 of the two disclinations are given by the relations

$$\begin{cases} \Theta_1 = -\vec{\Omega}_1 \vec{t} = -\vec{\Pi}\,\vec{m} \\ \Theta_2 = -\vec{\Omega}_2 \vec{t} = \vec{\Pi}\,\vec{m} = -\Theta_1 \end{cases} \qquad (9.71)$$

From this we deduce the existence of these two densities Θ_1 and Θ_2 on either side of the dislocation ribbon which generates a dipole field of flexion $\vec{\chi}_{dipolaire}$, located essentially in the vicinity of the two disclinations. This dipole field is shown in figure 9.39 in the case of a doublet of disclinations quantified at $\pm 90°$ in a cubic structure. It is easy to see the positive and negative flexion curvatures surrounding the two disclinations.

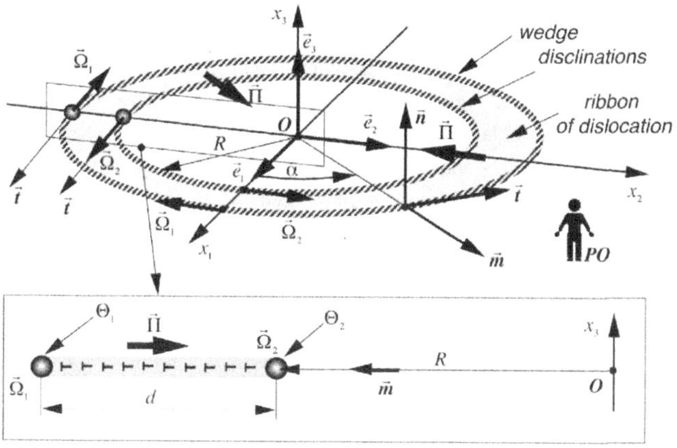

***Figure 9.38** - loop of doublet of wedge disclinations linked by a ribbon of dislocation*

It is interesting to see that the ribbon of dislocation of surface flexion charges $\vec{\Pi}$ can be reduced, by integration over the distance d separating the two disclinations, to a virtual linear charge $\vec{\Lambda}$ of edge dislocation distributed over a radius dislocation loop $R+d/2$. Thus, the disclinations of the doublet loop may be considered similar to a charge of edge dislocation loop with linear charge $\vec{\Lambda}$ such that

$$\vec{\Lambda} = \vec{\Pi} d = -d\,\Theta_1 \vec{m} \qquad (9.72)$$

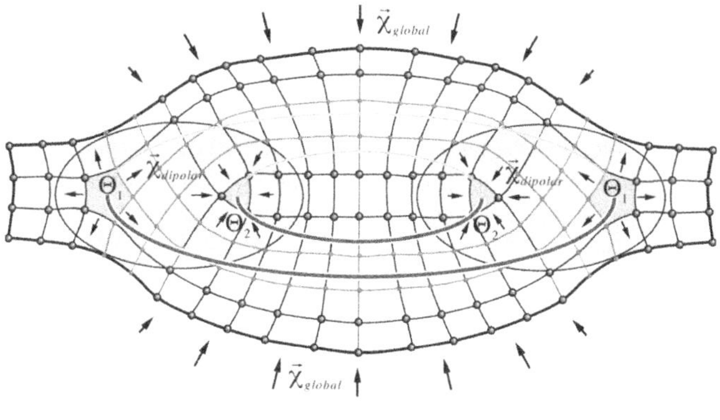

***Figure 9.39** - local dipolar field of flexion due to charges Θ_1 and Θ_2*
and divergent flexion field due to the charge Q_θ of a loop of doublet of wedge disclinations

The relationship *(9.69)* can be used to calculate the overall scalar charge of curvature q_θ of the virtual edge dislocation loop, and we obtain, knowing that $\vec{B} = \vec{t} \wedge \vec{\Lambda}$

$$q_\theta = 2\pi \vec{\Lambda}\vec{m} = 2\pi B_3 = 2\pi d \vec{\Pi}\vec{m} = -2\pi d\Theta_1 \tag{9.73}$$

One can also imagine that the diameter of the inner loop disclination goes to zero and there remains only the external disclination loop of linear charge Θ_1 and radius d. In this case, the overall charge of curvature q_θ remains unchanged and is always given by the same equation *(9.73)*. The global charge of curvature q_θ is the one due to the dislocation ribbon as a whole, and the one that is seen at large enough distance for the loop to be no longer possible to distinguish this loop from a simple edge dislocation loop. The charge q_θ is then responsible for the overall flexion $\vec{\chi}_{global}$ of the lattice at large scale, as figure 19.46 illustrates in the case of a doublet of quantified disclinations of $\pm 90°$ in a cubic structure.

The loops of twist disclinations

In figure 8.8, we have shown a macroscopic implementation of a loop of twist disclination[16]. We will give their mathematical description here. For this, consider a loop consisting of a twist disclination generated by a rotation $\vec{\Omega}_{twist}$ of the upper plane at an angle α relative to the lower plane, as shown in figure 9.40.

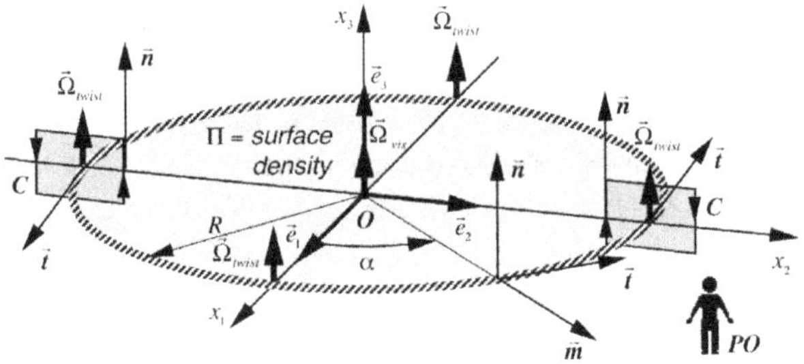

***Figure 9.40** - loop of twist disclination with its charged membrane*

The fact that one glues together two planes which have been moved relative to each other within the loop must create on the plane of the loop a surface charge Π of dislocation. By using relation *(9.35)*, one can write on the contour C surrounding the edge of the loop, and knowing that the surface charge Π comes from a volume density λ distributed in the thickness of the membrane with height h

$$\vec{\Omega}_{twist} = -\oint_C \lambda\, d\vec{r} = -\vec{n}\lambda h = -\vec{n}\Pi \tag{9.74}$$

in such a way that the angle of rotation α is written

$$\alpha = \Omega_{twist} = -\Pi \tag{9.75}$$

The surface charge Π of rotation can be integrated on the surface of the loop, giving us the overall charge of rotation q_λ of such a loop

[16] *W. Huang, T. Mura, Elastic J. of Applied Physics, vo. 41(13), p. 5175, 1970; A. Unzicker, arXiv:gr-qc/0011064, 2000; A. Unzicker, arXiv:gr-qc/9612061v2, 2010*

$$q_\lambda = \iint\limits_S \Pi dS = \pi R^2 \Pi \tag{9.76}$$

This global charge q_λ is actually the charge of rotation of the loop of twist disclination as seen at a great distance of the loop. This means that such a loop can act as the source of a divergent field of rotation $\vec{\omega}$ in the solid medium.

Note that it is possible to see a disclination loop somewhat differently. Indeed, the act of carrying out the rotation of the two planes with respect to the other causes a displacement along the string similar to that of a screw dislocation. The Burgers vector and the linear charge of this pseudo-dislocation is then worth

$$\vec{B}_{screw} = R\alpha\vec{t} = -R\Pi\vec{t} \quad \Rightarrow \quad \Lambda_{screw} = -\frac{1}{2}\vec{B}_{screw}\vec{t} = \frac{1}{2}R\Pi \tag{9.77}$$

so that the global charge of this pseudo-loop can be written

$$q_\lambda = \oint \Lambda_{screw}\, ds = 2\pi R \Lambda_{screw} = \pi R^2 \Pi \tag{9.78}$$

We therefore obtain much the same value of the overall charge q_λ that was obtained by considering the surface charge Π, which allows either consider this singularity as a twist disclination loop or as a screw pseudo-dislocation loop!

The effects of a variation of volume expansion on the torsional charge q_λ

Since the scalar charge of torsion q_λ of a twist disclination loop depends on R^2 and Π, and as the surface charge Π is independent of the local volume expansion of the medium τ *(9.30)*, the dependence of the overall charge of rotation q_λ of screw disinclination loops on the local volume expansion τ is easily deduced

$$q_\lambda = \Pi\pi R^2 = \Pi_0 \pi\, R_0^2\, e^{2\tau/3} = q_{\lambda 0}\, e^{2\tau/3} \tag{9.79}$$

On transforming a rectilinear string in a closed loop

A string of dislocation, disclination and/or rectilinear dispiration is fully characterized by its three linear charges Λ, $\vec{\Lambda}$ and Θ. Its linear scalar charge Λ is connected to the screw dislocation portion $\vec{B}_{screw} = -\Lambda\vec{t}$, its linear vectorial charge $\vec{\Lambda}$ is connected to its edge dislocation portion $\vec{B}_{edge} = \vec{t} \wedge \vec{\Lambda}$, and its linear scalar charge Θ is connected to its wedge disclination part $\vec{\Omega}_{wedge} = -\Theta\vec{t}$ (fig. 9.41a). From a straight string satisfying the conditions set out above, it is possible to construct a loop, by bending the string and closing it on itself to form a loop of radius R. For this, there are rules to follow, that we will now discuss.

The edge dislocation part of the string, represented by its Burgers vector $\vec{B}_{edge} = \vec{t} \wedge \vec{\Lambda}$ has to be conserved to meet the conservation equation $\operatorname{div} \vec{\lambda}_i = 0$. To satisfy this conservation necessarily involves bending the rope in the plane perpendicular to the Burgers vector $\vec{B}_{edge}$, which also implies that the vector density $\vec{\Lambda}$ is contained in the plane of the resulting loop (fig. 9.41b). Thus, the loop can only be a *prismatic loop* vis-à-vis its Burgers vector $\vec{B}_{edge}$, so it has a scalar charge of curvature $q_\theta^{disloc} = -2\pi\vec{n}\vec{B}_{edge} = 2\pi\vec{\Lambda}\vec{m}$ according to equation *(9.68)*.

If there is additionally a part of wedge disclinations on the linear string, its scalar charge $\Theta \neq 0$ will generate a virtual dislocation membrane of surface charge $\Theta \neq 0$ within the loop, as shown by the example in figure 9.37. Thus, the portion of the wedge disclination of the linear string leads to a loop of wedge disclination. Using the relation *(9.73)* in which d is replaced by R, one easily obtains the contribution $q_\theta^{disclin} = -2\pi R\Theta = 2\pi R\Pi\vec{m}$ of this disclination part to the *scalar charge of curvature*. We deduce that the overall charge of curvature q_θ of the loop due both to the edge dislocation and the wedge disclination of the initial string is worth

$$q_\theta = -2\pi\vec{n}\vec{B}_{edge} - 2\pi R\Theta = 2\pi\vec{\Lambda}\vec{m} + 2\pi R\vec{\Pi}\vec{m} \tag{9.80}$$

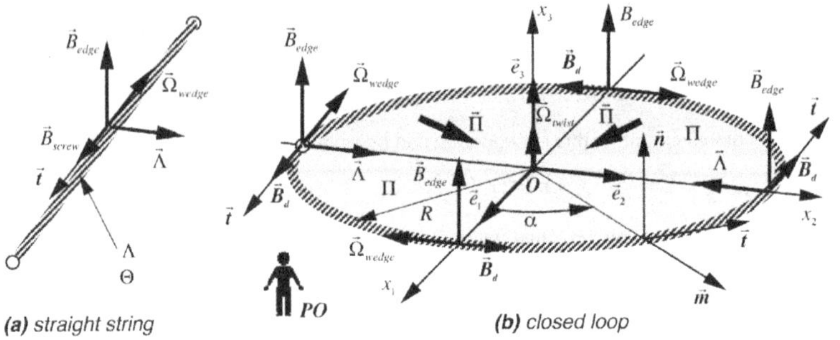

(a) *straight string* **(b)** *closed loop*

Figure 9.41 *- transformation of a rectilinear string in a closed loop string*

The fact of having closed on itself a string containing a linear charge, associated with a screw Burgers vector $\vec{B}_{screw} = -\Lambda\vec{t}$, implies that this screw Burgers vector $\vec{B}_{screw}$ can be a dislocation vector conserved along the loop, which appears quite paradoxical at first sight. But it is perfectly possible precisely because Λ is a scalar. By closing the string on itself, the screw dislocation $\vec{B}_{screw}$ loses its dislocation identity in favor of a disclination loop of the same type as the one described in figure 9.39. Then former Burgers vector $\vec{B}_{screw}$ becomes a simple displacement $\vec{B}_d$ along the loop, connected with the rotation $\vec{\Omega}_{twist}$ of the upper plane at an angle $\alpha = \vec{\Omega}_{twist} = -\Pi$ relative to the lower plane of the loop. There must therefore appear a surface charge of rotation Π on the plane of the loop, such that

$$\vec{B}_d = R\alpha\vec{t} = -R\Pi\vec{t} \tag{9.81}$$

But the conservation of the length $\left|\vec{B}_{screw}\right|$ which becomes $\left|\vec{B}_d\right|$ during the process of curving of the string to form a loop, implies that

$$\vec{B}_{screw} = -\Lambda\vec{t} \;\rightarrow\; \vec{B}_d = -R\Pi\vec{t} \;\Rightarrow\; \Pi = \Lambda / R \tag{9.82}$$

The initial linear charge Λ of rotation of the string disappears with the creation of a scalar surface charge Π of rotation of the loop, satisfying the simple transformation relation $\Pi = \Lambda / R$. The loop becomes then a loop of twist disclination vis-a-vis of the ex-vector of Burgers $\vec{B}_{screw}$, which carries then a global scalar charge q_λ *of rotation*, with

$$q_\lambda = \pi R\Lambda = \pi R^2\Pi \tag{9.83}$$

The transformation rules for converting a rectilinear loop in a closed loop are summarized in table 9.5.

Rectilinear string	$\Rightarrow$	Closed loop	Associated charge
screw dislocation $\Lambda \Rightarrow \vec{B}_{screw} = -\Lambda\vec{t}$	$\Rightarrow$	loop of twist disclination $\Pi = \dfrac{\Lambda}{R} \Rightarrow \vec{\Omega}_{twist} = -\Pi\vec{n} = -\dfrac{\Lambda}{R}\vec{n}$	$q_\lambda = \pi R\Lambda$ $= \pi R^2\Pi$ $= -\pi R^2\vec{\Omega}_{twist}\vec{n}$
edge dislocation $\vec{\Lambda} \Rightarrow \vec{B}_{edge} = \vec{t} \wedge \vec{\Lambda}$	$\Rightarrow$	loop of prismatic dislocation $\vec{\Lambda} \Rightarrow \vec{B}_{edge} = \vec{t} \wedge \vec{\Lambda}$	$q_\theta = 2\pi\vec{\Lambda}\vec{m}$ $= -2\pi\vec{n}\vec{B}_{edge}$
wedge disclination $\Theta \Rightarrow \vec{\Omega}_{wedge} = -\Theta\vec{t}$	$\Rightarrow$	loop of wedge disclination $\vec{\Pi} = -\Theta\vec{m} \Rightarrow \vec{\Omega}_{wedge} = (\vec{\Pi}\vec{m})\vec{t} = -\Theta\vec{t}$	$q_\theta = -2\pi R\Theta$ $= 2\pi R\vec{\Pi}\vec{m}$ $= 2\pi R\vec{\Omega}_{wedge}\vec{t}$

Table 9.5 - *rules of transformation of a rectilinear string into a closed loop string*

The quantification of the radius of a localized loop in an ordered lattice

A localized loop can contain, as we have seen, a global scalar charge q_θ of curvature and/or a global scalar charge q_λ of rotation. We can now show that the existence of a non null scalar charge of rotation q_λ may involve, under certain conditions, a quantization of the radius of the loop R if it is in an ordered array. Indeed, if we assume that, during the formation of the loop, the environment in question requires that the upper jaw closes again to the lower jaw in the orientation of the network structure, the angle of rotation $\vec{\Omega}_{twist}$ can then take quantized values $\Omega_{(q)}$ dependent on the lattice structure (for example, $\Omega_{(q)} = \pm n\pi / 2$ in a lattice having a cubic symmetry or $\Omega_{(q)} = \pm n\pi / 3$ in a lattice having a hexagonal symmetry). This implies that the surface charge $\Pi = -\Omega_{twist}$ also takes the quantized values $\Pi_{(q)} = -\Omega_{(q)}$. But since, on the other hand, the length of the Burgers vectors $\vec{B}_{screw}$ of the original string must also take quantized values in an ordered array, the linear charge Λ also has quantized values $\Lambda_{(k)}$.
The result of the quantization of Π and Λ is immediate, and results in a quantization of the radius of the loop R, which can only take eigenvalues $R_{(q,k)}$

$$R_{(q,k)} = \Lambda_{(k)} / \Pi_{(q)} \tag{9.84}$$

There are several interesting consequences of the quantification of the radius of twist disclinations. We will list them here:

- the first and very important result lies in the fact that the quantification of R automatically entails a *quantization of the overall scalar charge of torsion* q_λ of the loop, since in this case

$$q_\lambda = \pi R_{(q,k)}^2 \Pi_{(q)} = \pi R_{(q,k)} \Lambda_{(k)} = \pi \Lambda_{(k)}^2 / \Pi_{(q)} \tag{9.85}$$

- the second interesting result is linked to expressions of the quantized radius and the quantized torsional charge, i.e. $R_{(q,k)} = \Lambda_{(k)} / \Pi_{(q)}$ and $q_\lambda = \pi\Lambda_{(k)}^2 / \Pi_{(q)}$, as they clearly show that the radius and overall charge of rotation of the loop increase with the value of the charge of the string $\Lambda_{(k)}$ from which the loop is formed. This implies among other things that there cannot be a linear relationship between q_λ and $\Lambda_{(k)}$, for a series of given loops, unless the ratio $\Lambda_{(k)} / \Pi_{(q)}$ remains constant, which also implies that the radius does not change from one loop to another.

- the third important consequence is that the value of the global charge of curvature q_θ may depend on the quantification of R, but only its wedge disclination part $2\pi R \Pi \vec{m}$ depends on it, while its edge dislocation part does not depend on it.

- finally, for a loop whose charge of rotation q_λ is zero, there are no conditions requiring quantification of the radius R of the loop in an ordered array, so that the radius can take any value.

9.6 – Clusters of dislocations, disclinations and dispirations

Since the strings of dislocations, disclinations and dispirations containing nonzero tensorial densities of charges can be closed on themselves in loops, it is possible to imagine the existence of small clusters of very localized such loops within a solid medium, such as the cluster shown in the absolute frame of reference in figure 9.42.

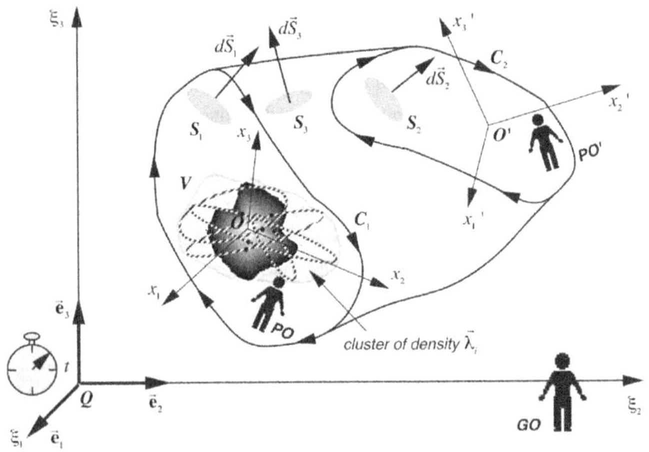

Figure 9.42 *- clusters of localized strings of dislocations and disclinations*

Such clusters are in principle fully characterized by their tensorial density $\vec{\lambda}_i$ of dislocation charges, which are non-zero in the strings, in the area represented by a hatched area in figure 9.42. By then defining a volume V such that its surface S_1 intersects the cluster of charges, the cutoff section is represented by the dark gray area in the figure 9.42, it is possible to break the surface surrounding the volume V into three parts S_1, S_2 and S_3. Then using the relationship *(9.5)*, we can show that, in this case, the components B_i of the Burgers vectors, and consequently also the components Ω_i of the Frank vectors obtained in the local referential $Ox_1x_2x_3$ by integration on the surface S_1 going through the cluster are identically zero

$$\boldsymbol{B}_i = -\iint_{S_1} \vec{\lambda}_i \, d\vec{S}_1 = -\iint_{S_2} \vec{\lambda}_i \, d\vec{S}_2 = \boldsymbol{B}_i' \tag{9.86}$$

This is due to the fact that each loop of strings or super-strings of the cluster cuts through the surface S_1 twice, once in one direction and the second time in the opposite direction, so that the contributions associated with each of these crossings exactly offset.

In the local referential $Ox_1x_2x_3$ the lack of vectors $\vec{B}$ and $\vec{\Omega}$ which are nonzero on all the contour surrounding the cluster without crossing it implies that there are no discontinuities in the virtual field of displacement $\vec{u}^{(\delta)}$ and no discontinuity in the rotation field in the solid surrounding the cluster, and that, therefore, the solid *remains perfect outside of the cluster.*

However, the presence of a cluster in the solid must certainly involve an elastic and inelastic distortion field of the *perfect solid surrounding the cluster* and this up to a certain distance thereof. This is analogous to the way the presence of a localized density ρ of electrical charges involves an electric field of displacement $\vec{D}$ at a distance of such charges.

To find this field, it is necessary to introduce here the fact that there is, apart from the tensorial charge conservation equation $\operatorname{div}\vec{\lambda}_i \equiv 0$, no restrictions on the scalar densities λ and θ of charges of rotation and flexion. Therefore, it is to possible, depending on the nature of the charges in the cluster, to have *non-zero global scalar charges Q_λ of rotation and Q_θ of curvature.* These global scalar charges are defined in the local referential frame by the sums on all closed loops of global charges $q_{\lambda(i)}$ and $q_{\theta(i)}$ as we have defined them for each individual loop in the previous section [17]. These considerations are then used to find the fields of elastic and anelastic distortions at long distances due to the presence of a localized cluster of charges, such as we are going to show in the following sections.

The divergent torsion field associated with a localized cluster of charges

The geometrocompatibility equation $\lambda = \operatorname{div}\vec{\omega}$ for the rotation vector $\vec{\omega}$ gives us the scalar density of torsional charges λ, linked to the trace of the tensorial density $\vec{\lambda}_i$ of distortion charges. It allows us to write a full relationship of compatibility in the following form in the local frame $Ox_1x_2x_3$

$$\oiint_S \vec{\omega} d\vec{S} = \iiint_V \operatorname{div}\vec{\omega} \, dV = \iiint_V \lambda \, dV = Q_\lambda \neq 0 \tag{9.87}$$

This relationship implies that a global scalar charge of rotation Q_λ, localized and non-zero, acts as the source of a divergent field of rotation $\vec{\omega}$ of the perfect solid in the neighborhood of the cluster of charges (fig. 9.43). The rotational field then has a *topological singularity* where is the cluster of charge Q_λ, and its norm $|\vec{\omega}|$ is a decreasing function of $1/R^2$ as a function of the distance R from the cluster.

Everything happens in fact exactly as in electromagnetism, where a local density ρ of electric

[17] *One can imagine for example that the curvature charge of a cluster can be due to prismatic dislocation loops of lacunar or interstitial nature (fig. 19.44) and/or to wedge disclination loops (fig. 9.39), and that the rotation charge can be due to twist disclination loops (fig. 9.40).*

charges leads to a localized macroscopic electric charge Q, which behaves like a singularity responsible for an electric displacement field $\vec{D}$ which is divergent in the neighboring space.

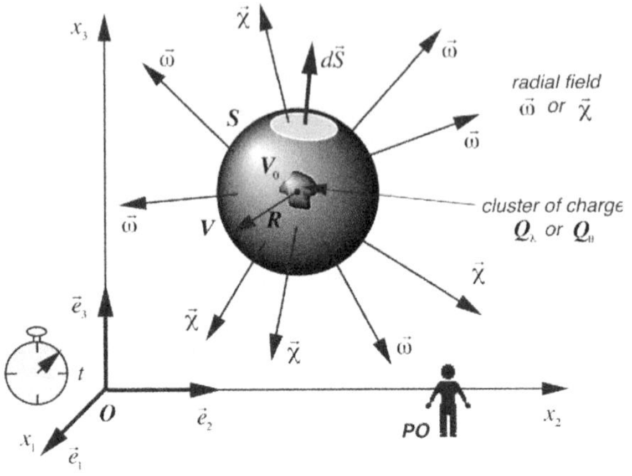

Figure 9.43 *- divergent fields of rotation and flexion in the neighborhood of a cluster of charges*

The divergent flexion field associated with a localized cluster of charges

The equation of geometro-compatibility $\theta = \operatorname{div} \vec{\chi}$ for the vector $\vec{\chi}$ of flexion shows the scalar charge density θ of curvature. It allows us to write a full relationship of compatibility in the following form in the local referential frame $Ox_1x_2x_3$

$$\oiint_S \vec{\chi} d\vec{S} = \iiint_V \operatorname{div} \vec{\chi} \, dV = \iiint_V \theta \, dV = Q_\theta \neq 0 \qquad (9.88)$$

This relationship implies that a macroscopic scalar charge of curvature Q_θ, localized and non-zero, acts as the source of a divergent flexion field $\vec{\chi}$ within the perfect solid surrounding the cluster of charges (fig. 9.43). The flexion field then has a *topological singularity* where is the charge Q_θ, and its norm $|\vec{\chi}|$ also shows a $1/R^2$ decrease as a function of the distance R from the cluster. In other words, in the vicinity of a localized overall charge of curvature Q_θ, the solid bends by flexion with a spherically symmetry around the singularity.

The singularity of the shear strains and the volume expansion in the vicinity of a cluster

As the flexion field $\vec{\chi}$ is directly related to the shear fields $\vec{\alpha}_i$ and the volume expansion τ by *(8.35)*, we can write the relation *(9.88)* in the following form in the local frame $Ox_1x_2x_3$

$$-\oiint_S \left(\sum_k \vec{e}_k \wedge \overrightarrow{\operatorname{rot}} \, \vec{\alpha}_k + \frac{2}{3} \overrightarrow{\operatorname{grad}} \, \tau \right) d\vec{S} = \iiint_V \theta \, dV = Q_\theta \neq 0 \qquad (9.89)$$

which implies that the presence of a localized global scalar charge Q_θ induces radial shear strain and volume expansion fields whose norms $|\vec{\alpha}_i|$ and $|\tau|$ must decrease as $1/R$ in the vicinity of the cluster. However, it is clear that equation *(9.89)* does not allow to determine unambiguously the fields $\vec{\alpha}_i$ and τ, unlike fields $\vec{\omega}$ and $\vec{\chi}$ that are defined unambiguously by the equations *(9.87)* and *(9.88)*. In fact, the exact determination of the distortion fields $\vec{\alpha}_i$ and τ will require an additional equation representing the equilibrium conditions of the elastic forces in the perfect solid surrounding the charges. These equation are necessarily linked to the various elastic moduli of the solid considered.

The simplified description of a solid in the presence of clusters of localized charge

Consider a hypothetical solid wherein the charges are confined in localized clusters, as illustrated for example in figure 9.44, and therefore in which there are *no strings of dislocations and disclinations propagating over large distances* compared to the scale at which the solid is studied. It is clear that, depending on the complexity of the internal structure of these clusters, i.e. the complexity of the intricate curls forming these clusters, the description of the fields of distortion and contortion within clusters can be very complex. But if these clusters have stable internal structures and that they can move individually within the solid, but not sufficiently interact with each other to change their internal structure, it is then possible to greatly simplify the description of the distortion fields prevailing in the solid.

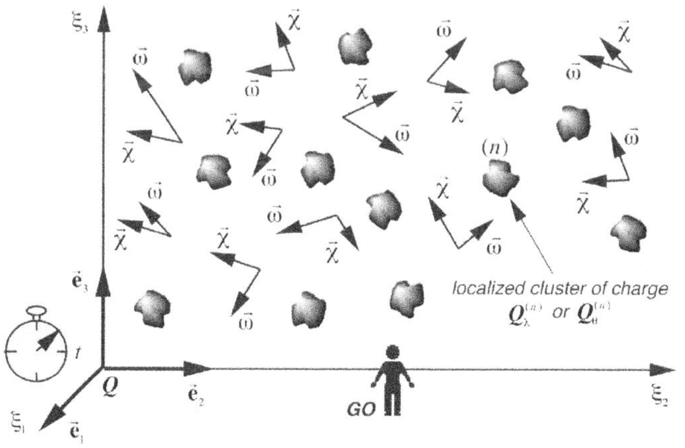

Figure 9.44 *- description of a solid containing localized clusters of charges*

In this case indeed, and so far as we are mainly concerned with describing the fields of distortion and contortion of elastic and anelastic nature in the perfect solid surrounding the clusters, that is to say at a certain distance external to the charged clusters, the problem can be solved much more simply by considering only the scalar charge densities λ and θ within the clusters, which can lead to the existence, at great distance, of two macroscopic scalar charges $Q_{\lambda(n)}$

and $Q_{\theta(n)}$ for each cluster number (n).
In fact, knowledge of the distribution of charge densities λ and θ within a cluster allows us to find purely topological conditions, and therefore independent of the elastic properties of the solid considered, which are imposed on the rotation and flexion fields $\vec{\omega}$ and $\vec{\chi}$ prevailing in the perfect solid outside of the cluster. These conditions are simply expressed by the scalar equations of geometro-compatibility $\lambda = \operatorname{div} \vec{\omega}$ and $\theta = \operatorname{div} \vec{\chi}$.
Depending on the inhomogeneity of the internal distribution of the charge densities λ and θ in the cluster, it may appear dipolar or multipolar fields $\vec{\omega}$ *and* $\vec{\chi}$ at short and medium distance of each cluster.
On the other hand, *at long distances* from the charge clusters, it is essentially the presence of non-null macroscopic scalar charges $Q_{\lambda(n)}$ and $Q_{\theta(n)}$ which will be responsible for the appearance of monopole radial fields of rotation $\vec{\omega}$ and flexion $\vec{\chi}$, as we have already shown in figure 9.43.
Thus, in this particular case of charges located in clusters, both invariant vector fields, namely the rotation $\vec{\omega}$ and flexion $\vec{\chi}$ fields, are affected at great distances from the clusters. And it is remarkable that each of these clusters can be individually and thoroughly characterized, concerning its long-range effects on the fields of distortion and contortion, from only the two macroscopic scalar charges $Q_{\lambda(n)}$ and $Q_{\theta(n)}$, although these clusters may have very complex structures, of tensorial nature and therefore highly dependent on their spatial orientation in the local referential frame.
In the analogy developed with electromagnetism, the rotation field $\vec{\omega}$ is analogous to the electric displacement $\vec{D}$, and the macroscopic charge of rotation $Q_{\lambda(n)}$ is the analog of the macroscopic electric charge Q of a particle in electromagnetism. But is there a similar analogy to the flexion field $\vec{\chi}$ and the overall charge of curvature $Q_{\theta(n)}$? A partially positive answer can be given here. Indeed, the presence of a macroscopic charge cluster $Q_{\theta(n)}$ is responsible for a nonzero vector field $\vec{\chi}$ of flexion, which is divergent in the neighborhood, and thus is also responsible for a spatial curvature of the solid lattice surrounding the cluster, which results in the appearance of nonzero fields of shear strains and volume expansion. Thus, the presence of a cluster of charges $Q_{\theta(n)} \neq 0$ implies vis-à-vis the solid lattice, a result analogous to that stipulated in the general theory of gravitation of Einstein vis-à-vis the spacetime in the presence of matter, i.e. that a matter accumulation (cluster) located in a volume of space is directly responsible for a curvature of the neighboring space-time. We will return later in detail to this analogy.

Chapter 10

Flows of dislocation charges and Orowan's relations

In this chapter, we try to interpret the concept of flow of dislocation charges, and link it to the concept of density of dislocation charges we described in the previous chapter.

We then recover the relationship between the *velocity of the macroscopic plastic deformation* described in section 7.9 and the existence of *flow of dislocation charges* in the solid medium in the form of *Orowan's relationships*.

10.1 – Interpretation of the flow of charges

The macroscopic interpretation of the tensorial density $\vec{\lambda}_i$ of dislocation charges and the conservation equation $\operatorname{div}\vec{\lambda}_i \equiv 0$ which these tensors satisfy, revealed the concepts of strings and loops of dislocation, disclination and dispiration. It was also shown that, at great distance from the clusters of plastic charges, it is essentially the two invariant vector fields, ie the fields of rotation $\vec{\omega}$ and of curvature $\vec{\chi}$, which are affected by the scalar components λ and θ of charge density. We still have to make the connection between these quantities and the flows of charges $\vec{J}_i$ and $\vec{J}$ introduced in chapter 9.

The continuity equation of the dislocation charges

Consider a tube filled with a dislocation density $\vec{\lambda}_i$, which moves with a *relative velocity* $\vec{\boldsymbol{v}}$ *with regard to the lattice*, which itself moves at velocity $\vec{\varphi}$ in the local frame $Ox_1x_2x_3$ of a **PO**. Define a mobile contour C surrounding the charge tube, with a movable surface S, which moves along with the tube at velocity $\vec{\boldsymbol{v}}+\vec{\varphi}$ with respect to $Ox_1x_2x_3$, as depicted in figure 10.1.

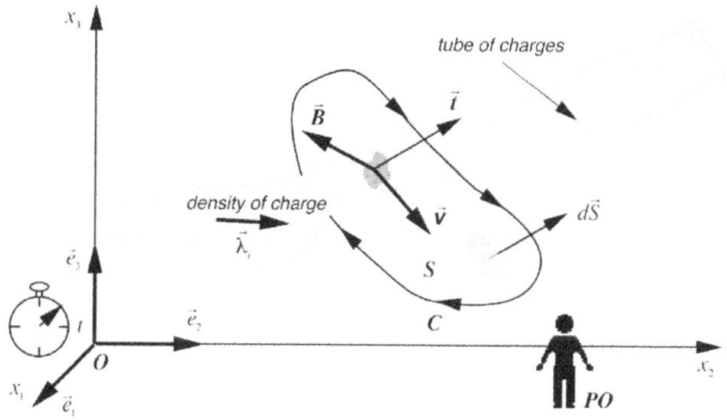

Figure 10.1 - *tube of charges* $\vec{\lambda}_i$ *moving with velocity* $\vec{\boldsymbol{v}}+\vec{\varphi}$ *in the frame* $Ox_1x_2x_3$

According to relation *(9.3)*, the global Burgers vector $\vec{B}$ defined by this contour is given by

$$\vec{B} = \sum_i B_i \vec{e}_i = -\sum_i \vec{e}_i \iint_S \vec{\lambda}_i d\vec{S} \tag{10.1}$$

so that the temporal variation of the Burgers vector on the tube of charges in movement can be written

$$\frac{d\vec{B}}{dt} = \sum_i \frac{dB_i}{dt}\vec{e}_i = -\sum_i \vec{e}_i \frac{d}{dt}\iint_S \vec{\lambda}_i\, d\vec{S} \tag{10.2}$$

It is possible to apply the formula of the derivative of an integral on a mobile surface *(see appendix F)* to relation *(10.2)* and we obtain

$$\frac{d\vec{B}}{dt} = -\sum_i \vec{e}_i \iint_S \left[\frac{\partial \vec{\lambda}_i}{\partial t} + (\vec{v} + \vec{\phi})\operatorname{div}\vec{\lambda}_i - \overrightarrow{\operatorname{rot}}\left[(\vec{v}+\vec{\phi})\wedge\vec{\lambda}_i\right]\right] d\vec{S} \tag{10.3}$$

As the components of the Burgers vector are locally conserved in the referential $Ox_1x_2x_3$, in order to be a change in the overall Burgers vector $\vec{B}$ defined by the movable contour, it is necessary that there is a non-zero source of charges $\vec{S}_i^{(\lambda_i)}$ in the lattice, so that

$$\frac{d\vec{B}}{dt} = \sum_i \frac{dB_i}{dt}\vec{e}_i = -\sum_i \vec{e}_i \iint_S \vec{S}_i^{(\vec{\lambda}_i)}\, d\vec{S} \tag{10.4}$$

For the equality between relations *(10.3)* and *(10.4)* to be satisfied for any velocity $\vec{v}+\vec{\phi}$, it is necessary that

$$\frac{\partial \vec{\lambda}_i}{\partial t} + (\vec{v} + \vec{\phi})\operatorname{div}\vec{\lambda}_i - \overrightarrow{\operatorname{rot}}\left[(\vec{v}+\vec{\phi})\wedge\vec{\lambda}_i\right] = \vec{S}_i^{(\vec{\lambda}_i)} \tag{10.5}$$

The condition of geometro-compatibility imposes that $\operatorname{div}\vec{\lambda}_i \equiv 0$, and we obtain the following relation, defined on a mobile lattice with velocity $\vec{\phi} \neq 0$

$$\begin{aligned}\frac{\partial \vec{\lambda}_i}{\partial t} &= \vec{S}_i^{(\vec{\lambda}_i)} + \overrightarrow{\operatorname{rot}}\left[(\vec{v}+\vec{\phi})\wedge\vec{\lambda}_i\right] = \vec{S}_i^{(\vec{\lambda}_i)} + \overrightarrow{\operatorname{rot}}\left[\vec{v}\wedge\vec{\lambda}_i\right] + \overrightarrow{\operatorname{rot}}\left[\vec{\phi}\wedge\vec{\lambda}_i\right] \\ &= \vec{S}_i^{(\vec{\lambda}_i)} + \overrightarrow{\operatorname{rot}}\left[\vec{v}\wedge\vec{\lambda}_i\right] + \left[\underbrace{\vec{\phi}\operatorname{div}\vec{\lambda}_i}_{\equiv 0} - \vec{\lambda}_i\operatorname{div}\vec{\phi} + (\vec{\lambda}_i\vec{\nabla})\vec{\phi} - (\vec{\phi}\vec{\nabla})\vec{\lambda}_i\right]\end{aligned} \tag{10.6}$$

It then follows that the equation of continuity describing the material derivative for the density of charges in a lattice which moves at velocity $\vec{\phi} \neq 0$ in the referential frame $Ox_1x_2x_3$

$$\frac{d\vec{\lambda}_i}{dt} = \left[\vec{S}_i^{(\vec{\lambda}_i)} - \vec{\lambda}_i\operatorname{div}\vec{\phi} + (\vec{\lambda}_i\vec{\nabla})\vec{\phi}\right] + \overrightarrow{\operatorname{rot}}\left[\vec{v}\wedge\vec{\lambda}_i\right] \tag{10.7}$$

The term of sources of dislocation charges

This continuity equation can now be compared to the equation of continuity for the charges $\vec{\lambda}_i$ *(8.14)* which states that

$$\frac{d\vec{\lambda}_i}{dt} = \left[\frac{d}{dt}\left(\overrightarrow{\operatorname{rot}}\,\vec{\beta}_i\right) - \overrightarrow{\operatorname{rot}}\left(\frac{d\vec{\beta}_i}{dt}\right)\right] - \overrightarrow{\operatorname{rot}}\,\vec{J}_i \tag{10.8}$$

Thus, in the case of a charge tube moving with velocity $\vec{v}$ relative to the lattice, the term of non-commutativity between brackets, which expresses the creation of local non-compatibility due to the charges may be connected to the existence of a non-zero source of charges $\vec{S}_i^{(\vec{\lambda}_i)}$ and the terms $\vec{\lambda}_i \operatorname{div}\vec{\varphi}$ and $\left(\vec{\lambda}_i\vec{\nabla}\right)\vec{\varphi}$

$$\frac{d}{dt}\left(\overrightarrow{\operatorname{rot}}\,\vec{\beta}_i\right)-\overrightarrow{\operatorname{rot}}\left(\frac{d\vec{\beta}_i}{dt}\right)=\vec{S}_i^{(\vec{\lambda}_i)}+\left(\vec{\lambda}_i\vec{\nabla}\right)\vec{\varphi}-\vec{\lambda}_i\operatorname{div}\vec{\varphi} \tag{10.9}$$

The term of charge sources $\vec{S}_i^{(\vec{\lambda}_i)}$ can only be due to the existence of a process of creating charges in the lattice, for example the *production of dislocations by a Frank-Read type mechanism* (see section 12.9).

The source term $\vec{\lambda}_i \operatorname{div}\vec{\varphi}$ is a decrease or an increase of the charge density $\vec{\lambda}_i$ due respectively to an *increase or a decrease of the local volume expansion* $\operatorname{div}\vec{\varphi}$. In the case where the local volume expansion *varies homogeneously*, this term ensures that charge density $\vec{\lambda}_i$ indeed changes according to the equation *(9.16)*.

Concerning the source term $\left(\vec{\lambda}_i\vec{\nabla}\right)\vec{\varphi}$, it provides an increase or a decrease of the density of charges $\vec{\lambda}_i$ due respectively to an *elongation or a shrinkage of the medium in the direction of the dislocation string* since the vectors $\vec{\lambda}_i$ are oriented in the direction of the string. In the event that the medium elongates only in the direction of the dislocation line, we can verify that this term $\left(\vec{\lambda}_i\vec{\nabla}\right)\vec{\varphi}$ compensates exactly the term $\vec{\lambda}_i \operatorname{div}\vec{\varphi}$, and thus ensures that the charge density $\vec{\lambda}_i$ does not change in this case.

But near the origin of the frame $Ox_1x_2x_3$, the velocity is, by definition, very low, and if we assume that $\vec{\boldsymbol{v}} > \vec{\varphi}$, the charge continuity equation can be written simply as follows, as we will adopt in the later parts of this chapter

$$\frac{\partial\vec{\lambda}_i}{\partial t}\cong\frac{d\vec{\lambda}_i}{dt}\cong\vec{S}_i^{(\vec{\lambda}_i)}+\overrightarrow{\operatorname{rot}}\left[\vec{\boldsymbol{v}}\wedge\vec{\lambda}_i\right] \tag{10.10}$$

The relations between flows and densities of charges

The comparison of equations *(10.7)* and *(10.8)* allows us to write a simple correspondence between the curl operators applied to $\vec{\boldsymbol{v}}\wedge\vec{\lambda}_i$ and $-\vec{J}_i$

$$\overrightarrow{\operatorname{rot}}\left(\vec{\boldsymbol{v}}\wedge\vec{\lambda}_i\right)=-\overrightarrow{\operatorname{rot}}\,\vec{J}_i \tag{10.11}$$

From which we deduce the relation that must exist between the charge density $\vec{\lambda}_i$ moving with velocity $\vec{\boldsymbol{v}}$ with respect to the lattice and flow of charges $\vec{J}_i$ associated with this movement

$$\vec{J}_i=\vec{\lambda}_i\wedge\vec{\boldsymbol{v}}+\overrightarrow{\operatorname{grad}}\,\boldsymbol{A}_i \tag{10.12}$$

In this relationship, we introduced the term $\overrightarrow{\operatorname{grad}}\,\boldsymbol{A}_i$, which is a vector field, gradient of an arbitrary scalar $\boldsymbol{A}_i$, which perfectly meets the relation *(10.10)*.

We can then deduce the relationship between the vector flow of charge $\vec{J}$ and the charge density $\vec{\lambda}_i$ by using equation *(8.20)*

$$\vec{J}=-\frac{1}{2}\sum_k\vec{e}_k\wedge\vec{J}_k=-\frac{1}{2}\sum_k\vec{e}_k\wedge\left(\vec{\lambda}_k\wedge\vec{\boldsymbol{v}}+\overrightarrow{\operatorname{grad}}\,\boldsymbol{A}_k\right)=\frac{1}{2}\sum_k\left(\vec{e}_k\vec{\lambda}_k\right)\vec{\boldsymbol{v}}-\frac{1}{2}\sum_k\left(\vec{e}_k\vec{\boldsymbol{v}}\right)\vec{\lambda}_k-\frac{1}{2}\sum_k\vec{e}_k\wedge\overrightarrow{\operatorname{grad}}\,\boldsymbol{A}_k \tag{10.13}$$

The first term of this relationship can be expressed as a function of the scalar torsion charge λ thanks to the relation *(8.16)*, and the third term is none other than the curl of the vector $\vec{\boldsymbol{A}}$ constructed with components $\boldsymbol{A}_i$

$$\vec{J} = \lambda\vec{v} - \frac{1}{2}\sum_k (\vec{e}_k\vec{v})\vec{\lambda}_k + \frac{1}{2}\overrightarrow{\mathrm{rot}}\,\vec{A} \qquad (10.14)$$

With regards to the second term we can use the following relation

$$\sum_k (\vec{e}_k\vec{v})\vec{\lambda}_k = \sum_k \underbrace{(\vec{\lambda}_k\vec{v})}_{=0}\vec{e}_k - \sum_k \vec{v}\wedge(\vec{e}_k\wedge\vec{\lambda}_k) = -\vec{v}\wedge\sum_k \vec{e}_k\wedge\vec{\lambda}_k = \vec{v}\wedge\vec{\lambda} \qquad (10.15)$$

in which the term $\vec{\lambda}_k\vec{v}$ is zero since $\vec{\lambda}_k$ is parallel to $\vec{t}$, while $\vec{v}$ is perpendicular to $\vec{t}$, and in which the sum of terms $\vec{e}_k \wedge \vec{\lambda}_k$ can be replaced by the relation *(8.36)*. The vectorial flow of charges $\vec{J}$ is therefore composed of two components $\vec{J}^{(\lambda)}$ and $\vec{J}^{(\vec{\lambda})}$ which depend, respectively, on the scalar charge density λ and the vectorial charge density $\vec{\lambda}$

$$\vec{J} = \vec{J}^{(\lambda)} + \vec{J}^{(\vec{\lambda})} + \frac{1}{2}\overrightarrow{\mathrm{rot}}\,\vec{A} = \lambda\vec{v} + \frac{1}{2}\left(\vec{\lambda}\wedge\vec{v}\right) + \frac{1}{2}\overrightarrow{\mathrm{rot}}\,\vec{A} \qquad (10.16)$$

Finally we can calculate the trace of the vectorial flow $\vec{J}_i$ by using relation *(8.25)*

$$\frac{S_n}{n} = \sum_k \vec{e}_k\vec{J}_k = \sum_k \vec{e}_k\left(\vec{\lambda}_k\wedge\vec{v} + \overrightarrow{\mathrm{grad}}\,A_k\right) = \sum_k \vec{v}\left(\vec{e}_k\wedge\vec{\lambda}_k\right) + \sum_k \vec{e}_k\,\overrightarrow{\mathrm{grad}}\,A_k \qquad (10.17)$$

By using once again relation *(8.36)*, and by remarking that the second term is the divergence of the vector $\vec{A}$ built with components A_i, we have

$$\frac{S_n}{n} = -\vec{\lambda}\vec{v} + \mathrm{div}\,\vec{A} \qquad (10.18)$$

The three main relationships we have obtained between the charge density tensors and the tensors of charge flows are given in table 10.1. As we do no have, for the moment, a direct interpretation for the vector $\vec{A}$ appearing in the continuity relations, other than the fact that this vector is the origin of flows $\vec{J}_i$ and $\vec{J}$ and source S_n / n which are not directly related to charges $\vec{\lambda}_i$, $\vec{\lambda}$ and λ, and since these flows and this source disappear in the continuity equation *(10.10)* of charges, we neglect them in table 10.1.

The continuity equations of the densities of dislocation charges

For the movements of charges at homogeneous velocity $\vec{v}$ such that $\vec{v} >> \vec{\varphi}$, it is easy to show thanks to the equation of conservation *(8.12)*, that

$$\overrightarrow{\mathrm{rot}}\left(\vec{v}\wedge\vec{\lambda}_i\right) = \vec{v}\underbrace{\mathrm{div}\,\vec{\lambda}_i}_{=0} - \left(\vec{v}\vec{\nabla}\right)\vec{\lambda}_i = -\left(\vec{v}\vec{\nabla}\right)\vec{\lambda}_i \qquad (10.19)$$

which allows to write the equation of continuity *(10.10)* for the tensorial density $\vec{\lambda}_i$ of charges in the following equivalent forms

$$\frac{d\vec{\lambda}_i}{dt} \underset{\vec{v}>>\vec{\varphi}}{\cong} \vec{S}_i^{(\vec{\lambda}_i)} + \overrightarrow{\mathrm{rot}}\left(\vec{v}\wedge\vec{\lambda}_i\right) = \vec{S}_i^{(\vec{\lambda}_i)} - \left(\vec{v}\vec{\nabla}\right)\vec{\lambda}_i = \vec{S}_i^{(\vec{\lambda}_i)} - \overrightarrow{\mathrm{rot}}\,\vec{J}_i \qquad (10.20)$$

By applying relation *(8.36)* to this last relation, we have the equation of continuity for the density $\vec{\lambda}$ of charges of flexion

$$\frac{d\vec{\lambda}}{dt} \cong -\sum_k \vec{e}_k\wedge\frac{d\vec{\lambda}_k}{dt} \underset{\vec{v}>>\vec{\varphi}}{\cong} -\sum_k \vec{e}_k\wedge\vec{S}_k^{(\vec{\lambda}_k)} + \sum_k \vec{e}_k\wedge\left(\vec{v}\vec{\nabla}\right)\vec{\lambda}_k = \vec{S}^{(\vec{\lambda})} - \left(\vec{v}\vec{\nabla}\right)\vec{\lambda} \qquad (10.21)$$

in which we have introduced the source $\vec{S}^{(\vec{\lambda})}$ of charges $\vec{\lambda}$, defined as

$$\vec{S}^{(\vec{\lambda})} = -\sum_k \vec{e}_k \wedge \vec{S}_k^{(\vec{\lambda}_i)} \tag{10.22}$$

By using again the following relation, with the density of charges of curvature $\theta = \operatorname{div} \vec{\lambda}$

$$\overrightarrow{\operatorname{rot}}\left(\vec{\boldsymbol{v}} \wedge \vec{\lambda}\right) = -2\ \overrightarrow{\operatorname{rot}}\, \vec{J}^{(\vec{\lambda})} = \vec{\boldsymbol{v}} \operatorname{div} \vec{\lambda} - \left(\vec{\boldsymbol{v}}\vec{\nabla}\right)\vec{\lambda} = \vec{\boldsymbol{v}}\theta - \left(\vec{\boldsymbol{v}}\vec{\nabla}\right)\vec{\lambda} \tag{10.23}$$

which calls upon the density of charges of curvature $\theta = \operatorname{div} \vec{\lambda}$, the vectorial flow of charges of flexion $\vec{J}^{(\vec{\lambda})} = (\vec{\lambda} \wedge \vec{\boldsymbol{v}}) / 2$ and the vectorial flow of charges of curvature $\theta\vec{\boldsymbol{v}}$, the equation of continuity for the charges $\vec{\lambda}$ can be written under the following forms

$$\frac{d\vec{\lambda}}{dt} \cong \vec{S}^{(\vec{\lambda})} - \left(\vec{\boldsymbol{v}}\vec{\nabla}\right)\vec{\lambda} = \vec{S}^{(\vec{\lambda})} + \overrightarrow{\operatorname{rot}}\left(\vec{\boldsymbol{v}} \wedge \vec{\lambda}\right) - \vec{\boldsymbol{v}} \operatorname{div} \vec{\lambda} = \vec{S}^{(\vec{\lambda})} - 2\ \overrightarrow{\operatorname{rot}}\, \vec{J}^{(\vec{\lambda})} - \vec{\boldsymbol{v}}\theta \tag{10.24}$$

Table 10.1 - Relations between tensors of charge densities and tensorial flows of charges

Relations between flows and densities
for charges at velocity $\vec{\boldsymbol{v}}$ with respect to the lattice

$$\begin{cases} \vec{J}_i = \vec{\lambda}_i \wedge \vec{\boldsymbol{v}} & (1) \\ \vec{J} = \vec{J}^{(\lambda)} + \vec{J}^{(\vec{\lambda})} = \lambda\vec{\boldsymbol{v}} + \dfrac{1}{2}\left(\vec{\lambda} \wedge \vec{\boldsymbol{v}}\right) & (2) \\ \dfrac{S_n}{n} = -\vec{\lambda}\vec{\boldsymbol{v}} & (3) \end{cases}$$

Continuity equations for the charges in the case $\vec{\boldsymbol{v}} >> \vec{\varphi}$

$$\begin{cases} \dfrac{d\vec{\lambda}_i}{dt} \cong \vec{S}_i^{(\vec{\lambda}_i)} - \left(\vec{\boldsymbol{v}}\vec{\nabla}\right)\vec{\lambda}_i = \vec{S}_i^{(\vec{\lambda}_i)} + \overrightarrow{\operatorname{rot}}\left(\vec{\boldsymbol{v}} \wedge \vec{\lambda}_i\right) = \vec{S}_i^{(\vec{\lambda}_i)} - \overrightarrow{\operatorname{rot}}\, \vec{J}_i & (4) \\ \dfrac{d\vec{\lambda}}{dt} \cong \vec{S}^{(\vec{\lambda})} - \left(\vec{\boldsymbol{v}}\vec{\nabla}\right)\vec{\lambda} = \vec{S}^{(\vec{\lambda})} + \overrightarrow{\operatorname{rot}}\left(\vec{\boldsymbol{v}} \wedge \vec{\lambda}\right) - \vec{\boldsymbol{v}} \operatorname{div} \vec{\lambda} = \vec{S}^{(\vec{\lambda})} - 2\ \overrightarrow{\operatorname{rot}}\, \vec{J}^{(\vec{\lambda})} - \vec{\boldsymbol{v}}\theta & (5) \\ \dfrac{d\lambda}{dt} \cong S^{(\lambda)} - \left(\vec{\boldsymbol{v}}\vec{\nabla}\right)\lambda = S^{(\lambda)} - \operatorname{div}\left(\lambda\vec{\boldsymbol{v}}\right) = S^{(\lambda)} - \operatorname{div} \vec{J}^{(\lambda)} & (6) \end{cases}$$

$$\text{with} \begin{cases} \vec{S}^{(\vec{\lambda})} = -\sum\limits_k \vec{e}_k \wedge \vec{S}_k^{(\vec{\lambda}_i)} & (7) \\ S^{(\lambda)} = \dfrac{1}{2}\sum\limits_k \vec{e}_k \vec{S}_k^{(\vec{\lambda}_i)} & (8) \end{cases}$$

By applying relation *(8.17)* to relation *(10.20)*, we obtain the continuity equation for the density λ of charges of rotation

$$\frac{d\lambda}{dt} = \frac{1}{2}\sum_k \vec{e}_k \frac{d\vec{\lambda}_k}{dt} \underset{\vec{\boldsymbol{v}}>>\vec{\varphi}}{\cong} \frac{1}{2}\sum_k \vec{e}_k \vec{S}_k^{(\vec{\lambda}_i)} - \frac{1}{2}\sum_k \vec{e}_k \left(\vec{\boldsymbol{v}}\vec{\nabla}\right)\vec{\lambda}_k = S^{(\lambda)} - \left(\vec{\boldsymbol{v}}\vec{\nabla}\right)\lambda \tag{10.25}$$

in which we have introduced the scalar source $S^{(\lambda)}$ of charges λ, defined as

$$S^{(\lambda)} = \frac{1}{2}\sum_k \vec{e}_k \vec{S}_k^{(\vec{\lambda}_i)} \qquad (10.26)$$

By using again the following relation

$$\operatorname{div}(\lambda\vec{\boldsymbol{v}}) = \vec{\boldsymbol{v}}\,\overrightarrow{\operatorname{grad}}\,\lambda = (\vec{\boldsymbol{v}}\vec{\nabla})\lambda \qquad (10.27)$$

which calls upon the vectorial flow $\vec{J}^{(\lambda)} = \lambda\vec{\boldsymbol{v}}$, the equation of continuity for the scalar charges λ can be written under the following forms

$$\frac{d\lambda}{dt} \cong S^{(\lambda)} - (\vec{\boldsymbol{v}}\vec{\nabla})\lambda = S^{(\lambda)} - \operatorname{div}(\lambda\vec{\boldsymbol{v}}) = S^{(\lambda)} - \operatorname{div}\vec{J}^{(\lambda)} \qquad (10.28)$$

The equations of continuity for the density of charges $\vec{\lambda}_i$, $\vec{\lambda}$ and λ are also shown in table 10.1, with the definitions of the sources of charges $\vec{S}^{(\vec{\lambda})}$ and $S^{(\lambda)}$.

10.2 – Charges and linear flows for lines of dislocations

To interpret the relationships of table 10.1, it is useful to apply them to the case of dislocation lines. So consider a line like that reported in figure 10.2, which moves relative to the lattice with velocity $\vec{\boldsymbol{v}}$.

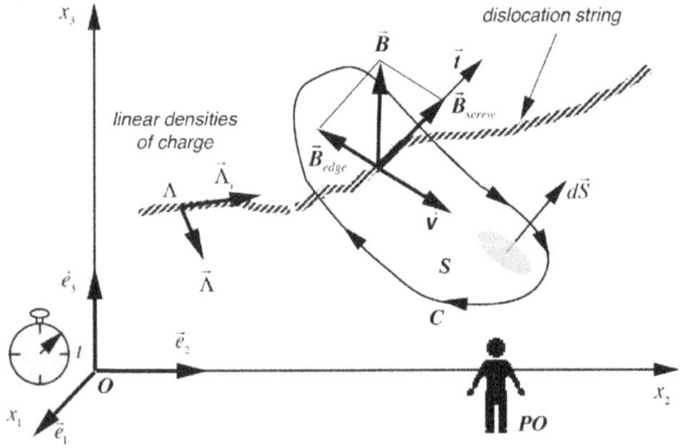

Figure 10.2 - *lines of dislocation moving with velocity* $\vec{\boldsymbol{v}}$ *in the referential* $Ox_1x_2x_3$

Clearly, in the case of a line, the velocity $\vec{\boldsymbol{v}}$ can only be perpendicular at all points to the direction $\vec{t}$ of the line. In the case of a line, one can integrate relationships of table 10.1 on the mobile contour C. Integrations of the contour's surface of charge flows $\vec{J}_i$ and $\vec{J}$ will give the *linear flow associated with the movable dislocation,* which we represent by the symbols $\vec{\Upsilon}_i$ and $\vec{\Upsilon}$. As flows $\vec{J}_i$ and $\vec{J}$ have dimension of the inverse of a time (1/s), the linear flows $\vec{\Upsilon}_i$ and $\vec{\Upsilon}$ have as dimension a surface per time unit (m²/s). As for the source of lattice sites S_n / n, its integral also represents a surface per unit of time (m²/s), and we will write it Υ. We have, taking into account relationships *(9.5)*, *(9.8)* and *(9.9)* for the linear charges $\vec{\Lambda}_i$, $\vec{\Lambda}$ and Λ of the dislocation

$$\begin{cases} \vec{\Upsilon}_i = \iint\limits_S \vec{J}_i\, dS = \iint\limits_S \vec{\lambda}_i\, dS \wedge \vec{\boldsymbol{v}} = \vec{\Lambda}_i \wedge \vec{\boldsymbol{v}} \\ \vec{\Upsilon} = \iint\limits_S \vec{J}\, dS = \Lambda \vec{\boldsymbol{v}} + \frac{1}{2}\left(\vec{\Lambda} \wedge \vec{\boldsymbol{v}}\right) \\ \Upsilon = \iint\limits_S \frac{S_n}{n}\, dS = -\vec{\Lambda}\vec{\boldsymbol{v}} \end{cases} \tag{10.30}$$

We may now introduce the relations which link the linear charge of a dislocation with he Burgers vector of said dislocation, namely

$$\begin{cases} \vec{B} = -\sum_i \left(\vec{t}\vec{\Lambda}_i\right)\vec{e}_i & \Rightarrow \quad \vec{\Lambda}_i = -\vec{t}B_i \\ \vec{B}_{screw} = \left(\vec{B}\vec{t}\right)\vec{t} = -2\,\Lambda\vec{t} & \Rightarrow \quad \Lambda = -\vec{B}\vec{t}\,/\,2 \\ \vec{B}_{edge} = \vec{t} \wedge \left(\vec{B} \wedge \vec{t}\right)t = \vec{t} \wedge \vec{\Lambda} & \Rightarrow \quad \vec{\Lambda} = \vec{B} \wedge \vec{t} \end{cases} \tag{10.31}$$

in relations *(10.30)* and we obtain the following expressions for the linear flows

$$\begin{cases} \vec{\Upsilon}_i = \vec{\Lambda}_i \wedge \vec{\boldsymbol{v}} = -\left(\vec{t} \wedge \vec{\boldsymbol{v}}\right)B_i = \left(\vec{\boldsymbol{v}} \wedge \vec{t}\right)\left(\vec{B}\vec{e}_i\right) \\ \vec{\Upsilon} = \Lambda\vec{\boldsymbol{v}} + \frac{1}{2}\left(\vec{\Lambda} \wedge \vec{\boldsymbol{v}}\right) = -\frac{1}{2}\left(\vec{B}\vec{t}\right)\vec{\boldsymbol{v}} - \frac{1}{2}\left[\vec{\boldsymbol{v}} \wedge \left(\vec{B} \wedge \vec{t}\right)\right] = -\frac{1}{2}\left(\vec{B}\vec{t}\right)\vec{\boldsymbol{v}} + \frac{1}{2}\left(\vec{\boldsymbol{v}}\vec{B}\right)\vec{t} \\ \Upsilon = -\vec{\Lambda}\vec{\boldsymbol{v}} = \left(\vec{t} \wedge \vec{B}\right)\vec{\boldsymbol{v}} \end{cases} \tag{10.32}$$

Thanks to these relations, we will be able to interpret the different terms of the flow of charges associated with the movement of dislocations.

The flow associated with the slipping of a screw dislocation

If the dislocation contains only a screw component, the flux $\vec{\Upsilon}$ will be given by relation

$$\vec{\Upsilon}_{screw} = \Lambda_{screw}\vec{\boldsymbol{v}} = -\frac{1}{2}\left(\vec{B}_{screw}\vec{t}\right)\vec{\boldsymbol{v}} \tag{10.33}$$

This relationship shows that, as the velocity $\vec{\boldsymbol{v}}$ is always perpendicular to the direction $\vec{t}$ of the line, the pure screw dislocations can move in all directions perpendicular to the direction $\vec{t}$. One speaks in this case of a slipping movement of screw dislocations, and the planes on which the screw dislocation moves are called *slip planes*.

The flows associated with the slipping or the climbing of an edge dislocation

In the case of a pure edge dislocation, the flows $\vec{\Upsilon}$ and Υ may appear, since the edge dislocation is completely represented by the charge $\vec{\Lambda}$, so that

$$\begin{cases} \vec{\Upsilon}^{(slip)}_{edge} = \frac{1}{2}\left(\vec{\Lambda}_{edge} \wedge \vec{\boldsymbol{v}}\right) = \frac{1}{2}\left[\left(\vec{B}_{edge} \wedge \vec{t}\right) \wedge \vec{\boldsymbol{v}}\right] = \frac{1}{2}\left(\vec{\boldsymbol{v}}\,\vec{B}_{edge}\right)\vec{t} \\ \Upsilon^{(climb)}_{edge} = -\vec{\boldsymbol{v}}\vec{\Lambda}_{edge} = \left(\vec{t} \wedge \vec{B}_{edge}\right)\vec{\boldsymbol{v}} \end{cases} \tag{10.34}$$

With relations *(10.33)* and *(10.34)*, one understands the dimensions (m²/s) of the linear flows Υ and $\vec{\Upsilon}$ since these flows are the product of the length of the Burgers vector with the relative speed of the line relative to lattice.

According to *(10.34)*, there appears two types of movements of the edge dislocation:

- the movement for which $\vec{v}$ is perpendicular to $\vec{\Lambda}$, and which is responsible for a vectorial flow of charges $\vec{\Upsilon}^{(slip)}_{edge}$. This movement is shown in figure 10.3a. It corresponds to a conservative *slipping* movement of the edge dislocation on its *slip plane*, which is defined as the plane perpendicular to $\vec{\Lambda}$, so the plane that contains both the Burgers vector $\vec{B}_{edge}$, the direction $\vec{t}$ of the line and the velocity vector $\vec{v}$.

- the movement for which $\vec{v}$ is parallel to $\vec{\Lambda}$ and which is responsible for a scalar flow of charges $\Upsilon^{(climb)}_{edge}$. This movement is shown in figure 10.3b. It corresponds to a non-conservative upward movement of the edge dislocation perpendicular to its slip plane. The dislocation goes "up" in the lattice, it climbs, creating or destroying a plane of the lattice. This movement is not conservative in the sense that it destroys or builds the lattice, and it is this movement that is responsible for the existence of a source of lattice sites S_n in the equations of geometro-kinetic volume expansion *(8.25)*, and which is given by

$$S_n = -n\vec{v}\vec{\Lambda}_{edge} = n\left(\vec{t} \wedge \vec{B}_{edge}\right)\vec{v} \qquad (10.35)$$

(a)

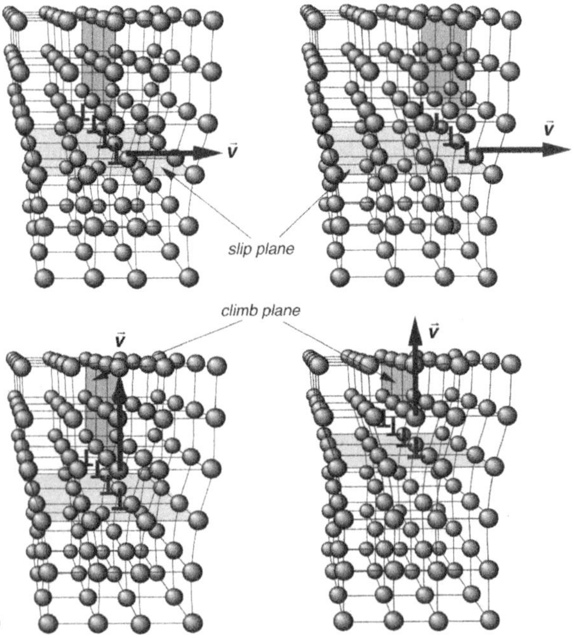

(b)

Figure 10.3 *- movement of slipping* ***(a)*** *and climbing* ***(b)*** *of an edge dislocation*

The various flows associated with the slipping and climbing movements of dislocations are summarized in table 10.2.

Table 10.2 - Relations between tensors of density and flow for a line dislocation

$\vec{v}$ = *relative velocity of the string with respect to the lattice* $\Rightarrow$

$$\begin{cases} \vec{\Upsilon}_i = \vec{\Lambda}_i \wedge \vec{v} = -(\vec{t} \wedge \vec{v}) B_i = (\vec{v} \wedge \vec{t})(\vec{B}\vec{e}_i) & (1) \\ \vec{\Upsilon} = \vec{\Upsilon}_{screw} + \vec{\Upsilon}_{edge}^{(slip)} = \Lambda_{screw}\vec{v} + \frac{1}{2}(\vec{\Lambda}_{edge} \wedge \vec{v}) = -\frac{1}{2}(\vec{B}_{screw}\vec{t})\vec{v} + \frac{1}{2}(\vec{v}\,\vec{B}_{edge})\vec{t} & (2) \\ \Upsilon = \Upsilon_{edge}^{(climb)} = -\vec{v}\vec{\Lambda}_{edge} = (\vec{t} \wedge \vec{B}_{edge})\vec{v} & (3) \end{cases}$$

The flow associated with the movement of a local charge of rotation

If we consider a localized charge of rotation Q_λ moving with velocity $\vec{v}$ relative to the solid medium, we obtain the relation between charge and flow by integrating the equation *(10.13)* on the volume of the charge, by asking that $\vec{\lambda} = 0$. It then comes, by introducing a current $\vec{I}$ which has a dimension of volume per time unit (m³/s)

$$\vec{I} = \iiint_V \vec{J}\, dV = \iiint_V \lambda\vec{v}\, dV = Q_\lambda \vec{v} \qquad (10.36)$$

This relationship implies that a localized scalar charge Q_λ, whose analogy is a localized electric charge becomes, as it moves in the solid medium, the source of a current $\vec{I}$ which is an analog current to the electric current.

10.3 – Relations of Orowan

From the knowledge of the flows $\vec{J}_i$ of dislocation charges, it is possible to retrieve the *macroscopic plastic distortion of the solid* $\vec{\beta}_i^{pl}$, as defined in chapter 8, thanks to the relation *(8.8)*. Total derivatives $\dot{\vec{\beta}}_i^{pl}$, $\dot{\vec{\varepsilon}}_i^{pl}$, $\dot{\vec{\alpha}}_i^{pl}$, $\dot{\vec{\omega}}^{pl}$ and $\dot{\tau}^{pl}$ can be written

$$\begin{cases} \dot{\vec{\beta}}_i^{pl} = \vec{J}_i = \vec{\lambda}_i \wedge \vec{v} \\ \dot{\vec{\omega}}^{pl} = \vec{J} = \lambda\vec{v} + \frac{1}{2}(\vec{\lambda} \wedge \vec{v}) \\ \dot{\vec{\varepsilon}}_i^{pl} = \vec{J}_i - \vec{e}_i \wedge \vec{J} = \vec{\lambda}_i \wedge \vec{v} - (\vec{e}_i \wedge \vec{v})\lambda - \frac{1}{2}\vec{e}_i \wedge (\vec{\lambda} \wedge \vec{v}) \\ \dot{\tau}^{pl} = \frac{S_n}{n} = -\vec{\lambda}\vec{v} \\ \dot{\vec{\alpha}}_i^{pl} = \vec{J}_i - \vec{e}_i \wedge \vec{J} - \frac{1}{3}\vec{e}_i \frac{S_n}{n} = \vec{\lambda}_i \wedge \vec{v} - (\vec{e}_i \wedge \vec{v})\lambda - \frac{1}{2}\vec{e}_i \wedge (\vec{\lambda} \wedge \vec{v}) + \frac{1}{3}\vec{e}_i(\vec{\lambda}\vec{v}) \end{cases} \qquad (10.37)$$

Similar relationships can be established in the case of many dislocations moving with velocity $\vec{v}$ relative to the solid medium. For example, imagine the same dislocations, all parallel to one direction and moving all at the same velocity $\vec{v}$. The integral $\vec{\beta}_i^{pl}$ on a surface S perpendicular to the direction of the lines is written

$$\iint\limits_S \dot{\vec{\beta}}_i^{pl}\, dS = \dot{\vec{\beta}}_i^{pl} S = \iint\limits_S \left(\vec{\lambda}_i \wedge \vec{\boldsymbol{v}}\right) dS = \iint\limits_S \vec{\lambda}_i\, dS \wedge \vec{\boldsymbol{v}} = \boldsymbol{N}\left(\vec{\Lambda}_i \wedge \vec{\boldsymbol{v}}\right) \tag{10.38}$$

where $\boldsymbol{N}$ is the number of lines which cut perpendicularly the surface S.
We obtain

$$\dot{\vec{\beta}}_i^{pl} = \frac{\boldsymbol{N}}{\boldsymbol{S}}\left(\vec{\Lambda}_i \wedge \vec{\boldsymbol{v}}\right) = \boldsymbol{\Lambda}\left(\vec{\Lambda}_i \wedge \vec{\boldsymbol{v}}\right) \tag{10.39}$$

wherein $\boldsymbol{\Lambda} = \boldsymbol{N} / \boldsymbol{S}$ is the density of *active dislocations*, that is to say the number of dislocation lines crossing perpendicularly the surface unit. It can be ensured that the dimension of the total derivative $\dot{\vec{\beta}}_i^{pl}$ corresponds to the inverse of a time (1/s), because the density $\boldsymbol{\Lambda}$ has dimension of the inverse of a surface (1/m^2) and the product $\vec{\Lambda}_i \wedge \vec{\boldsymbol{v}}$ has dimension (m^2/s). The quantity $\boldsymbol{\Lambda}$ can also be interpreted as *the length of lines of active dislocations per unit volume of lattice* since its size can also be written as (m/m^3).
By applying the same treatment to all relations *(10.36)*, we obtain the following equations set, in which we can introduce the linear flow of charges $\vec{\Upsilon}_i$, $\vec{\Upsilon}$ and Υ associated with the movement of the dislocation lines

$$\begin{cases}
\dot{\vec{\beta}}_i^{pl} = \boldsymbol{\Lambda}\left[\vec{\Lambda}_i \wedge \vec{\boldsymbol{v}}\right] = \boldsymbol{\Lambda}\vec{\Upsilon}_i \\
\dot{\vec{\omega}}^{pl} = \boldsymbol{\Lambda}\left[\Lambda_{screw}\vec{\boldsymbol{v}} + \dfrac{1}{2}\left(\vec{\Lambda}_{edge} \wedge \vec{\boldsymbol{v}}\right)\right] = \boldsymbol{\Lambda}\vec{\Upsilon} \\
\dot{\vec{\varepsilon}}_i^{pl} = \boldsymbol{\Lambda}\left[\vec{\Lambda}_i \wedge \vec{\boldsymbol{v}} + \left(\vec{\boldsymbol{v}} \wedge \vec{e}_i\right)\Lambda_{screw} + \dfrac{1}{2}\vec{e}_i \wedge \left(\vec{\boldsymbol{v}} \wedge \vec{\Lambda}_{edge}\right)\right] = \boldsymbol{\Lambda}\left(\vec{\Upsilon}_i - \vec{e}_i \wedge \vec{\Upsilon}\right) \\
\dot{\tau}^{pl} = \boldsymbol{\Lambda}\left[-\vec{\boldsymbol{v}}\vec{\Lambda}_{edge}\right] = \boldsymbol{\Lambda}\Upsilon \\
\dot{\vec{\alpha}}_i^{pl} = \boldsymbol{\Lambda}\left[\vec{\Lambda}_i \wedge \vec{\boldsymbol{v}} + \left(\vec{\boldsymbol{v}} \wedge \vec{e}_i\right)\Lambda_{screw} + \dfrac{1}{2}\vec{e}_i \wedge \left(\vec{\boldsymbol{v}} \wedge \vec{\Lambda}_{edge}\right) + \dfrac{1}{3}\vec{e}_i\left(\vec{\boldsymbol{v}}\vec{\Lambda}_{edge}\right)\right] = \boldsymbol{\Lambda}\left(\vec{\Upsilon}_i - \vec{e}_i \wedge \vec{\Upsilon} - \vec{e}_i \Upsilon / 3\right)
\end{cases} \tag{10.40}$$

These relationships between the macroscopic deformation velocity and the dislocation movement velocities at the microscopic level are usually called *relations of Orowan*[1]. We can also express these relationships of Orowan as functions of the Burgers vectors of the dislocations instead of the linear charge densities of the dislocation lines

$$\begin{cases}
\dot{\vec{\beta}}_i^{pl} = \boldsymbol{\Lambda}\left[\left(\vec{\boldsymbol{v}} \wedge \vec{\boldsymbol{t}}\right)\left(\vec{\boldsymbol{B}}\vec{e}_i\right)\right] \\
\dot{\vec{\omega}}^{pl} = \boldsymbol{\Lambda}\left[-\dfrac{1}{2}\left(\vec{\boldsymbol{B}}_{vis}\vec{\boldsymbol{t}}\right)\vec{\boldsymbol{v}} + \dfrac{1}{2}\left(\vec{\boldsymbol{v}}\vec{\boldsymbol{B}}_{coin}\right)\vec{\boldsymbol{t}}\right] \\
\dot{\vec{\varepsilon}}_i^{pl} = \boldsymbol{\Lambda}\left[\left(\vec{\boldsymbol{v}} \wedge \vec{\boldsymbol{t}}\right)\left(\vec{\boldsymbol{B}}\vec{e}_i\right) + \dfrac{1}{2}\left(\vec{e}_i \wedge \vec{\boldsymbol{v}}\right)\left(\vec{\boldsymbol{B}}_{vis}\vec{\boldsymbol{t}}\right) - \dfrac{1}{2}\left(\vec{e}_i \wedge \vec{\boldsymbol{t}}\right)\left(\vec{\boldsymbol{v}}\vec{\boldsymbol{B}}_{coin}\right)\right] \\
\dot{\tau}^{pl} = \boldsymbol{\Lambda}\left[\left(\vec{\boldsymbol{t}} \wedge \vec{\boldsymbol{B}}_{coin}\right)\vec{\boldsymbol{v}}\right] \\
\dot{\vec{\alpha}}_i^{pl} = \boldsymbol{\Lambda}\left[\left(\vec{\boldsymbol{v}} \wedge \vec{\boldsymbol{t}}\right)\left(\vec{\boldsymbol{B}}\vec{e}_i\right) + \dfrac{1}{2}\left(\vec{e}_i \wedge \vec{\boldsymbol{v}}\right)\left(\vec{\boldsymbol{B}}_{vis}\vec{\boldsymbol{t}}\right) - \dfrac{1}{2}\left(\vec{e}_i \wedge \vec{\boldsymbol{t}}\right)\left(\vec{\boldsymbol{v}}\vec{\boldsymbol{B}}_{coin}\right) - \dfrac{1}{3}\vec{e}_i\left[\left(\vec{\boldsymbol{t}} \wedge \vec{\boldsymbol{B}}_{coin}\right)\vec{\boldsymbol{v}}\right]\right]
\end{cases} \tag{10.41}$$

---

[1] *E. Orowan, Z. Phys., vol. 89, p. 605, 614, 634, 1934*

Chapter 11

Evolution equations of a charged lattice and the force of Peach and Koehler

The tensor of density $\vec{\lambda}_i$ and the tensor of flow $\vec{J}_i$ of dislocation charges, defined in the previous chapters, allow us to find in this chapter all the fundamental and phenomenological equations of spatiotemporal evolution to be met by a lattice presenting elastic, anelastic and self-diffusive properties and containing dislocation charges. Furthermore, from the energy balance equation of charged solids, it is possible to find the forces acting on the dislocation charges in the presence of non-zero stress fields. These forces are called *forces of Peach and Koehler.*

11.1 - Replacement of the tensor of plastic distortion

To write the spatiotemporal evolution equations of an elastic and anelastic solid containing dislocation charges, we will simply consider all evolution equations obtained in chapter 6, and reported in table 6.1, and we will replace in them all the terms showing the tensor of plastic distortions $\vec{\beta}_i^{pl}$ by the concepts of density and flow of dislocation charges that we developed in the previous chapters. In other words, we perform the following systematic replacements of the plastic distortion tensors

$$\begin{cases} \vec{\beta}_i \to \vec{\beta}_i + \vec{\beta}_i^{pl} \\ d\vec{\beta}_i^{pl} / dt \to \vec{J}_i \\ \overrightarrow{\text{rot}}\,\vec{\beta}_i^{pl} \to -\vec{\lambda}_i \end{cases} \qquad \begin{cases} \vec{\omega} \to \vec{\omega} + \vec{\omega}^{pl} \\ d\vec{\omega}^{pl} / dt \to \vec{J} \\ \text{div}\,\vec{\omega}^{pl} \to -\vec{\lambda} \end{cases} \qquad \begin{cases} \tau \to \tau + \tau^{pl} \\ d\tau^{pl} / dt \to S_n / n \end{cases} \tag{11.1}$$

In reality, the number of transformations to be performed is relatively low. Indeed, we find that only the topological equations, the heat equations and the energy balance equation are affected by these replacements. While the transformation of topological equations are immediate since we established them in detail in the previous chapter, we need to discuss in more detail in this chapter the changes in the thermal equations and the energy balance equation.

The equation of heat and the source of entropy

Considering again equations *(6.23)* and *(6.30)*, and doing all the necessary changes, we have

$$\begin{aligned} nT\frac{ds}{dt} &= -\left(\mu_L^* + \mu_I^*\right)S_{I-L} - \left(\mu_L^* + h^*\right)S_L^{pl} - \left(\mu_I^* - h^*\right)S_I^{pl} \\ &+\left(-\,\overrightarrow{\text{grad}}\mu_L^* + m\frac{d}{dt}\left(\vec{\phi}_L - 2\Delta\vec{\varphi}_L\right) - m\vec{g}\right)\vec{J}_L + \left(-\,\overrightarrow{\text{grad}}\mu_I^* - m\frac{d}{dt}\left(\vec{\phi}_I\right) + m\vec{g}\right)\vec{J}_I \\ &+\vec{s}_k^{dis}\frac{d\vec{\beta}_k^{an}}{dt} + \vec{m}^{dis}\frac{d\vec{\omega}^{an}}{dt} + \vec{s}_k\vec{J}_k + \vec{m}\vec{J} - \text{div}\,\vec{J}_q \end{aligned} \tag{11.2}$$

$$S_e = -\frac{1}{T}\left(\mu_L^* + \mu_I^*\right)S_{I-L} - \frac{1}{T}\left(\mu_L^* + g^*\right)S_L^{pl} - \frac{1}{T}\left(\mu_I^* - g^*\right)S_I^{pl}$$
$$+\frac{\vec{J}_L}{T}\left(-\overrightarrow{\text{grad}}\mu_L^* + m\frac{d}{dt}\left(\vec{\phi}_L - 2\Delta\vec{\varphi}_L\right) - m\vec{g}\right) + \frac{\vec{J}_I}{T}\left(-\overrightarrow{\text{grad}}\mu_I^* - m\frac{d}{dt}\left(\vec{\phi}_I\right) + m\vec{g}\right) \quad (11.3)$$
$$+\frac{1}{T}\left(\vec{s}_k^{dis}\frac{d\vec{\beta}_k^{an}}{dt} + \vec{m}^{dis}\frac{d\vec{\omega}^{an}}{dt} + \vec{s}_k\vec{J}_k + \vec{m}\vec{J}\right) + \vec{J}_q\overrightarrow{\text{grad}}\left(\frac{1}{T}\right)$$

The dissipative equations of plasticity

The terms of source of entropy associated with plastic deformation generated by the movement of the plastic charges in equation *(11.3)* are the following

$$S_e^{pl} = -\frac{1}{T}\left(\mu_L^* + g^*\right)S_L^{pl} - \frac{1}{T}\left(\mu_I^* - g^*\right)S_I^{pl} + \frac{1}{T}\left(\vec{s}_k\vec{J}_k + \vec{m}\vec{J}\right) \quad (11.4)$$

From these sources of entropy, it is possible to define the dissipative equations of plasticity that characterize the tensorial flow of dislocation charges. The first of these concerns the trace of $\vec{J}_k$, which is associated with the source of creation or annihilation of lattice sites $S_n / n = -\vec{\lambda}\vec{\boldsymbol{v}}$ via *the phenomenon of climbing of edge dislocations* with velocity $\vec{\boldsymbol{v}}$. This climbing is controlled by the source of vacancies and interstitials, and depends essentially on creation-annihilation energies $\mu_L^* + g^*$ and $\mu_I^* - g^*$, but certainly also on the local pressure within the medium and the local dislocation density charge $\vec{\lambda}$ of edge dislocation which are able to leave a *"trace"* of their passage

$$\begin{cases} \dfrac{S_n}{n} = \sum_k \vec{e}_k\vec{J}_k = -\vec{\lambda}\vec{\boldsymbol{v}} = \dfrac{1}{n}\left(S_L^{pl} - S_I^{pl}\right) \\ S_L^{pl} = S_L^{pl}\left[\left(\mu_L^* + g^*\right), p, \vec{\lambda}, v, T, C_L, C_I, ...\right] \\ S_I^{pl} = S_I^{pl}\left[\left(\mu_I^* - g^*\right), p, \vec{\lambda}, v, T, C_L, C_I, ...\right] \end{cases} \quad (11.5)$$

The other two relationships are deduced from the product of charge flows $\vec{J}_k$ and $\vec{J}$ and thermodynamic forces $\vec{s}_k / T$ and $\vec{m} / T$. These products show that the tensorial flow of charges $\vec{J}_k$ must depend essentially on the tensor of shear stress $\vec{s}_i$, while the vector flow of charges $\vec{J}$ should depend on the torsor of moment $\vec{m}$, so that

$$\begin{cases} \vec{J}_k = \vec{\lambda}_k \wedge \vec{\boldsymbol{v}} = \vec{J}_k\left(\vec{s}_m, \vec{\lambda}_n, v, T, ...\right) \\ \vec{J} = -\dfrac{1}{2}\sum_k \vec{e}_k \wedge \vec{J}_k = \lambda\vec{\boldsymbol{v}} + \dfrac{1}{2}\left(\vec{\lambda} \wedge \vec{\boldsymbol{v}}\right) = \vec{J}\left(\vec{m}, \lambda, v, T, ...\right) \end{cases} \quad (11.6)$$

These two relationships are the *dynamic equations of motion of charges* $\vec{\lambda}_k$ and λ within the medium. Indeed, we will see in section 11.2 that forces dependent on $\vec{s}_k$ and $\vec{m}$ apply to the charges of dislocation $\vec{\lambda}_k$ and λ. In this way, the associated flows $\vec{J}_k = \vec{\lambda}_k \wedge \vec{\boldsymbol{v}}$ and $J = \lambda\vec{\boldsymbol{v}} + \left(\vec{\lambda} \wedge \vec{\boldsymbol{v}}\right) / 2$ depend on $\vec{s}_k$ and $\vec{m}$ via a dynamic equation which should allow us to calculate the velocity $\vec{\boldsymbol{v}}$ of the charges by knowing the forces that depend on $\vec{s}_k$ and $\vec{m}$. We will return later to the problem of the dynamic equation of dislocation movement, which will lead us to *the string model of the dislocation.*

The dissipative equations of sources of plastic charges

By taking the continuity equations for dislocation charges of table 10.1, there appears the sources of charges $\vec{S}_i^{\vec{\lambda}}$, $\vec{S}^{(\vec{\lambda})}$ and S^{λ}, which are partially responsible for the non-commutative nature of the operators of time and space on distortion tensors $\vec{\beta}_i$ and $\vec{\omega}$. The expression of these sources $\vec{S}_i^{\vec{\lambda}}$, $\vec{S}^{(\vec{\lambda})}$ and S^{λ} will certainly depend on the variables $\vec{s}_m, \vec{\lambda}_n, v, T, \ldots$, but can be obtained through the description of the dynamics of dislocation charges at the level of the lattice itself. We cannot fully understand the nature of this until we develop the *string model of dislocation*, the dynamic equation of motion of the dislocations, which will allow us to describe the existence of dislocation sources (see section 12.9).

The equation of energy balance

The main change to the energy balance equation is the emergence of new dissipative terms associated with the flows $\vec{J}_i$ and $\vec{J}$, as well as the source of lattice sites S_n

$$n\vec{\phi}\left(\frac{d\vec{p}}{dt}-m\vec{\phi}_I\frac{dC_I}{dt}+m\vec{\phi}_L\frac{dC_L}{dt}\right)+\left(\vec{s}_k\frac{d\vec{\beta}_k}{dt}+\vec{m}\frac{d\vec{\omega}}{dt}-p\frac{d\tau}{dt}\right)$$
$$=-\left(\vec{s}_k\vec{J}_k+\vec{m}\vec{J}-p\frac{S_n}{n}\right)+\rho\vec{g}\vec{\phi}-\operatorname{div}\left[-\phi_k\vec{s}_k-\frac{1}{2}\left(\vec{\phi}\wedge\vec{m}\right)+p\vec{\phi}\right] \tag{11.7}$$

These terms $\vec{s}_k\vec{J}_k+\vec{m}\vec{J}-pS_n/n$ will allow us to find, in the following section, the forces acting on the dislocation charges in the presence of non-zero stress fields.

11.2 - Force of Peach and Koehler acting on the dislocation charges

In the energy balance equations, the power $\boldsymbol{P}_{ch}$ supplied to the dislocation charges by the stress fields corresponds to the balance of the term containing the flow of charges

$$\boldsymbol{P}_{ch}=\vec{s}_k\vec{J}_k+\vec{m}\vec{J}-p\frac{S_n}{n} \tag{11.8}$$

In this term of power, it is possible to replace flows $\vec{J}_k$, $\vec{J}$ and S_n/n by their expression as a function of the velocity $\vec{\boldsymbol{v}}$ of the charges, from table 10.1. One obtains an expression of $\boldsymbol{P}_{ch}$ containing the charge densities $\vec{\lambda}_k$, $\vec{\lambda}$ and λ

$$P_{ch}=\sum_k\vec{s}_k\left(\vec{\lambda}_k\wedge\vec{\boldsymbol{v}}\right)+\vec{m}\lambda\vec{\boldsymbol{v}}+\frac{1}{2}\vec{m}\left(\vec{\lambda}\wedge\vec{\boldsymbol{v}}\right)+p\vec{\lambda}\vec{\boldsymbol{v}}=\left[\sum_k\left(\vec{s}_k\wedge\vec{\lambda}_k\right)+\lambda\vec{m}+\frac{1}{2}\left(\vec{m}\wedge\vec{\lambda}\right)+\vec{\lambda}p\right]\vec{\boldsymbol{v}} \tag{11.9}$$

The power $\boldsymbol{P}_{ch}$ given to the charges is thus the product of a velocity by a term that is a force $\vec{f}_{PK}$ acting by unit of volume on the charge densities $\vec{\lambda}_k$, $\vec{\lambda}$, λ

$$\boldsymbol{P}_{ch}=\vec{f}_{PK}\vec{\boldsymbol{v}}\quad\Rightarrow\quad\vec{f}_{PK}=\sum_k\left(\vec{s}_k\wedge\vec{\lambda}_k\right)+\lambda\vec{m}+\frac{1}{2}\left(\vec{m}\wedge\vec{\lambda}\right)+\vec{\lambda}p+\vec{\boldsymbol{v}}\wedge\vec{\boldsymbol{A}} \tag{11.10}$$

This force depends on the stress tensors $\vec{s}_k$, $\vec{m}$ and/or p, and is usually called the *force of Peach and Koehler*. The use of the stress tensors $\vec{s}_k$, $\vec{m}$ and/or p depends of course, as we saw in chapter 5, on the choice of how we write the free energy state function of the solid medium considered.

As the dimension is an inverse of a length (1/m) and the dimension of the torsor of moments is a moment per unit volume, namely (Nm/ m³), the force $\vec{f}_{PK}$ has the dimension of a force per unit volume (N/m³).
The last term containing a vector $\vec{\boldsymbol{A}}$ is added here because it is a term for a possible force which would not produce work, so that it does not appear in the expression *(11.9)* of power. We will also later see that this term actually corresponds to a relativistic force similar to the *Lorentz force* in electromagnetism.

The force of Peach and Kohler acting on a dislocation line

The force of Peach and Koehler can also be written in the case of a dislocation, integrating equation *(11.10)* on the surface of a section of the string. It then comes, neglecting the term containing the vector $\vec{\boldsymbol{A}}$

$$\vec{F}_{PK} = \iint_S \vec{f}_{PK}\, dS = \iint_S \left[\sum_k \left(\vec{s}_k \wedge \vec{\lambda}_k \right) + \lambda \vec{m} + \frac{1}{2}\left(\vec{m} \wedge \vec{\lambda} \right) + \vec{\lambda} p \right] dS \tag{11.11}$$

The integral of $\vec{\lambda}_k$, $\vec{\lambda}$, λ on the section of a dislocation string gives linear densities, so that the force of Peach and Kohler acting on the dislocation line is given by

$$\vec{F}_{PK} = \sum_k \left(\vec{s}_k \wedge \vec{\Lambda}_k \right) + \Lambda \vec{m} + \frac{1}{2}\left(\vec{m} \wedge \vec{\Lambda} \right) + \vec{\Lambda} p \tag{11.12}$$

The dimension of the force $\vec{F}_{PK}$ acting on the dislocation is a force per unit length (N/m). This is in fact *the force per unit length of the string* in the presence of the stress fields $\vec{s}_k, \vec{m}, p$. Writing the force of Peach and Koehler using the constraint tensors $\vec{s}_k, \vec{m}, p$ is in fact very interesting because it gives a much clearer picture than by using the notation $\vec{F}_{PK} = \sum \left(\vec{\sigma}_k \wedge \vec{\Lambda}_k \right)$ usually used with the symmetric stress tensor $\vec{\sigma}_k$. Indeed, imagine a solid in which the volume expansion is zero, so that one would have a negligible pressure p. In this case, it is already known that we can replace the shear tensor $\vec{\alpha}_k$ by the vector of rotation $\vec{\omega}$, so that the force becomes a *sliding force*, which is written $\vec{F}_{PK} = \Lambda \vec{m} + \left(\vec{m} \wedge \vec{\Lambda} \right)/2$, wherein the term $\Lambda \vec{m}$ is the force acting on the screw component Λ of the dislocation and the term $\left(\vec{m} \wedge \vec{\Lambda} \right)/2$ is the force acting on the edge component $\vec{\Lambda}$ of the dislocation. As the component m_k of torque is associated with the shear components σ_{ij} and σ_{ji} of the stress tensor, it can give a very clear representation of the forces acting on a dislocation. The same here goes for the pressure force $\vec{\Lambda} p$, which acts on the edge component $\vec{\Lambda}$ of dislocations and which matches, given its direction (figure 10.3b), a *climbing force for the dislocations.*
We can again use relations *(10.31)* to express the Peach and Koehler force directly from the Burgers vector of the dislocation

$$\vec{F}_{PK} = \sum_k \left(\vec{t} \wedge \vec{s}_k \right)\left(\vec{B} \vec{e}_k \right) - \frac{1}{2}\left(\vec{B} \vec{t} \right) \vec{m} - \frac{1}{2}\left[\left(\vec{B} \wedge \vec{t} \right) \wedge \vec{m} \right] + \left(\vec{B} \wedge \vec{t} \right) p \tag{11.13}$$

The Peach and Kohler force acting on a localized charge of rotation

In the case of a localized charge of rotation Q_λ, we obtain the force of Peach and Koehler by integrating equation *(11.10)* on the volume of the localized charge. It gives

$$\vec{F}_{PK} = \iiint_V \vec{f}_{PK}\, dV = \vec{m} \iiint_V \lambda\, dV = Q_\lambda \vec{m} \tag{11.14}$$

In this case, the dimension of $\vec{F}_{PK}$ is that of a pure force, expressed in (N), and which is analogous to the electrical force acting on a localized electrical charge ($\vec{F}_{electrical} = q\vec{E}$).
The possible expressions of the strength of Peach and Koehler are shown in Table 11.1.

Table 11.1 - Forces of Peach et Koehler

Volume force on volume density of dislocation charges

$$\vec{f}_{PK} = \sum_k \left(\vec{s}_k \wedge \vec{\lambda}_k\right) + \lambda\vec{m} + \frac{1}{2}\left(\vec{m} \wedge \vec{\lambda}\right) + \vec{\lambda}p + \vec{\boldsymbol{v}} \wedge \vec{\boldsymbol{A}} \qquad (1)$$

Linear forces on a line of dislocation

$$\vec{F}_{PK} = \sum_k \left(\vec{s}_k \wedge \vec{\Lambda}_k\right) + \Lambda\vec{m} + \frac{1}{2}\left(\vec{m} \wedge \vec{\Lambda}\right) + \vec{\Lambda}p \qquad (2)$$

$$\vec{F}_{PK} = \sum_k \left(\vec{t} \wedge \vec{s}_k\right)\left(\vec{B}\vec{e}_k\right) - \frac{1}{2}\left(\vec{B}\vec{t}\right)\vec{m} - \frac{1}{2}\left[\left(\vec{B} \wedge \vec{t}\right) \wedge \vec{m}\right] + \left(\vec{B} \wedge \vec{t}\right)p \qquad (3)$$

Global force on a localized charge of rotation

$$\vec{F}_{PK} = \iiint_V \vec{f}_{PK}\, dV = \vec{m}\iiint_V \lambda\, dV = Q_\lambda \vec{m} \qquad (4)$$

11.3 - Equations of spatio-temporal evolution in charged solids

It is now easy to combine all of the results obtained in this chapter with the equations of tables 6.1 and 10.1 to obtain the complete equations of spatio-temporal evolution of a solid self-diffusive lattice, with phenomenological behaviors of elasticity and anelasticity, and containing densities and flows of dislocation charges. As shown in Table 11.2, this system of equations is quite complex, especially in the high number of phenomenological equations of state and phenomenological equations of dissipation required for a complete description of all possible phenomena in such an environment .
We can already imagine that the new concepts of density and flow of dislocation charges are expected to describe the plasticity phenomena and the anelasticity at the microscopic scale in a discrete solid lattice by introducing a scalable microstructure of plastics charges which should help us to encode the non-Markovian behavior of plasticity. Moreover, the approach by plastic charges of a microscopically discrete solid lattice should also permit, in principle, to find exact local expressions for the dissipative equations *(11.5)* to *(11.9)* associated with the plastic charges.

Table 11.2 - Fundamental equations of evolution of self-diffusive solids with elasticity, anelasticity and charges of dislocations

Topological equations

$$\vec{J}_i = -\frac{d\vec{\beta}_i}{dt} + \overrightarrow{\text{grad}}\,\phi_i \quad (1)$$

$$\vec{J} = -\frac{1}{2}\sum_k \vec{e}_k \wedge \vec{J}_k = -\frac{d\vec{\omega}}{dt} + \frac{1}{2}\overrightarrow{\text{rot}}\,\vec{\phi} \quad (2)$$

$$\frac{S_n}{n} = \sum_k \vec{e}_k \vec{J}_k = -\frac{d\tau}{dt} + \text{div}\,\vec{\phi} \quad (3)$$

$$\vec{\lambda}_i = \overrightarrow{\text{rot}}\,\vec{\beta}_i \qquad avec \qquad \text{div}\,\vec{\lambda}_i = 0 \quad (4)$$

$$\vec{\lambda} = -\sum_k \vec{e}_k \wedge \vec{\lambda}_k = -\sum_k \vec{e}_k \wedge \overrightarrow{\text{rot}}\vec{\beta}_k \quad (5)$$

$$\lambda = \frac{1}{2}\sum_k \vec{e}_k \vec{\lambda}_k = \text{div}\,\vec{\omega} \quad (6)$$

$$\vec{\beta}_i = \vec{\beta}_i^{(\delta)} + \vec{e}_i \wedge \vec{\omega}_O(t) = \vec{\beta}_i^{él} + \vec{\beta}_i^{an} + \vec{e}_i \wedge \vec{\omega}_O(t) \quad (7)$$

$$\vec{\omega} = -\frac{1}{2}\sum_k \vec{e}_k \wedge \vec{\beta}_k = \vec{\omega}^{(\delta)} + \vec{\omega}_0(t) = \vec{\omega}^{él} + \vec{\omega}^{an} + \vec{\omega}_0(t) \quad (8)$$

$$\tau = \sum_k \vec{\beta}_k \vec{e}_k = \sum_k \vec{\beta}_k^{(\delta)} \vec{e}_k = \tau^{él} \quad (\tau^{an} \equiv 0 \text{ par hypothèse}) \quad (9)$$

$$\vec{\varepsilon}_i = \vec{\beta}_i - \vec{e}_i \wedge \vec{\omega} = \vec{\beta}_i^{(\delta)} - \vec{e}_i \wedge \vec{\omega}^{(\delta)} = \vec{\varepsilon}_i^{él} + \vec{\varepsilon}_i^{an} \quad (10)$$

$$\vec{\alpha}_i = \vec{\varepsilon}_i - \frac{1}{3}\tau \vec{e}_i = \vec{\alpha}_i^{él} + \vec{\alpha}_i^{an} \quad (11)$$

$$d/dt = \partial/\partial t + (\vec{\varphi}\vec{\nabla}) \quad (12)$$

$$\vec{\varphi} = \vec{\phi} - \vec{\phi}_O(t) - \dot{\vec{\omega}}_O(t) \wedge \vec{r} \quad (13)$$

Dynamic equations

$$n\frac{d\vec{p}}{dt} = \rho\vec{g} + \sum_k \vec{e}_k\,\text{div}\,\vec{s}_k - \frac{1}{2}\overrightarrow{\text{rot}}\,\vec{m} - \overrightarrow{\text{grad}}p + nm\vec{\phi}_I\frac{dC_I}{dt} - nm\vec{\phi}_L\frac{dC_L}{dt} \quad (14)$$

$$n = 1/v = n_0 e^{-\tau} \quad (15)$$

$$\vec{p} = m\left(\vec{\phi} + C_I\vec{\phi}_I - C_L\vec{\phi}_L\right) = m\vec{\phi} + m\left(C_I - C_L\right)\vec{\phi} + \frac{m}{n}\left(\vec{J}_I - \vec{J}_L\right) = \frac{1}{n}\left[\rho\vec{\phi} + m\left(\vec{J}_I - \vec{J}_L\right)\right] \quad (16)$$

$$\rho = mn\left(1 + C_I - C_L\right) \quad (17)$$

Diffusion equations

$$n\frac{dC_L}{dt} = \left(S_{I-L} + S_L^{pl}\right) - C_L\left(S_L^{pl} - S_I^{pl}\right) - \text{div}\,\vec{J}_L \quad (18)$$

$$n\frac{dC_I}{dt} = \left(S_{I-L} + S_I^{pl}\right) - C_I\left(S_L^{pl} - S_I^{pl}\right) - \text{div}\,\vec{J}_I \quad (19)$$

$$\vec{J}_L = nC_L\Delta\vec{\varphi}_L = nC_L\left(\vec{\phi}_L - \vec{\phi}\right) = nC_L\left(\vec{\varphi}_L - \vec{\varphi}\right) \quad (20)$$

$$\vec{J}_I = nC_I\Delta\vec{\varphi}_I = nC_I\left(\vec{\phi}_I - \vec{\phi}\right) = nC_I\left(\vec{\varphi}_I - \vec{\varphi}\right) \quad (21)$$

Thermal equations

$$nT\frac{ds}{dt} = -\left(\mu_L^* + \mu_I^*\right)S_{I-L} - \left(\mu_L^* + h^*\right)S_L^{pl} - \left(\mu_I^* - h^*\right)S_I^{pl} + T\vec{J}_L\vec{X}_L$$
$$+T\vec{J}_I\vec{X}_I + \vec{s}_k^{dis}\frac{d\vec{\beta}_k^{an}}{dt} + \vec{m}^{dis}\frac{d\vec{\omega}^{an}}{dt} + \vec{s}_k\vec{J}_k + \vec{m}\vec{J} - \operatorname{div}\vec{J}_q \qquad (22)$$

$$\begin{cases} \mu_L^* = \mu_L - \dfrac{1}{2}m\left(\vec{\phi}_L^2 - 2\Delta\vec{\varphi}_L^2\right) & (23) \\ \mu_I^* = \mu_I + \dfrac{1}{2}m\vec{\phi}_I^2 & (24) \end{cases}$$

$$\begin{cases} \vec{X}_q = \overrightarrow{\operatorname{grad}}\dfrac{1}{T} & (25) \\ \vec{X}_L = \dfrac{1}{T}\left(-\overrightarrow{\operatorname{grad}}\ \mu_L^* + m\dfrac{d}{dt}\left(\vec{\phi}_L - 2\Delta\vec{\varphi}_L\right) - m\vec{g}\right) & (26) \\ \vec{X}_I = \dfrac{1}{T}\left(-\overrightarrow{\operatorname{grad}}\ \mu_I^* - m\dfrac{d}{dt}\left(\vec{\phi}_I\right) + m\vec{g}\right) & (27) \end{cases}$$

$$h^* = f + Ts + pv + \frac{1}{2}m\vec{\phi}^2 - \mu_L C_L - \mu_I C_I \qquad (28)$$

Charge equations

$$\begin{cases} \dfrac{d\vec{\lambda}_i}{dt} \cong \vec{S}_i^{(\vec{\lambda}_i)} - \left(\boldsymbol{\vec{v}}\vec{\nabla}\right)\vec{\lambda}_i = \vec{S}_i^{(\vec{\lambda}_i)} + \overrightarrow{\operatorname{rot}}\left(\boldsymbol{\vec{v}} \wedge \vec{\lambda}_i\right) = \vec{S}_i^{(\vec{\lambda}_i)} - \overrightarrow{\operatorname{rot}}\,\vec{J}_i & (29) \\ \dfrac{d\vec{\lambda}}{dt} \cong \vec{S}^{(\vec{\lambda})} - \left(\boldsymbol{\vec{v}}\vec{\nabla}\right)\vec{\lambda} = \vec{S}^{(\vec{\lambda})} + \overrightarrow{\operatorname{rot}}\left(\boldsymbol{\vec{v}} \wedge \vec{\lambda}\right) - \boldsymbol{\vec{v}}\operatorname{div}\vec{\lambda} = \vec{S}^{(\vec{\lambda})} - 2\ \overrightarrow{\operatorname{rot}}\,\vec{J}^{(\vec{\lambda})} - \boldsymbol{\vec{v}}\theta & (30) \\ \dfrac{d\lambda}{dt} \cong S^{(\lambda)} - \left(\boldsymbol{\vec{v}}\vec{\nabla}\right)\lambda = S^{(\lambda)} - \operatorname{div}\left(\lambda\boldsymbol{\vec{v}}\right) = S^{(\lambda)} - \operatorname{div}\vec{J}^{(\lambda)} & (31) \end{cases}$$

$$\begin{cases} \vec{J}_i = \vec{\lambda}_i \wedge \boldsymbol{\vec{v}} & (32) \\ \vec{J} = \vec{J}^{(\lambda)} + \vec{J}^{(\vec{\lambda})} = \lambda\boldsymbol{\vec{v}} + \left(\vec{\lambda} \wedge \boldsymbol{\vec{v}}\right)/2 & (33) \\ S_n / n = -\vec{\lambda}\boldsymbol{\vec{v}} & (34) \end{cases}$$

$$\begin{cases} \vec{S}^{(\vec{\lambda})} = -\displaystyle\sum_k \vec{e}_k \wedge \vec{S}_k^{(\vec{\lambda}_k)} & (35) \\ S^{(\lambda)} = \dfrac{1}{2}\displaystyle\sum_k \vec{e}_k\vec{S}_k^{(\vec{\lambda}_k)} & (36) \end{cases}$$

$$\vec{f}_{PK} = \sum_k\left(\vec{s}_k \wedge \vec{\lambda}_k\right) + \lambda\vec{m} + \frac{1}{2}\left(\vec{m} \wedge \vec{\lambda}\right) + \vec{\lambda}p + \boldsymbol{\vec{v}} \wedge \boldsymbol{\vec{A}} \qquad (37)$$

Phenomenological equations of evolution of self-diffusive solids with elasticity, anelasticity and charges of dislocations

Functions and equations of state

$$f = f\left(\alpha_{ij}^{\acute{e}l}, \alpha_{ij}^{an}, \omega_k^{\acute{e}l}, \omega_k^{an}, \tau^{\acute{e}l}, C_L, C_I, T\right) \quad (38)$$

$$\begin{cases} s_{ij} = \dfrac{n}{2}\left(\dfrac{\partial f}{\partial \alpha_{ij}^{\acute{e}l}} + \dfrac{\partial f}{\partial \alpha_{ji}^{\acute{e}l}}\right) - \dfrac{n}{3}\delta_{ij}\sum_k \dfrac{\partial f}{\partial \alpha_{kk}^{\acute{e}l}} = s_{ij}(\alpha_{lm}^{\acute{e}l}, \omega_n^{\acute{e}l}, \tau^{\acute{e}l}, \alpha_{lm}^{an}, \omega_n^{an}, C_L, C_I, T) & (39) \\ m_k = n\dfrac{\partial f}{\partial \omega_k^{\acute{e}l}} = m_k(\alpha_{lm}^{\acute{e}l}, \omega_n^{\acute{e}l}, \tau^{\acute{e}l}, \alpha_{lm}^{an}, \omega_n^{an}, C_L, C_I, T) & (40) \\ p = -n\dfrac{\partial f}{\partial \tau^{\acute{e}l}} = p(\alpha_{lm}^{\acute{e}l}, \omega_n^{\acute{e}l}, \tau^{\acute{e}l}, \alpha_{lm}^{an}, \omega_n^{an}, C_L, C_I, T) & (41) \end{cases}$$

$$\begin{cases} s_{ij}^{dis} = \dfrac{n}{2}\left(\dfrac{\partial f}{\partial \alpha_{ij}^{an}} + \dfrac{\partial f}{\partial \alpha_{ji}^{an}}\right) - \dfrac{n}{3}\delta_{ij}\sum_k \dfrac{\partial f}{\partial \alpha_{kk}^{an}} = s_{ij}^{dis}(\alpha_{lm}^{\acute{e}l}, \omega_n^{\acute{e}l}, \tau^{\acute{e}l}, \alpha_{lm}^{an}, \omega_n^{an}, C_L, C_I, T) & (42) \\ m_k^{dis} = n\dfrac{\partial f}{\partial \omega_k^{an}} = m_k^{dis}(\alpha_{lm}^{\acute{e}l}, \omega_n^{\acute{e}l}, \tau^{\acute{e}l}, \alpha_{lm}^{an}, \omega_n^{an}, C_L, C_I, T) & (43) \end{cases}$$

$$\begin{cases} s = -\dfrac{\partial f}{\partial T} = s(\alpha_{lm}^{\acute{e}l}, \omega_n^{\acute{e}l}, \tau^{\acute{e}l}, \alpha_{lm}^{an}, \omega_n^{an}, C_L, C_I, T) & (44) \\ \mu_L = \dfrac{\partial f}{\partial C_L} = \mu_L(\alpha_{lm}^{\acute{e}l}, \omega_n^{\acute{e}l}, \tau^{\acute{e}l}, \alpha_{lm}^{an}, \omega_n^{an}, C_L, C_I, T) & (45) \\ \mu_I = \dfrac{\partial f}{\partial C_I} = \mu_I(\alpha_{lm}^{\acute{e}l}, \omega_n^{\acute{e}l}, \tau^{\acute{e}l}, \alpha_{lm}^{an}, \omega_n^{an}, C_L, C_I, T) & (46) \end{cases}$$

Dissipation equations: self-diffusion and creation-annihilation of pairs

$$\begin{cases} \vec{J}_q = \vec{J}_q(\vec{X}_q, \vec{X}_L, \vec{X}_I, n, T, C_L, C_I, \ldots) & (47) \\ \vec{J}_L = \vec{J}_L(\vec{X}_q, \vec{X}_L, \vec{X}_I, n, T, C_L, C_I, \ldots) & (48) \\ \vec{J}_I = \vec{J}_I(\vec{X}_q, \vec{X}_L, \vec{X}_I, n, T, C_L, C_I, \ldots) & (49) \end{cases}$$

$$S_{I-L} = S_{I-L}\left(\mu_L^* + \mu_I^*, n, T, C_L, C_I, \ldots\right) \quad (50)$$

Dissipation equations: anelasticity

$$\begin{cases} \vec{s}_i = \vec{s}_i^{\,cons}\left(\vec{\alpha}_m^{an}, v, T, \ldots\right) + \vec{s}_i^{\,dis}\left(\dfrac{d\vec{\alpha}_m^{an}}{dt}, v, T, \ldots\right) & (51) \\ \vec{m} = \vec{m}^{cons}\left(\vec{\omega}^{an}, v, T, \ldots\right) + \vec{m}^{dis}\left(\dfrac{d\vec{\omega}^{an}}{dt}, v, T, \ldots\right) & (52) \end{cases}$$

Dissipation equations: flows of dislocation charges

$$\begin{cases} \vec{J}_k = \vec{\lambda}_k \wedge \vec{\boldsymbol{v}} = \vec{J}_k\left(\vec{s}_m, \vec{\lambda}_n, v, T, \ldots\right) & (53) \\ \vec{J} = -\frac{1}{2}\sum_k \vec{e}_k \wedge \vec{J}_k = \lambda\vec{\boldsymbol{v}} + \frac{1}{2}\left(\vec{\lambda} \wedge \vec{\boldsymbol{v}}\right) = \vec{J}\left(\vec{m}, \lambda, v, T, \ldots\right) & (54) \\ \frac{S_n}{n} = \sum_k \vec{e}_k \vec{J}_k = -\vec{\lambda}\vec{\boldsymbol{v}} = \frac{1}{n}\left(S_L^{pl} - S_I^{pl}\right) & (55) \end{cases}$$

$$\begin{cases} S_L^{pl} = S_L^{pl}\left[\left(\mu_L^* + g^*\right), p, \vec{\lambda}, v, T, C_L, C_I, \ldots\right] & (56) \\ S_I^{pl} = S_I^{pl}\left[\left(\mu_I^* - g^*\right), p, \vec{\lambda}, v, T, C_L, C_I, \ldots\right] & (57) \end{cases}$$

$$g^* = f + pv + m\vec{\phi}^2 / 2 - \mu_L C_L - \mu_I C_I \qquad (58)$$

Dissipation equations: sources of dislocation charges

$$\vec{S}_i^{\vec{\lambda}} = \vec{S}_i^{\vec{\lambda}}(\vec{s}_m, \vec{\lambda}_n, v, T, \ldots) \quad (59) \quad \Rightarrow \quad \begin{cases} \vec{S}^{(\lambda)} = -\sum_k \vec{e}_k \wedge \vec{S}_k^{(\vec{\lambda}_i)} = \vec{S}^{(\lambda)}(\vec{s}_m, \vec{\lambda}_n, v, T, \ldots) & (60) \\ S^{(\lambda)} = \frac{1}{2}\sum_k \vec{e}_k \vec{S}_k^{(\vec{\lambda}_i)} = S^{(\lambda)}(\vec{s}_m, \vec{\lambda}_n, v, T, \ldots) & (61) \end{cases}$$

Additional equations of evolution

Mass continuity

$$\frac{\partial \rho}{\partial t} = -\operatorname{div}\left[\rho\vec{\phi} + m\left(\vec{J}_I - \vec{J}_L\right)\right] = -\operatorname{div}\left(n\vec{p}\right) \quad \text{dans } Q\xi_1\xi_2\xi_3 \qquad (62)$$

Flux of work and surface forces

$$\vec{J}_w = \mu_L^*\vec{J}_L + \mu_I^*\vec{J}_I - \phi_k\vec{s}_k - \frac{1}{2}\left(\vec{\phi} \wedge \vec{m}\right) + p\vec{\phi} \qquad (63)$$

$$\vec{F}_S = \sum_k \vec{e}_k\left(\vec{s}_k\vec{\boldsymbol{n}}\right) + \frac{1}{2}\left(\vec{m} \wedge \vec{\boldsymbol{n}}\right) - \vec{\boldsymbol{n}}p \qquad (64)$$

Entropy source

$$\begin{aligned} S_e = & -\frac{1}{T}\left(\mu_L^* + \mu_I^*\right)S_{I-L} - \frac{1}{T}\left(\mu_L^* + g^*\right)S_L^{pl} - \frac{1}{T}\left(\mu_I^* - g^*\right)S_I^{pl} + \vec{J}_L\vec{X}_L \\ & + \vec{J}_I\vec{X}_I + \frac{1}{T}\left(\vec{s}_k^{dis}\frac{d\vec{\beta}_k^{an}}{dt} + \vec{m}^{dis}\frac{d\vec{\omega}^{an}}{dt} + \vec{s}_k\vec{J}_k + \vec{m}\vec{J}\right) + \vec{J}_q\overrightarrow{\operatorname{grad}}\left(\frac{1}{T}\right) \end{aligned} \qquad (65)$$

Energy balance

$$\begin{aligned} & n\vec{\phi}\left(\frac{d\vec{p}}{dt} - m\vec{\phi}_I\frac{dC_I}{dt} + m\vec{\phi}_L\frac{dC_L}{dt}\right) + \left(\vec{s}_k\frac{d\vec{\beta}_k}{dt} + \vec{m}\frac{d\vec{\omega}}{dt} - p\frac{d\tau}{dt}\right) \\ & + \left(\vec{s}_k\vec{J}_k + \vec{m}\vec{J} - p\frac{S_n}{n}\right) = \rho\vec{g}\vec{\phi} - \operatorname{div}\left[-\phi_k\vec{s}_k - \frac{1}{2}\left(\vec{\phi} \wedge \vec{m}\right) + p\vec{\phi}\right] \end{aligned} \qquad (66)$$

PART D

Application: elements of the dislocation theory in the usual solids

Maxwell's equations at constant volume expansion

Fields and energies of the screw and edge dislocations

Interactions between dislocations

String model of dislocation

Chapter 12

Elements of dislocation theory in usual solids

In this chapter, we introduce the simplest solid that can be considered, namely the perfect isotropic solid, and discuss Newton's equation of this solid. We can then show that this solid is perfectly described by equations analogous to the Maxwell equations when the volume expansion is homogeneous in the solid.

Then we calculate the distortion fields, energies and interactions of dislocations in that perfect solid. In the case of immobile dislocations in the solid lattice, the *static lattice distortions* induced by these dislocation store elastic energy within the lattice. This stored energy can then be considered as *the rest energy* of these immobile dislocations. In the case where dislocations are mobile within the lattice, *the movement of lattice itself* induced by the movement of dislocations are associated with a *kinetic energy.* At low speed, the kinetic energy is directly linked to the rest energy of the dislocations through relationships similar to the famous expression of Einstein $E_0 = M_0 c^2$. This allows us to introduce, in a *completely classical way* the notion of *inertial mass of dislocations.*

From the distortion fields induced by the dislocations and the force of Peach and Koehler, we then describe some interactions that may occur between dislocations. Finally, we introduce *the string model of dislocation*, which can process the dynamics of a dislocation line that moves and deforms in the lattice.

12.1 – The perfect solid and its equation of Newton

We will call *perfect lattice*, the isotropic lattice as it has been defined in section 7.1, with the state function *(7.7)*. It is further assumed that, very generally, only the modulus k_2 is strictly positive in this state function, while the moduli k_1 and k_0 can be either positive, negative or zero. Furthermore we will assume that these moduli can be temperature dependent (section 7.3). The purely elastic part of the state function of the perfect lattice depends only on the values of τ, τ^2 and $(\vec{\alpha}_i^{el})^2$

$$f^{el} = -k_0(T)\tau + k_1\tau^2 + k_2\sum_i \vec{\alpha}_i^{\,2} \qquad (12.1)$$

so that the equations of state which completely characterize the elasticity of the *perfect lattice* are limited to the scalar pressure p and the symmetric tensor of transverse shear stresses $\vec{s}_i$

$$\vec{s}_i = n\frac{\partial f}{\partial \alpha_{ik}^{el}}\vec{e}_k = 2nk_2\,\vec{\alpha}_i^{el} \quad \text{and} \quad p = -n\frac{\partial f}{\partial \tau^{el}} = n\left(k_0(T) - 2k_1\tau^{el}\right) \qquad (12.2)$$

It is found that the pressure p depends on scalar modulus $k_0(T)$ which may contain the effects of temperature, and the volume expansion τ through the modulus k_1.

The equation of Newton of the perfect solid

By neglecting the force of gravity, knowing that the torsor of moment $\vec{m}$ is null, and by using state equations *(12.2)*, the equation of Newton *(6.11)* is written as

$$n\frac{d\vec{p}}{dt} = \sum_k \vec{e}_k \operatorname{div} \vec{s}_k - \overrightarrow{\operatorname{grad}}\, p + nm\vec{\phi}_I \frac{dC_I}{dt} - nm\vec{\phi}_L \frac{dC_L}{dt}$$
$$= 2k_2 \sum_k \vec{e}_k \operatorname{div}\left(n\vec{\alpha}_i^{el}\right) - \overrightarrow{\operatorname{grad}}\left(nk_0(T) - 2nk_1\tau^{el}\right) + nm\vec{\phi}_I \frac{dC_I}{dt} - nm\vec{\phi}_L \frac{dC_L}{dt} \qquad (12.3)$$

By taking into account the fact that $\overrightarrow{\operatorname{grad}}\, n = -n\, \overrightarrow{\operatorname{grad}}\, \tau$, we obtain the following equation of Newton

$$\frac{d\vec{p}}{dt} = 2k_2 \sum_k \vec{e}_k \operatorname{div} \vec{\alpha}_k - 2k_2 \sum_k \left(\vec{e}_k \overrightarrow{\operatorname{grad}}\, \tau\right) \vec{\alpha}_k + \overrightarrow{\operatorname{grad}}\left[\left(2k_1(1-\tau) + k_0\right)\tau\right] + m\vec{\phi}_I \frac{dC_I}{dt} - m\vec{\phi}_L \frac{dC_L}{dt} \qquad (12.4)$$

Written in this form, the Newton's equation does not allow to separate the effects leading to a "rotational" quantity of movement, which would be related to pure shears without volume expansion of the lattice, from those leading to a "divergent" quantity of movement, which would be linked to the local volume expansions of the lattice.
But it is possible to find another formulation of the Newton's equation to separate these effects. Indeed, one can use the equations giving the *vector of flexion* to connect the space derivatives of the distortion tensors

$$\vec{\chi} = -\sum_k \vec{e}_k \wedge \overrightarrow{\operatorname{rot}}\, \vec{\alpha}_k - \frac{2}{3}\overrightarrow{\operatorname{grad}}\, \tau = \sum_k \vec{e}_k \operatorname{div} \vec{\alpha}_k - \frac{2}{3}\overrightarrow{\operatorname{grad}}\, \tau = -\overrightarrow{\operatorname{rot}}\, \vec{\omega} + \vec{\lambda} \qquad (12.5)$$

If we assume the existence of elasticity in the lattice but *only by shear*, the deformations of the lattice are entirely characterized by $\vec{\alpha}_i$ and τ so that

Hypothesis 1: $\vec{\alpha}_i = \vec{\alpha}_i^{el} + \vec{\alpha}_i^{an}$ *and* $\tau = \tau^{el}$ (12.6)

Expression *(12.5)* can be written

$$\sum_k \vec{e}_k \operatorname{div}\left(\vec{\alpha}_k^{el} + \vec{\alpha}_k^{an}\right) = \vec{\lambda} - \overrightarrow{\operatorname{rot}}\left(\vec{\omega}^{el} + \vec{\omega}^{an}\right) + \frac{2}{3}\overrightarrow{\operatorname{grad}}\, \tau^{el} \qquad (12.7)$$

This relationship combines elasticity effects with anelasticity effects, which cannot be separated, given the presence of the density $\vec{\lambda}$ of flexion charges, which greatly limits the applicability of this relationship. But we can assume that the presence of a vector density $\vec{\lambda}$ of flexion charges in the lattice should primarily be related to the behavior of elastic distortions, so that we will make the simplifying assumption that the relation *(12.7)* can be split into two separate relationships, one for the elastic part and the other for the anelastic part of the distortions in the following form

Hypothesis 2:
$$\begin{cases} \sum_k \vec{e}_k \operatorname{div} \vec{\alpha}_k^{el} = \vec{\lambda} - \overrightarrow{\operatorname{rot}}\, \vec{\omega}^{el} + \frac{2}{3}\overrightarrow{\operatorname{grad}}\, \tau^{el} \\ \sum_k \vec{e}_k \operatorname{div} \vec{\alpha}_k^{an} = -\overrightarrow{\operatorname{rot}}\, \vec{\omega}^{an} \end{cases} \qquad (12.8)$$

Using then the first relation *(12.8)*, the equation of Newton *(12.4)* can also be written

$$\frac{d\vec{p}}{dt} = -2k_2 \overrightarrow{\mathrm{rot}}\,\vec{\omega} - 2k_2 \sum_k \left(\vec{e}_k \overrightarrow{\mathrm{grad}}\,\tau\right)\vec{\alpha}_k + \overrightarrow{\mathrm{grad}}\left[\left(\frac{4}{3}k_2 + 2k_1(1-\tau) + k_0\right)\tau\right]$$
$$+ m\vec{\phi}_I \frac{dC_I}{dt} - m\vec{\phi}_L \frac{dC_L}{dt} + 2k_2\vec{\lambda} \tag{12.9}$$

In this new form, Newton's equation becomes really interesting because it is now possible to separate the effects of pure shear in the lattice without volume expansion from the effects of pure volume expansion of the lattice. The only term of this equation which is not linear is the coupling term $2k_2 \sum \left(\vec{e}_k \overrightarrow{\mathrm{grad}}\,\tau\right)\vec{\alpha}_k$. For small deformations, as in the common solids, this second order term may be usually neglected.

12.2 - Analogy with Maxwell's equations under homogeneous expansion

By supposing a solid in which the volume expansion is homogeneous, and which therefore satisfies the following

Hypothesis 3: $\quad \tau = \tau(t) \neq \tau(\vec{r}, t)$ *(12.10)*

and by assuming further that the concentrations of vacancies and interstitials are independent of time

Hypothesis 4: $\quad \frac{dC_I}{dt} = \frac{dC_L}{dt} = 0$ *(12.11)*

the Newton equation becomes

$$\frac{d\vec{p}}{dt} = -2k_2 \overrightarrow{\mathrm{rot}}\,\vec{\omega} + 2k_2\vec{\lambda} \tag{12.12}$$

In this case, it is possible to introduce in equation *(12.12)* a new quantity, namely the torsor of moment $\vec{m}$ copied after the torsor of moment which appears in the equation of Newton *(6.11)*, by writing

$$\frac{d\vec{p}}{dt} = -2k_2 \overrightarrow{\mathrm{rot}}\,\vec{\omega} + 2k_2\vec{\lambda} = -\frac{1}{2n}\overrightarrow{\mathrm{rot}}\,\vec{m} + 2k_2\vec{\lambda} \tag{12.13}$$

Thus, in the *perfect lattice* with homogeneous volume expansion, the torsor of moment $\vec{m}$ is linked to the vector of elastic rotation $\vec{\omega}^{el}$ by a very simple relation

$$\vec{m} = 4nk_2\,\vec{\omega}^{el} \tag{12.14}$$

As a consequence, if we imagine that the torsor of moment $\vec{m}$ derives from a state equation, we obtain the *volume density of free energy of elastic rotation* by lattice site, at homogenous expansion, under the form

$$m_k = n\frac{\partial f^{el}_{rotation}}{\partial \omega_k^{el}} = 4nk_2\vec{\omega}^{el} \Rightarrow f^{el}_{rotation} = 2k_2\left(\vec{\omega}^{el}\right)^2 \tag{12.15}$$

Thus, the *volume density of free energy of elastic rotation*, linked to the deformation by pure elastic rotations at homogenous volume expansion is equal to the elastic energy of deformation by pure shear at homogenous volume expansion

$$f^{el}_{rotation}\Big|_{\tau=\tau_0} = 2k_2\left(\vec{\omega}^{el}\right)^2 = k_2\sum_i(\vec{\alpha}_i^{el})^2 = f^{el}_{shear}\Big|_{\tau=\tau_0} \tag{12.16}$$

Finally, by using the second relation of hypothesis *(12.8)*, namely that there exists a one-to-one relationship $\sum \vec{e}_k \operatorname{div} \vec{\alpha}_k^{an} = -\overrightarrow{\operatorname{rot}}\,\vec{\omega}^{an}$ between spatial derivatives of the anelastic shears and anelastic rotations, we deduce that the anelasticity of the lattice by pure anelastic shear without changes in the volume expansion can also be reduced, at homogenous volume expansion, to a simple *lattice anelasticity by pure anelastic rotation*. This anelasticity can then be represented by a global rotation vector of the lattice $\vec{\omega}$, by writing the following relationship

$$\vec{\omega} = \vec{\omega}^{el} + \vec{\omega}^{an} + \vec{\omega}_0(t) = \frac{\vec{m}}{4nk_2} + \vec{\omega}^{an} + \vec{\omega}_0(t) \qquad (12.17)$$

where $\vec{\omega}_0(t)$ represents the uniform global rotation of the lattice.
It is concluded that, in the lattice where the volume expansion may be regarded as homogeneous, it is possible to treat univocally the elastic and anelastic pure lattice shear using only rotation vectors $\vec{\omega}^{el}$ and $\vec{\omega}^{an}$, which is very useful for dealing with the problem of the fields, energies and interactions of dislocations in a perfect lattice.
As the density $\vec{\lambda}$ of flexion charges associated with the edge dislocation charges, which are likely to bring locally a solid volume expansion variation, it will be assumed a-priori here a solid in which this type of charge is negligible, posing

Hypothesis 5: $\vec{\lambda} \approx 0$ *(12.18)*

As n is a constant in this case, we can rewrite the equations of Newton *(12.13)* as

$$\frac{d\vec{p}}{dt} = -2k_2 \overrightarrow{\operatorname{rot}}\,\vec{\omega} = -\frac{1}{2n} \overrightarrow{\operatorname{rot}}\,\vec{m} \qquad (12.19)$$

The equations needed for a complete description of the pure shear of a perfect solid still need to incorporate the topological equations for the rotation vector $\vec{\omega}$, i.e. the geometrokinetic equations and the equation of geometrocompatibility in the presence of dislocation charges

$$\vec{J} = -\frac{1}{2}\sum_k \vec{e}_k \wedge \vec{J}_k = -\frac{d\vec{\omega}}{dt} + \frac{1}{2}\overrightarrow{\operatorname{rot}}\,\vec{\phi} \quad \text{and} \quad \lambda = \frac{1}{2}\sum_k \vec{e}_k \vec{\lambda}_k = \operatorname{div}\vec{\omega} \qquad (12.20)$$

With hypothesis 4, we have that $C_L = cste$ and $C_I = cste$, so that the density of mass ρ becomes a constant

$$\rho = m\left(n + n_I - n_L\right) = mn\left(1 + C_I - C_L\right) = cste \qquad (12.21)$$

We also have that the equation of evolution of this density in the local referential $Ox_1x_2x_3$ allows us to deduce that the divergence of $n\vec{p}$ is null

$$\partial\rho / \partial t = 0 = -\operatorname{div}(n\vec{p}) \quad \Rightarrow \quad \operatorname{div}\vec{p} = 0 \qquad (12.22)$$

The quantity $\vec{p}$ can then be deduced thanks to the following relationships

$$\vec{p} = \frac{1}{n}\left(\rho\vec{\phi} + \vec{J}_m\right) = \frac{1}{n}\left[\rho\vec{\phi} + m\left(\vec{J}_I - \vec{J}_L\right)\right] = m\left[\vec{\phi} + \left(C_I - C_L\right)\vec{\phi} + \frac{1}{n}\left(\vec{J}_I - \vec{J}_L\right)\right] \qquad (12.23)$$

Furthermore, by imagining that there are no sources S^λ of charges of rotation

Hypothesis 6: $S^\lambda = 0$ *(12.24)*

we also obtain an equation for the continuity of charges of rotation

$$\frac{d\lambda}{dt} = -\operatorname{div}\vec{J} \qquad (12.25)$$

Finally, we can establish a continuity equation of the energy from equations *(12.19)* and *(12.20)*

$$-\vec{m}\vec{J} = \vec{m}\frac{d\vec{\omega}}{dt} + \vec{\phi}\,\frac{d(n\vec{p})}{dt} - \operatorname{div}\left(\frac{1}{2}\vec{\phi} \wedge \vec{m}\right) \tag{12.26}$$

The relations thus obtained for the pure shears of a perfect solid in the local framework $Ox_1x_2x_3$ in translation $\vec{\phi}_O(t)$ and in rotation $\vec{\omega}_O(t)$ in the absolute referential frame are written in table 12.1, and compared to the *equations of Maxwell for electromagnetism*

Table 12.1 - "Maxwellian" formulation of the equations of a perfect solid presenting homogeneous expansion τ in the local frame $Ox_1x_2x_3$

$\begin{cases} -\dfrac{d(2\vec{\omega})}{dt} + \overrightarrow{\mathrm{rot}}\,\vec{\phi} = (2\vec{J}) \\ \operatorname{div}(2\vec{\omega}) = (2\lambda) \end{cases}$	$\Leftrightarrow$	$\begin{cases} -\dfrac{\partial \vec{D}}{\partial t} + \overrightarrow{\mathrm{rot}}\,\vec{H} = \vec{j} \\ \operatorname{div}\vec{D} = \rho \end{cases}$
$\begin{cases} \dfrac{d(n\vec{p})}{dt} = -\overrightarrow{\mathrm{rot}}\left(\dfrac{\vec{m}}{2}\right) \\ \operatorname{div}(n\vec{p}) = 0 \end{cases}$	$\Leftrightarrow$	$\begin{cases} \dfrac{\partial \vec{B}}{\partial t} = -\overrightarrow{\mathrm{rot}}\,\vec{E} \\ \operatorname{div}\vec{B} = 0 \end{cases}$
$\begin{cases} (2\vec{\omega}) = \left(\dfrac{1}{nk_2}\right)\left(\dfrac{\vec{m}}{2}\right) + (2\vec{\omega}^{an}) + \left(2\vec{\omega}_0(t)\right) \\ (n\vec{p}) = (nm)\left[\vec{\phi} + (C_I - C_L)\vec{\phi} + \left(\dfrac{1}{n}\left(\vec{J}_I - \vec{J}_L\right)\right)\right] \end{cases}$	$\Leftrightarrow$	$\begin{cases} \vec{D} = \varepsilon_0\,\vec{E} + \vec{P} + \vec{P}_0(t) \\ \vec{B} = \mu_0\left[\vec{H} + \left(\chi^{para} + \chi^{dia}\right)\vec{H} + \vec{M}\right] \end{cases}$
$\left\{ \dfrac{d(2\lambda)}{dt} = -\operatorname{div}(2\vec{J}) \right.$	$\Leftrightarrow$	$\left\{ \dfrac{\partial \rho}{\partial t} = -\operatorname{div}\vec{j} \right.$
$\begin{cases} -\left(\dfrac{\vec{m}}{2}\right)(2\vec{J}) = \\ \vec{\phi}\,\dfrac{d(n\vec{p})}{dt} + \left(\dfrac{\vec{m}}{2}\right)\dfrac{d(2\vec{\omega})}{dt} - \operatorname{div}\left(\vec{\phi} \wedge \left(\dfrac{\vec{m}}{2}\right)\right) \end{cases}$	$\Leftrightarrow$	$\begin{cases} -\vec{E}\vec{j} = \\ \vec{H}\dfrac{\partial \vec{B}}{\partial t} + \vec{E}\dfrac{\partial \vec{D}}{\partial t} - \operatorname{div}\left(\vec{H} \wedge \vec{E}\right) \end{cases}$
$\left\{ c_t = \sqrt{\dfrac{nk_2}{nm}} = \sqrt{\dfrac{k_2}{m}} \right.$	$\Leftrightarrow$	$\left\{ c = \sqrt{\dfrac{1}{\varepsilon_0 \mu_0}} \right.$

It is noted that there is a remarkable and absolutely complete analogy between these two sets of equations, except for the fact that the evolution equations involve the total derivative, while the Maxwell equations only involve the partial derivative with respect to time. However, it is recalled here that the material derivative in the local frame $Ox_1x_2x_3$ can be replaced without problems by the partial derivative with respect to time if the strains are sufficiently small and/or slow in the vicinity of the origin of the local frame $Ox_1x_2x_3$!

This analogy will be discussed in detail in the third part of the book. For now, let's enumerate the following analogies:

- the *rotational field* $2\vec{\omega}$ is analogous to the *electrical field of displacement* $\vec{D}$,
- the *quantity of movement and the flow of mass* $n\vec{p}$ is analogous of the *magnetic induction field* $\vec{B}$,
- the *torsor of moment* $\vec{m}/2$ is analogous to the *electric field* $\vec{E}$,
- the *velocity field* $\vec{\phi}$ is analogous to the *magnetic field* $\vec{H}$,
- the *density* 2λ *of rotational charges* is analogous to the *density* ρ *of electric charges*
- the *flux* $2\vec{J}$ *of rotational charges* is analogous to the *density* $\vec{j}$ *of electric current*
- the anelasticity $2\vec{\omega}^{an}$ is analogous to the *dielectric polarisation* $\vec{P}$ *of matter,*
- the *mass transport* $nm\vec{\phi}$ *by the movement of the lattice is analogous to the term* $\mu_0\vec{H}$ *of magnetic induction of vacuum,*
- the *mass transport* $nm(C_I - C_L)\vec{\phi}$ *by the drag-along movement of point defects of the lattice* is analogous to the term $\mu_0(\chi^{para}+\chi^{dia})\vec{H}$ *of paramagnetism and diamagnetism of matter,*
- the *mass transport* $m(\vec{J}_I - \vec{J}_L)$ *due to self-diffusion of vacancies and interstitials* is analogous to the *magnetization* $\vec{M}$ *of matter,*
- the *inverse of elastic modulus* $1/K_2$ is analogous to the *dielectric permittivity of vacuum* ε_0,
- the *density of mass* $nm\vec{\phi}$ is analogous to the *magnetic permeability of vacuum* μ_0,
- the *speed of transversal waves* c_t is analogous to the *speed of light* c,
- the *flow of elastic energy* $\vec{\phi}\wedge\vec{m}/2$ is analogous to the *vector of Poynting* $\vec{H}\wedge\vec{E}$.

12.3 – Fields and energies of a screw dislocation

The rotational field of a screw dislocation

Let's consider a *screw dislocation string,* in the form of a straight cylinder of infinite length and radius, containing a density λ of rotational charges (figure 12.1).
Apply the compatibility relationship $\operatorname{div}\vec{\omega}^{el} = \lambda$ and integrate on a cylindrical volume of radius r and length unit. We have

$$\iiint\limits_V \operatorname{div}\vec{\omega}^{screw}\,dV = \oint\!\!\oint\limits_S \vec{\omega}^{screw}\vec{n}\,dS = \vec{\omega}^{screw}\vec{n}\,2\pi r = \iiint\limits_V \lambda\,dV \qquad (12.27)$$

in which $\vec{n}$ is the vector normal to the surface cylinder.
Outside the charge, meaning for $r>R$, this relation gives us the following $\vec{\omega}$ field

$$\vec{\omega}_{ext}^{screw}\vec{n} = \frac{1}{2\pi r}\left(\lambda\pi R^2\right) = \frac{\Lambda}{2\pi r} \quad \Rightarrow \quad \vec{\omega}_{ext}^{screw} = \frac{\Lambda}{2\pi}\frac{\vec{r}}{r^2} \quad (r>R) \qquad (12.28)$$

where Λ is the *linear charge of rotation of the string* given by the integration of density λ on the unit length of the string

$$\Lambda = \pi R^2 \lambda \qquad (12.29)$$

Inside the charge, meaning for $r<R$, we then have

$$\vec{\omega}_{int}^{screw}\vec{n} = \frac{1}{2\pi r}\left(\pi\lambda r^2\right) = \frac{\Lambda}{2\pi R^2}r \quad \Rightarrow \quad \vec{\omega}_{int}^{screw} = \frac{\Lambda}{2\pi R^2}\vec{r} \quad (r<R) \qquad (12.30)$$

The behavior of the module of $\vec{\omega}^{screw}$ as a function of distance r from the center of the cylinder is shown in figure 12.2.

The displacement field by rotation of a screw dislocation

Knowing the field of external rotations $\vec{\omega}_{ext}^{screw}$ outside the string, it is possible to derive easily the elastic displacement field $\vec{u}_{ext}^{screw}$ associated with the rotation due to the presence of the charged string since outside of the string λ is zero. We have

$$\vec{\omega}_{ext}^{screw} = -\frac{1}{2}\overrightarrow{\mathrm{rot}}\,\vec{u}_{ext}^{screw} = \frac{\Lambda}{2\pi}\frac{\vec{r}}{r^2} = \frac{\Lambda}{2\pi}\frac{x_1\vec{e}_1 + x_3\vec{e}_3}{x_1^{\ 2} + x_3^{\ 2}} \qquad (r > R) \tag{12.31}$$

It is clear that the field $\vec{u}_{ext}^{screw}$ must be parallel to $\vec{e}_2$ to satisfy this relationship, so that

$$\vec{u}_{ext}^{screw} = f\vec{e}_2 \quad \Rightarrow \quad \overrightarrow{\mathrm{rot}}\,\vec{u}_{ext}^{screw} = -\frac{\partial f}{\partial x_3}\vec{e}_1 + \frac{\partial f}{\partial x_1}\vec{e}_3 \tag{12.32}$$

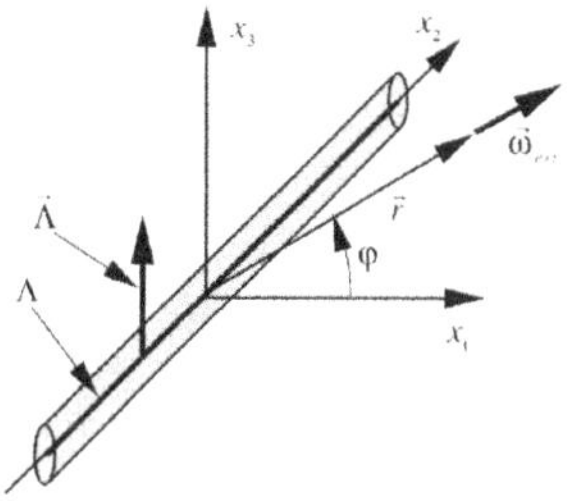

Figure 12.1 *- cylindrical charge of radius R and density λ*

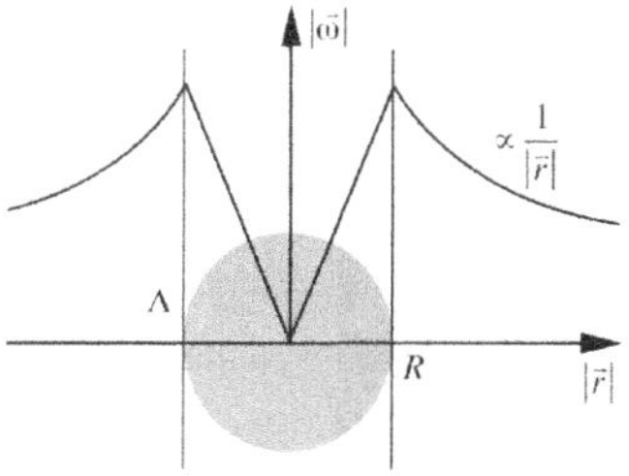

Figure 12.2 *- the norm of the field $\vec{\omega}$ inside and outside of the linearly charged string Λ*

so that we obtain the following system of partial differential equations

$$\frac{\partial f}{\partial x_3} = +\frac{\Lambda}{\pi}\frac{x_1}{x_1^2 + x_3^2} \quad \text{et} \quad \frac{\partial f}{\partial x_1} = -\frac{\Lambda}{\pi}\frac{x_3}{x_1^2 + x_3^2} \tag{12.33}$$

for which the solution is

$$\vec{u}_{ext}^{screw} = -\vec{e}_2\frac{\Lambda}{\pi}\arctan\frac{x_1}{x_3} \tag{12.34}$$

by introducing the polar coordinates $x_1 = r\cos\varphi$ and $x_2 = r\sin\varphi$ in the plane of the cylinder, the displacement field can be written

$$\vec{u}_{ext}^{screw} = \vec{e}_2\frac{\Lambda}{\pi}\left(\varphi - \frac{\pi}{2}\right) \tag{12.35}$$

In polar coordinates, this field does not depend on r and increases with angle φ. It does indeed correspond to a displacement of type 'screw' along the axis Ox_2.

The rotational rest energy of a screw dislocation in a perfect lattice

As the rotational plane of the screw dislocation is not dependent on the volume expansion field, we can use without problem equation *(12.15)*, and write the stored elastic energy of rotation stored outside of the string per unit length in a "quasi-infinite" environment, as such

$$E_0^{screw\,ext} = 2nk_2 \int_R^{R_\infty} \frac{\Lambda^2}{4\pi^2 r^2} 2\pi r\, dr = \frac{nk_2\Lambda^2}{\pi} \ln\frac{R_\infty}{R} \tag{12.36}$$

and the energy stored inside the string as

$$E_0^{screw\,int} = 2nk_2 \int_0^{R} \frac{\Lambda^2 r^2}{4\pi^2 R^4} 2\pi r\, dr = \frac{nk_2\Lambda^2}{4\pi} \tag{12.37}$$

It is remarkable that the energy of the interior of the dislocation $E_0^{screw\,int}$ thus obtained does not depend on the radius of the string R.

The sum of the two energies stored in the lattice can be considered as the rest energy E_0^{screw} of the dislocation string, expressed by unit length of the string or radius R, and thus is worth

$$E_0^{screw} = E_0^{screw\,int} + E_0^{screw\,ext} = \frac{nk_2\Lambda^2}{\pi}\left(\frac{1}{4} + \ln\frac{R_\infty}{R}\right) \tag{12.38}$$

We notice that the internal energy of the string, which does not depend on radius R, is a lot smaller than the energy stored outside the string, which depends on the radius R of the string and the size of the lattice $R_\infty >> R$. Since the internal energy by unit of length does not depend on R, we can neglect it compared to $E_0^{screw\,ext}$ and write approximatively the rest energy of dislocation under the following forms, by using the relation $\Lambda = -\vec{B}_{screw}\vec{t}\,/\,2$ and the *Lamé coefficient* $\mu = K_2$

$$E_0^{screw} \cong \frac{nk_2\Lambda^2}{\pi} \ln\frac{R_\infty}{R} = \frac{nk_2\,\vec{B}_{screw}^2}{4\pi} \ln\frac{R_\infty}{R} = \frac{\mu\,\vec{B}_{screw}^2}{4\pi} \ln\frac{R_\infty}{R} \tag{12.39}$$

In the expressions of the rest energy of a screw dislocation appears the external dimension of the lattice in question. Thus, the rest energy of a single line of dislocation $R_\infty >> R$ in an infinite network should be infinite. However, apart from the fact that a real solid is not of infinite size and the presence of a single dislocation in a solid is an exceptional fact, the size R_∞ is generally limited to smaller values than the actual size R_∞ of the solid . Indeed, if there exists in the lattice a large number of dislocations separated by an average distance d, and these dislocations have random orientations in the solid, we have a screening effect of the individual fields of displacement of each lines by the opposite sign lines located in the immediate vicinity, so that long-distance energy stored by an individual line becomes negligible (one can show that it actually decreases as $\sim 1/r^2$ for distances r greater than d). In this case, a fairly good estimate of the rest energy of the dislocation is obtained by choosing, for R_∞, the average distance d between two adjacent lines of opposite sign. Note that R_∞ appears in a logarithm, which greatly mitigates the effects of large variations of d.

It should also be noted that the *energy associated with a line segment of length L of screw dislocation does not depend on the volume expansion of the solid lattice*. Indeed, it is easy to verify that

$$E_0^{screw}\Big|_L = L\frac{nk_2\vec{B}_{screw}^2}{4\pi}\ln\frac{R_\infty}{R} = L_0\frac{n_0k_2\,\vec{B}_{0screw}^2}{4\pi}\ln\frac{R_{\infty 0}}{R_0} = E_{00}^{screw}\Big|_{L_0} \tag{12.40}$$

The kinetic energy of rotation at low velocity of a screw dislocation

In case a screw dislocation is moving in the direction Ox_1 at low velocity $\vec{\mathbf{v}}$ compared to the speed of transverse waves c_t, the dynamic displacement field associated with rotations is obtained by substituting x_1 by $x_1(t) = x_1 - \mathbf{v}t$ in the expression *(12.34)*

$$\vec{u}_{ext}^{screw} = -\vec{e}_2 \frac{\Lambda}{\pi} \arctan \frac{x_1 - \mathbf{v}t}{x_3} \tag{12.41}$$

The velocity field is then obtain by temporal derivation of this expression

$$\vec{\phi}_{ext}^{screw} = -\frac{d\vec{u}}{dt} = \vec{e}_2 \frac{\Lambda}{\pi} \frac{\partial}{\partial t} \left(\arctan \frac{x_1 - \mathbf{v}t}{x_3} \right) = \vec{e}_2 \frac{\Lambda}{\pi} \frac{-x_3}{x_3^2 + (x_1 - \mathbf{v}t)^2} \mathbf{v} \tag{12.42}$$

The kinetic energy per lattice particle can then be deduced. For this, it is convenient to express it in the referential frame $Ox_1'x_2'x_3'$, which is in translation with velocity $\vec{\mathbf{v}}$ with the dislocation line and in which $x_1 - \mathbf{v}t$ is x_1', and then switch to polar coordinates in this referential frame with $x_1' = r'\cos\varphi'$ and $x_3' = r'\sin\varphi'$ in the plane of the moving cylinder. We have

$$e_{kin}^{screw} = \frac{1}{2} m \left(\vec{\phi}_{ext}^{screw} \right)^2 = \frac{1}{2} m \left(\frac{\Lambda}{\pi} \right)^2 \left(\frac{x_3'}{x_3'^2 + x_1'^2} \right)^2 \mathbf{v}^2 = \frac{1}{2} m \left(\frac{\Lambda}{\pi} \right)^2 \left(\frac{\sin^2\varphi'}{r'} \right)^2 \mathbf{v}^2 \tag{12.43}$$

The kinetic energy stored in the lattice by the dynamic field of rotation, expressed in unit of length of the screw dislocation that is in movement with velocity $\vec{\mathbf{v}}$ is obtained by the integration of energy density $e_{kin}^{\vec{\omega}}$ on a cylinder of radius R_∞ and of unit length

$$E_{kin}^{screw} = \frac{1}{2} m \int_R^{R_\infty} \int_0^{2\pi} n \left(\frac{\Lambda}{\pi} \right)^2 \frac{\sin^2\varphi'}{r'^2} \mathbf{v}^2 r' d\varphi' dr' = \frac{1}{2} mn\mathbf{v}^2 \frac{\Lambda^2}{\pi} \ln \frac{R_\infty}{R} \tag{12.44}$$

Using the fact that $\Lambda = -\vec{B}_{screw}\vec{t} / 2$ and that the specific mass of the medium is $\rho = mn$, we can write the kinetic energy associated with the movement at low velocity of the screw dislocation under the following forms

$$E_{kin}^{screw} = \frac{mn\Lambda^2}{2\pi} \left(\ln \frac{R_\infty}{R} \right) \mathbf{v}^2 = \frac{mn\vec{B}_{screw}^2}{8\pi} \left(\ln \frac{R_\infty}{R} \right) \mathbf{v}^2 = \frac{\rho \vec{B}_{screw}^2}{8\pi} \left(\ln \frac{R_\infty}{R} \right) \mathbf{v}^2 \tag{12.45}$$

By comparing the kinetic energy E_{kin}^{screw} stored in the medium by the movement of the string with the elastic potential energy E_0^{screw} stored in the medium by the presence of this line, using the expression *(12.39)* of the rest energy of the screw dislocation and the relationship $c_t^2 = k_2 / m$ of the speed of the transverse waves in the medium *(7.59)*, we find the following relationship between the rest energy and the kinetic energy of a screw dislocation

$$E_{kin}^{screw} = \frac{1}{2} \frac{m}{k_2} \frac{nk_2\Lambda^2}{\pi} \ln \frac{R_\infty}{R} \mathbf{v}^2 = \frac{1}{2} \frac{E_0^{screw}}{c_t^2} \mathbf{v}^2 \tag{12.50}$$

This kinetic energy E_{kin}^{screw} stored in the solid lattice by the mobile dislocation can be considered as the kinetic energy of the dislocation in motion, and we can therefore introduce an *inertial mass M_0^{screw} of the dislocation* by writing relations

$$E_{kin}^{screw} = \frac{1}{2} M_0^{screw} \mathbf{v}^2 \quad \Rightarrow \quad M_0^{screw} = \frac{1}{c_t^2} E_0^{screw} \quad \Rightarrow \quad E_0^{screw} = M_0^{screw} c_t^2 \tag{12.51}$$

We thus find for the dislocation line the famous *expression of Einstein* connecting the inertial energy to the rest energy via the speed of transverse waves. But here, this relationship is found without *any appeal to a relativistic dynamics of the string*, because it is due to the fact that the rest energy and the kinetic energy of the line are the elastic potential energy and the kinetic energy stored in the lattice by the dynamic distortions imposed on the lattice by the presence of the moving screw dislocation!

12.4 – Fields and energies of an edge dislocation

The rotational field and the volume expansion field of an edge dislocation

Consider now an *edge dislocation string* in the form of a straight cylinder of infinite length and radius R, as shown in figure 12.1, containing this time a vector density of flexion charges $\vec{\lambda}$ perpendicular to the direction of the string, which points in the direction of the axis Ox_3. In conventional solid media, the volume expansion is still very low, and the modulus k_0 associated with thermal effects is negligible vis-à-vis the moduli k_1 and k_2. So we can make three simplifying assumptions: the volume expansion τ is still very small ($\tau_0 \cong 0$) and can therefore be treated as a disturbance $\tau^{(p)} << 1$, and the modulus k_0 can be neglected

Hypothesis: $\tau = \tau^{(p)} << 1$ and $k_0 \cong 0$ and $\tau_0 \cong 0$ (12.52)

One can then use the linearized equation of Newton *(12.9)*, expressed in terms of $\tau^{(p)}$ and $\vec{\omega}^{el}$, to find the static fields of elastic torsion $\vec{\omega}^{el}$ and perturbation of volume expansion $\tau^{(p)}$ associated with the presence of the vector charge density $\vec{\lambda}$. We thus get the following equilibrium equation around the charge

$$\overrightarrow{\mathrm{rot}}\,\vec{\omega}^{el} - \frac{2k_2+3k_1}{3k_2}\,\overrightarrow{\mathrm{grad}}\,\tau^{(p)} \cong \begin{cases} \vec{\lambda} & (r<R) \\ 0 & (r>R) \end{cases} \tag{12.53}$$

To avoid manipulating expressions containing the elastic moduli k_1 and k_2 we can use *the Poisson's modulus (7.25 and 7.30)* of the solid, and write the equilibrium equation

$$\overrightarrow{\mathrm{rot}}\,\vec{\omega}^{el} - \frac{1-\nu}{1-2\nu}\,\overrightarrow{\mathrm{grad}}\,\tau^{(p)} \cong \begin{cases} \vec{\lambda}\cdot & (r<R) \\ 0 & (r>R) \end{cases} \tag{12.54}$$

Outside of the dislocation string $(r>R)$, the vector of rotation $\vec{\omega}^{el}$ should have a single-axis component along Ox_2 which must depend on $\cos\varphi$, while the expansion scalar $\tau^{(p)}$ should depend on $\sin\varphi$. On the other hand, the two variables $\vec{\omega}^{el}$ and $\tau^{(p)}$ should decrease with the distance as $1/r$ in the same way as for the screw dislocation string treated before, so the solution must surely be written in the following form

$$\begin{cases} \vec{\omega}^{el} = A\dfrac{\cos\varphi}{r}\vec{e}_2 = A\dfrac{x_1}{x_1^2+x_3^2}\vec{e}_2 \\ \tau^{(p)} = B\dfrac{\sin\varphi}{r} = B\dfrac{x_3}{x_1^2+x_3^2} \end{cases} \quad (r>R) \tag{12.55}$$

where A and B are constants of integration.

We can then calculate $\overrightarrow{\mathrm{rot}}\,\vec{\omega}^{el}$ and $\overrightarrow{\mathrm{grad}}\,\tau^{(p)}$

$$\begin{cases} \overrightarrow{\mathrm{rot}}\,\vec{\omega}^{el} = A\left[\dfrac{2x_1x_3}{\left(x_1^2+x_3^2\right)^2}\vec{e}_1 + \dfrac{x_3^2-x_1^2}{\left(x_1^2+x_3^2\right)^2}\vec{e}_3\right] \\ \overrightarrow{\mathrm{grad}}\,\tau^{(p)} = -B\left[\dfrac{2x_1x_3}{\left(x_1^2+x_3^2\right)^2}\vec{e}_1 + \dfrac{x_3^2-x_1^2}{\left(x_1^2+x_3^2\right)^2}\vec{e}_3\right] \end{cases} \tag{12.56}$$

which, when introduced in the equation of equilibrium, give us the relation existing between A and B

$$A + \frac{1-\nu}{1-2\nu}B = 0 \tag{12.57}$$

Inside the string, we can integrate the equilibrium equation on a cylinder containing the density of charge $\vec{\lambda}$, of radius R and of unit length. We then have the integral relation

$$\iiint_V \overrightarrow{\mathrm{rot}}\,\vec{\omega}^{el}\,dV - \frac{1-\nu}{1-2\nu}\iiint_V \overrightarrow{\mathrm{grad}}\,\tau^{(p)}\,dV \cong \iiint_V \vec{\lambda}\,dV \tag{12.58}$$

As the integral on $\vec{\lambda}$ on the cylinder of unit length gives us the linear charge $\vec{\Lambda}$ of the string, the previous relation is easily transformed in

$$\oiint_{cylinder} d\vec{S}\wedge\vec{\omega}^{el} - \frac{1-\nu}{1-2\nu}\oiint_{cylinder} \tau^{(p)}\,d\vec{S} \cong \vec{\Lambda} \tag{12.59}$$

By symmetry, the integrals on the two sections (lateral faces) of the cylinder cancel each other, so that with $d\vec{S} = R\left(\vec{e}_1\cos\varphi + \vec{e}_3\sin\varphi\right)d\varphi$

$$\begin{cases} \displaystyle\iint_S d\vec{S}\wedge\vec{\omega}^{el} = \iint_S A\frac{\cos\varphi}{R}\left(d\vec{S}\wedge\vec{e}_2\right) = \int_0^{2\pi} A\frac{\cos\varphi}{R}R\left(\vec{e}_3\cos\varphi - \vec{e}_1\sin\varphi\right)d\varphi = A\pi\vec{e}_3 \\ \displaystyle\iint_S \tau^{(p)}\,d\vec{S} = \iint_S B\frac{\sin\varphi}{R}d\vec{S} = \int_0^{2\pi} B\frac{\sin\varphi}{R}R\left(\vec{e}_1\cos\varphi + \vec{e}_3\sin\varphi\right)d\varphi = B\pi\vec{e}_3 \end{cases} \tag{12.60}$$

We then have as a result of the equation of equilibrium inside the string

$$A\pi\vec{e}_3 - \frac{1-\nu}{1-2\nu}B\pi\vec{e}_3 \cong \vec{\Lambda} \quad\Rightarrow\quad A - \frac{1-\nu}{1-2\nu}B \cong \frac{\vec{\Lambda}\vec{e}_3}{\pi} \tag{12.61a}$$

From relations *(12.57)* and *(12.61)*, we deduce the constants A and B

$$A = \frac{\vec{\Lambda}\vec{e}_3}{2\pi} \quad\text{and}\quad B = -\left(\frac{1-2\nu}{1-\nu}\right)\frac{\vec{\Lambda}\vec{e}_3}{2\pi} \tag{12.61b}$$

With the help of relations *(12.55)*, and recalling that the linear charge $\vec{\Lambda}$ is worth $\vec{\Lambda} = \vec{\boldsymbol{B}}_{edge}\wedge\vec{\boldsymbol{t}}$, the fields $\vec{\omega}^{el}$ and $\tau^{(p)}$ outside the dislocation charge are then completely known

$$\begin{cases} \vec{\omega}^{el} = \dfrac{\vec{\Lambda}\vec{e}_3}{2\pi}\dfrac{\cos\varphi}{r}\vec{e}_2 = \dfrac{\vec{\Lambda}\vec{e}_3}{2\pi}\dfrac{x_1}{x_1^2+x_3^2}\vec{e}_2 \\ \tau^{(p)} = -\dfrac{\vec{\Lambda}\vec{e}_3}{2\pi}\dfrac{1-2\nu}{1-\nu}\dfrac{\sin\varphi}{r} = -\dfrac{\vec{\Lambda}\vec{e}_3}{2\pi}\dfrac{1-2\nu}{1-\nu}\dfrac{x_3}{x_1^2+x_3^2} \end{cases} \quad (r>R) \tag{12.62}$$

The displacement field of an edge dislocation

It is possible to find the displacement field $\vec{u}$ outside an edge dislocation string. Indeed, as the field $\vec{u}$ can only have components along Ox_1 and Ox_3, and that $\vec{u}$ must not depend on x_2, we have

$$\vec{u} = u_1(x_1, x_3)\vec{e}_1 + u_3(x_1, x_3)\vec{e}_3 \tag{12.63}$$

So that we have two differential equations

$$\begin{cases} \vec{\omega}^{el} = -\dfrac{1}{2}\overrightarrow{\text{rot}}\,\vec{u} = \dfrac{1}{2}\left(\dfrac{\partial u_3}{\partial x_1} - \dfrac{\partial u_1}{\partial x_3}\right)\vec{e}_2 \;\Rightarrow\; \dfrac{\partial u_3}{\partial x_1} - \dfrac{\partial u_1}{\partial x_3} = \dfrac{\vec{\Lambda}\vec{e}_3}{\pi}\dfrac{x_1}{x_1^2 + x_3^2} \\ \tau^{(p)} = -\text{div}\,\vec{u} = -\left(\dfrac{\partial u_1}{\partial x_1} + \dfrac{\partial u_3}{\partial x_3}\right) \;\Rightarrow\; \dfrac{\partial u_1}{\partial x_1} + \dfrac{\partial u_3}{\partial x_3} = \dfrac{\vec{\Lambda}\vec{e}_3}{2\pi}\dfrac{1-2\nu}{1-\nu}\dfrac{x_3}{x_1^2 + x_3^2} \end{cases} \tag{12.64}$$

We can try the following classic solution[1] for the displacement field $\vec{u}$

$$\begin{cases} u_1(x_1, x_3) = \alpha \arctan\left(\dfrac{x_3}{x_1}\right) + \beta\dfrac{x_1 x_3}{x_1^2 + x_3^2} \\ u_3(x_1, x_3) = \gamma \ln\left(x_1^2 + x_3^2\right) + \delta\dfrac{x_1^2 - x_3^2}{x_1^2 + x_3^2} \end{cases} \tag{12.65}$$

where $\alpha, \beta, \delta, \gamma$ are constants to be determined. The fields $\vec{\omega}^{el}$ and $\tau^{(p)}$ are then written

$$\begin{cases} \vec{\omega}^{el} = -\dfrac{1}{2}\overrightarrow{\text{rot}}\,\vec{u} = \left[-\dfrac{1}{2}(\alpha + \beta - 2\gamma)\dfrac{x_1}{x_1^2 + x_3^2} + (\beta + 2\delta)\dfrac{x_1 x_3^2}{\left(x_1^2 + x_3^2\right)^2}\right]\vec{e}_2 \\ \tau^{(p)} = -\text{div}\,\vec{u} = \left[(\alpha - \beta - 2\gamma)\dfrac{x_3}{x_1^2 + x_3^2} + 2(\beta + 2\delta)\dfrac{x_1^2 x_3}{\left(x_1^2 + x_3^2\right)^2}\right] \end{cases} \tag{12.66}$$

Comparing these relations with *(12.62)*, we deduce three relations

$$\alpha - 2\gamma + \beta = -\frac{\vec{\Lambda}\vec{e}_3}{\pi} \quad ; \quad \alpha - 2\gamma - \beta = -\frac{\vec{\Lambda}\vec{e}_3}{2\pi}\frac{1-2\nu}{1-\nu} \quad ; \quad \beta + 2\delta = 0 \tag{12.67}$$

whose solutions are

$$\alpha - 2\gamma = -\frac{\vec{\Lambda}\vec{e}_3}{2\pi}\frac{3-4\nu}{2(1-\nu)} \quad ; \quad \beta = -\frac{\vec{\Lambda}\vec{e}_3}{2\pi}\frac{1}{2(1-\nu)} \quad ; \quad \delta = \frac{\vec{\Lambda}\vec{e}_3}{2\pi}\frac{1}{4(1-\nu)} \tag{12.68}$$

With this system, the constants α and γ are still undetermined and there exists an infinite number of possible equilibrium solutions. The determination of these constants is actually a balance energy minimization problem in the case of a solid free of all external constraints, or with boundary conditions in the case of a solid subjected to restrictive external constraints. We can solve for the moment the general problem without making a prior choice as to the respective values of α and γ. Therefore introduce *an adjustable parameter* ς such that we can write

[1] *Hirth and Lothe, theory of dislocations, second edition, John Wiley and Sons, 1982, p.78*

$$\begin{cases} \alpha = -\dfrac{\vec{\Lambda}\vec{e}_3}{2\pi}\varsigma \\ \gamma = \dfrac{\vec{\Lambda}\vec{e}_3}{2\pi}\left[\dfrac{3-4\nu}{4(1-\nu)} - \dfrac{\varsigma}{2}\right] \end{cases} \quad et \quad \begin{cases} \beta = -\dfrac{\vec{\Lambda}\vec{e}_3}{2\pi}\dfrac{1}{2(1-\nu)} \\ \delta = \dfrac{\vec{\Lambda}\vec{e}_3}{2\pi}\dfrac{1}{4(1-\nu)} \end{cases} \tag{12.69}$$

The displacement field is then written

$$\begin{cases} u_1(x_1, x_3) = -\dfrac{\vec{\Lambda}\vec{e}_3}{2\pi}\varsigma \arctan\left(\dfrac{x_3}{x_1}\right) - \dfrac{\vec{\Lambda}\vec{e}_3}{2\pi}\dfrac{1}{2(1-\nu)}\dfrac{x_1 x_3}{x_1^2 + x_3^2} \\ u_3(x_1, x_3) = \dfrac{\vec{\Lambda}\vec{e}_3}{2\pi}\left[\dfrac{3-4\nu}{4(1-\nu)} - \dfrac{\varsigma}{2}\right]\ln\left(x_1^2 + x_3^2\right) + \dfrac{\vec{\Lambda}\vec{e}_3}{2\pi}\dfrac{1}{4(1-\nu)}\dfrac{x_1^2 - x_3^2}{x_1^2 + x_3^2} \end{cases} \tag{12.70}$$

How we got here the displacement field associated with an edge dislocation is new. Indeed, it is based on finding the equilibrium of the solid conditions on the basis of Newton's equation *(12.9)* containing a vector density $\vec{\lambda}$ of flexion charges, whereas in the classical approach[2], the field displacement is obtained from a particular solution of the differential equation for the *Airy function* of plane stress, which is much more complicated, and involves a strong a priori of not having restrictive external constraints!

The deformation and shear fields of an edge dislocation

From these expressions, using the relationship *(2.48)*, we can deduce the elastic deformation tensor $\vec{\varepsilon}_i^{el}$ and the tensor of elastic shear strains $\vec{\alpha}_i^{el}$ outside of the dislocation line

$$\begin{cases} \vec{\varepsilon}_1^{\,el} = -\dfrac{\partial u_1}{\partial x_1}\vec{e}_1 - \dfrac{1}{2}\left(\dfrac{\partial u_1}{\partial x_3} + \dfrac{\partial u_3}{\partial x_1}\right)\vec{e}_3 \\ \vec{\varepsilon}_2^{\,el} = 0 \\ \vec{\varepsilon}_3^{\,el} = -\dfrac{1}{2}\left(\dfrac{\partial u_1}{\partial x_3} + \dfrac{\partial u_3}{\partial x_1}\right)\vec{e}_1 - \dfrac{\partial u_3}{\partial x_3}\vec{e}_3 \end{cases} \tag{12.71}$$

$$\begin{cases} \vec{\alpha}_1^{\,el} = \dfrac{1}{3}\left(\dfrac{\partial u_3}{\partial x_3} - 2\dfrac{\partial u_1}{\partial x_1}\right)\vec{e}_1 - \dfrac{1}{2}\left(\dfrac{\partial u_1}{\partial x_3} + \dfrac{\partial u_3}{\partial x_1}\right)\vec{e}_3 \\ \vec{\alpha}_2^{\,el} = \dfrac{1}{3}\left(\dfrac{\partial u_1}{\partial x_1} + \dfrac{\partial u_3}{\partial x_3}\right)\vec{e}_2 \\ \vec{\alpha}_3^{\,el} = -\dfrac{1}{2}\left(\dfrac{\partial u_1}{\partial x_3} + \dfrac{\partial u_3}{\partial x_1}\right)\vec{e}_1 + \dfrac{1}{3}\left(\dfrac{\partial u_1}{\partial x_1} - 2\dfrac{\partial u_3}{\partial x_3}\right)\vec{e}_3 \end{cases} \tag{12.72}$$

The derivatives of the components of the displacement field that appear in these relations are deduced from *(23.56)*

$$\begin{cases} \dfrac{\partial u_1}{\partial x_1} = \dfrac{\vec{\Lambda}\vec{e}_3}{2\pi}\dfrac{1}{2(1-\nu)}\left\{\left[2(1-\nu)\varsigma - 1\right]\dfrac{x_3}{x_1^2 + x_3^2} + \dfrac{2x_1^2 x_3}{\left(x_1^2 + x_3^2\right)^2}\right\} \\ \dfrac{\partial u_1}{\partial x_3} = \dfrac{\vec{\Lambda}\vec{e}_3}{2\pi}\dfrac{1}{2(1-\nu)}\left\{-\left[2(1-\nu)\varsigma + 1\right]\dfrac{x_1}{x_1^2 + x_3^2} + \dfrac{2x_1 x_3^2}{\left(x_1^2 + x_3^2\right)^2}\right\} \end{cases} \tag{12.73a}$$

[2] *Hirth and Lothe, theory of dislocations, second edition, John Wiley and Sons, 1982, p.75*

$$\begin{cases} \dfrac{\partial u_3}{\partial x_1}=\dfrac{\vec{\Lambda}\vec{e}_3}{2\pi}\dfrac{1}{2(1-v)}\left\{\left[4(1-v)-2(1-v)\varsigma\right]\dfrac{x_1}{x_1^2+x_3^2}+\dfrac{x_1x_3^2-x_1^3}{\left(x_1^2+x_3^2\right)^2}\right\} \\ \dfrac{\partial u_3}{\partial x_3}=\dfrac{\vec{\Lambda}\vec{e}_3}{2\pi}\dfrac{1}{2(1-v)}\left\{\left[2(1-2v)-2(1-v)\varsigma\right]\dfrac{x_3}{x_1^2+x_3^2}+\dfrac{x_3^3-x_1^2x_3}{\left(x_1^2+x_3^2\right)^2}\right\} \end{cases} \qquad (12.73b)$$

As will be preferable thereafter, for the sake of simplicity, to work in polar coordinates $x_1 = r\cos\varphi$ and $x_2 = r\sin\varphi$ in the plane of the cylinder (fig. 12.1), the relationships

$$\frac{\cos\varphi}{r}=\frac{x_1}{x_1^2+x_3^2}\Rightarrow x_1=r\cos\varphi \quad \text{and} \quad \frac{\sin\varphi}{r}=\frac{x_3}{x_1^2+x_3^2}\Rightarrow x_3=r\sin\varphi \qquad (12.74)$$

allow us to obtain the derivatives as function of coordinates $[r,\varphi]$

$$\begin{cases} \dfrac{\partial u_1}{\partial x_1}=\dfrac{\vec{\Lambda}\vec{e}_3}{2\pi}\dfrac{1}{2(1-v)}\left\{\left[2(1-v)\varsigma-1\right]+2\cos^2\varphi\right\}\dfrac{\sin\varphi}{r} \\ \dfrac{\partial u_1}{\partial x_3}=\dfrac{\vec{\Lambda}\vec{e}_3}{2\pi}\dfrac{1}{2(1-v)}\left\{-\left[2(1-v)\varsigma+1\right]+2\sin^2\varphi\right\}\dfrac{\cos\varphi}{r} \\ \dfrac{\partial u_3}{\partial x_1}=\dfrac{\vec{\Lambda}\vec{e}_3}{2\pi}\dfrac{1}{2(1-v)}\left\{\left[4(1-v)-2(1-v)\varsigma\right]-\cos2\varphi\right\}\dfrac{\cos\varphi}{r} \\ \dfrac{\partial u_3}{\partial x_3}=\dfrac{\vec{\Lambda}\vec{e}_3}{2\pi}\dfrac{1}{2(1-v)}\left\{\left[2(1-2v)-2(1-v)\varsigma\right]-\cos2\varphi\right\}\dfrac{\sin\varphi}{r} \end{cases} \qquad (12.75)$$

The tensor of elastic deformation $\vec{\varepsilon}_i^{\,el}$ and the tensor of elastic shear $\vec{\alpha}_i^{\,el}$ can then be expressed in polar coordinates by introducing these derivatives in *(23.58)* and *(23.59)*

$$\begin{cases} \vec{\varepsilon}_1^{\,el}=-\dfrac{\vec{\Lambda}\vec{e}_3}{4\pi(1-v)}\left\{\left[2(1-v)\varsigma+\cos2\varphi\right]\dfrac{\sin\varphi}{r}\vec{e}_1+\left[2(1-v)(1-\varsigma)-\cos2\varphi\right]\dfrac{\cos\varphi}{r}\vec{e}_3\right\} \\ \vec{\varepsilon}_2^{\,el}=0 \\ \vec{\varepsilon}_3^{\,el}=-\dfrac{\vec{\Lambda}\vec{e}_3}{4\pi(1-v)}\left\{\left[2(1-v)(1-\varsigma)-\cos2\varphi\right]\dfrac{\cos\varphi}{r}\vec{e}_1+\left[2(1-2v)-2(1-v)\varsigma-\cos2\varphi\right]\dfrac{\sin\varphi}{r}\vec{e}_3\right\} \end{cases} \qquad (12.76)$$

$$\begin{cases} \vec{\alpha}_1^{\,el}=-\dfrac{\vec{\Lambda}\vec{e}_3}{4\pi(1-v)}\left\{\left[2(1-v)\varsigma-\dfrac{2}{3}(1-2v)+\cos2\varphi\right]\dfrac{\sin\varphi}{r}\vec{e}_1+\left[2(1-v)(1-\varsigma)-\cos2\varphi\right]\dfrac{\cos\varphi}{r}\vec{e}_3\right\} \\ \vec{\alpha}_2^{\,el}=\dfrac{\vec{\Lambda}\vec{e}_3}{4\pi(1-v)}\left[\dfrac{2}{3}(1-2v)\right]\dfrac{\sin\varphi}{r}\vec{e}_2 \\ \vec{\alpha}_3^{\,el}=-\dfrac{\vec{\Lambda}\vec{e}_3}{4\pi(1-v)}\left\{\left[2(1-v)(1-\varsigma)-\cos2\varphi\right]\dfrac{\cos\varphi}{r}\vec{e}_1+\left[-2(1-v)\varsigma+\dfrac{4}{3}(1-2v)-\cos2\varphi\right]\dfrac{\sin\varphi}{r}\vec{e}_3\right\} \end{cases} \qquad (12.77)$$

The rest energy of an edge dislocation

The rest energy of an edge dislocation is obtained by integrating the elastic potential energy stored outside of the string by the lattice

$$e_0 = k_2 \sum_i \left(\vec{\alpha}_i^{el}\right)^2 + k_1 \left(\tau^{(p)}\right)^2 \tag{12.78}$$

We deduce with the following integration

$$E_0^{edge} = \int_R^{R_\infty} \int_0^{2\pi} n e_0(r,\varphi)\, r\, d\varphi\, dr = \int_R^{R_\infty} \int_0^{2\pi} n \left[k_2 \sum_i \left(\vec{\alpha}_i^{el}\right)^2 + k_1 \left(\tau^{(p)}\right)^2 \right] r\, d\varphi\, dr \tag{12.79}$$

the rest energy E_0^{edge} by unit of length of the edge dislocation by using relations *(12.77)* and *(12.62)*

$$E_0^{edge} = \frac{n\vec{\Lambda}^2}{4\pi} \left\{ k_2 \left[4\varsigma^2 - 8\varsigma + \frac{2}{3} \left(\frac{7\nu^2 - 13\nu + 7}{(1-\nu)^2} \right) \right] + k_1 \left(\frac{1-2\nu}{1-\nu} \right)^2 \right\} \ln \frac{R_\infty}{R} \tag{12.80}$$

It is found that this rest energy contains a term dependent on modulus k_2 associated to shear deformations, and a term dependent on modulus k_1 associated to deformation by volume expansion of the medium. The rest energy per unit length of the edge dislocation line can also be expressed more simply by using the relation *(7.30)*

$$E_0^{edge} = \frac{nk_2\vec{\Lambda}^2}{4\pi} \left[4\varsigma^2 - 8\varsigma + \frac{4\nu^2 - 9\nu + 5}{(1-\nu)^2} \right] \ln \frac{R_\infty}{R} \tag{12.81}$$

It should be noted here that *the energy associated to a certain segment of length* L *of a line of edge dislocation does not depend on the state of volume expansion of the solid lattice.* Indeed it can easily be verified that

$$E_0^{edge}\Big|_L = L_0 \frac{n_0 k_2 \left(\vec{\Lambda}_0\right)^2}{4\pi} \left[4\varsigma^2 - 8\varsigma + \frac{4\nu^2 - 9\nu + 5}{(1-\nu)^2} \right] \ln \frac{R_{\infty 0}}{R_0} = E_{00}^{edge}\Big|_{L_0} \tag{12.82}$$

To compare the value of the rest energy of an edge dislocation with that of a screw dislocation, it is enough to remember that the relationship between the linear charge $\vec{\Lambda}$ and the Burgers vector $\vec{B}_{edge}$ is simply written $\vec{\Lambda}^2 = \vec{B}_{edge}^2$ and the modulus k_2 is connected to the *Lamé coefficient* by the relationship $\mu = nk_2$. It is then found that for Burgers vectors of equal length, the edge dislocation has a different rest energy than the screw dislocation by a factor that depends on both the tunable parameter ς and the Poisson's ratio ν of the material

$$\begin{cases} E_0^{screws} = \dfrac{\mu \vec{B}_{screws}^2}{4\pi} \ln \dfrac{R_\infty}{R} \\ E_0^{edge} = \dfrac{\mu \vec{B}_{edge}^2}{4\pi} \ln \dfrac{R_\infty}{R} \left[4\varsigma^2 - 8\varsigma + \dfrac{4\nu^2 - 9\nu + 5}{(1-\nu)^2} \right] \end{cases} \tag{12.83}$$

The kinetic energy of a low velocity edge dislocation

In the case where a string of edge dislocation is moving in direction Ox_1, with velocity $\vec{\boldsymbol{V}}$ which is small compared with the speed of transversal waves c_t, the components of the velocity field are obtained by replacing x_1 by $x_1(t) = x_1 - \boldsymbol{V}t$ in expressions *(12.70)* of the components of the field of displacement, and then by taking the derivative $\vec{\phi} = -d\vec{u}/dt$

$$\begin{cases} \phi_1(x_1,x_3,t) = +\dfrac{\vec{\Lambda}\vec{e}_3}{2\pi}\dfrac{\partial}{\partial t}\left\{\varsigma \arctan\left(\dfrac{x_3}{x_1-\mathbf{v}t}\right)+\dfrac{1}{2(1-\nu)}\dfrac{(x_1-\mathbf{v}t)x_3}{(x_1-\mathbf{v}t)^2+x_3^2}\right\} \\ \phi_3(x_1,x_3,t) = -\dfrac{\vec{\Lambda}\vec{e}_3}{2\pi}\dfrac{\partial}{\partial t}\left\{\left(\dfrac{3-4\nu}{4(1-\nu)}-\dfrac{\varsigma}{2}\right)\ln\left[(x_1-\mathbf{v}t)^2+x_3^2\right]+\dfrac{1}{4(1-\nu)}\dfrac{(x_1-\mathbf{v}t)^2-x_3^2}{(x_1-\mathbf{v}t)^2+x_3^2}\right\} \end{cases} \tag{12.84}$$

After derivation, it is useful to jump into the coordinate system $Ox_1'x_2'x_3'$ which is in translation with velocity $\vec{\mathbf{v}}$ with the line of dislocation, in which $x_1-\mathbf{v}t$ becomes x_1'

$$\begin{cases} \phi_1(x_1',x_3') = +\dfrac{\vec{\Lambda}\vec{e}_3}{2\pi}\left[\left(\varsigma+\dfrac{1}{2(1-\nu)}\right)\dfrac{x_1'^2x_3'}{(x_1'^2+x_3'^2)^2}+\left(\varsigma-\dfrac{1}{2(1-\nu)}\right)\dfrac{x_3'^3}{(x_1'^2+x_3'^2)^2}\right]\mathbf{v} \\ \phi_3(x_1',x_3') = -\dfrac{\vec{\Lambda}\vec{e}_3}{2\pi}\left[\left(\varsigma-\dfrac{5-4\nu}{2(1-\nu)}\right)\dfrac{x_3'^2x_1'}{(x_1'^2+x_3'^2)^2}+\left(\varsigma-\dfrac{3-4\nu}{2(1-\nu)}\right)\dfrac{x_1'^3}{(x_1'^2+x_3'^2)^2}\right]\mathbf{v} \end{cases} \tag{12.85}$$

and to use in $Ox_1'x_2'x_3'$ the polar coordinates $x_1'=r'\cos\varphi'$ and $x_3'=r'\sin\varphi'$ expressed in the plane of the moving cylinder

$$\begin{cases} \phi_1(r',\varphi') = +\dfrac{\vec{\Lambda}\vec{e}_3}{2\pi}\dfrac{1}{r'}\left[\left(\varsigma+\dfrac{1}{2(1-\nu)}\right)\cos^2\varphi'\sin\varphi'+\left(\varsigma-\dfrac{1}{2(1-\nu)}\right)\sin^3\varphi'\right]\mathbf{v} \\ \phi_3(r',\varphi') = -\dfrac{\vec{\Lambda}\vec{e}_3}{2\pi}\dfrac{1}{r'}\left[\left(\varsigma-\dfrac{5-4\nu}{2(1-\nu)}\right)\sin^2\varphi'\cos\varphi'+\left(\varsigma-\dfrac{3-4\nu}{2(1-\nu)}\right)\cos^3\varphi'\right]\mathbf{v} \end{cases} \tag{12.86}$$

The kinetic energy of the edge dislocation is then obtain by operating the following integral

$$E_{kin}^{edge} = \frac{1}{2}m\int_R^{R_\infty}\int_0^{2\pi} n\vec{\phi}^2(r',\varphi')\,r'd\varphi'\,dr' \tag{12.87}$$

It is a rather long computation that leads finally to the kinetic energy of an edge dislocation. This energy is given by unit of length, and it depends on the parameter ς and the Poisson modulus ν, just as with the rest energy of a screw dislocation *(12.81)*

$$E_{kin}^{edge} = \frac{mn\vec{\Lambda}^2}{64\pi}\ln\frac{R_\infty}{R}\left[16\varsigma^2-32\varsigma+\frac{26-56\nu+32\nu^2}{(1-\nu)^2}\right]\mathbf{v} \tag{12.88}$$

12.5 – Effects of the boundary conditions and of the nature of the lattice

The edge dislocation in a usual, finite and free medium

In the case of a *finite and free* lattice, that is to say not subjected to any boundary stress, the presence of an edge dislocation may bend the solid, as illustrated by the model in figure 9.11, which is taken into account by a non-zero value of the constant γ in the relations *(12.65)*. To determine the constants α and γ completely, look for the value of the adjustable parameter ς, so it minimizes the rest energy of the edge dislocation. From equation *(12.81)* we have

$$\frac{\partial E_0^{edge}}{\partial \varsigma}=\frac{nk_2\vec{\Lambda}^2}{\pi}2(\varsigma-1)\ln\frac{R_\infty}{R}=0 \quad \Rightarrow \quad \varsigma=1 \tag{12.89}$$

Thus, the value of ς that minimizes the rest energy of the edge dislocation string is 1. It can be verified that if the string moves relative to the lattice with velocity $\vec{\boldsymbol{v}}$, the value that minimizes the sum of potential energy and kinetic energy is always equal to 1. Indeed, as we have, from the relation *(12.88)*

$$\partial E_{kin}^{edge} / \partial \varsigma = 0 \quad \Rightarrow \quad \varsigma=1 \tag{12.90}$$

Thus in the case of a *usual medium, free and finite,* the energies of the edge dislocation are written

$$\begin{cases} E_0^{edge}=\dfrac{nk_2\vec{\Lambda}^2}{4\pi(1-v)}\ln\dfrac{R_\infty}{R}=\left[\dfrac{4(1-v)}{8v^2-12v+5}\right]M_0^{edge}c_t^2 \\ E_{kin}^{edge}=\dfrac{mn\vec{\Lambda}^2}{8\pi}\left[\dfrac{8v^2-12v+5}{4(1-v)^2}\right]\left(\ln\dfrac{R_\infty}{R}\right)\mathbf{v}^2=\dfrac{1}{2}M_0^{edge}\mathbf{v}^2 \end{cases} \tag{12.91}$$

The rest energy then corresponds to the classic value obtained in the literature[3].
Since the usual solids have a Poisson modulus close to 1/3, the Einstein relation is modified and is approximatively worth $E_0^{edge} \cong 1.41\, M_0^{edge}c_t^2$.
It is interesting to realize that the field of displacement is in this case written as

$$\begin{cases} u_1^{edge}(x_1,x_3)=\dfrac{\vec{\Lambda}\vec{e}_3}{2\pi}\left[-\arctan\dfrac{x_3}{x_1}-\dfrac{1}{2(1-v)}\dfrac{x_1x_3}{x_1^2+x_3^2}\right] \\ u_3^{edge}(x_1,x_3)=\dfrac{\vec{\Lambda}\vec{e}_3}{2\pi}\left[\dfrac{1-2v}{4(1-v)}\ln\left(x_1^2+x_3^2\right)+\dfrac{1}{4(1-v)}\dfrac{x_1^2-x_3^2}{x_1^2+x_3^2}\right] \end{cases} \tag{12.92}$$

and that it corresponds exactly with what is found by the classical theories of dislocations[3]. Only the sign of the expressions $u_1^{edge}(x_1,x_3)$ and $u_3^{edge}(x_1,x_3)$ are inverted, which is a result of the fact that the displacement field expressed in Euler coordinates has a sign opposite to the displacement field expressed in Lagrange coordinates. The tensor of elastic deformation $\vec{\varepsilon}_i^{\,edge}$ can be written

$$\begin{cases} \vec{\varepsilon}_1^{\,edge}=\dfrac{\vec{\Lambda}\vec{e}_3}{4\pi(1-v)}\dfrac{1}{r}\left[-(2-2v+\cos 2\varphi)\sin\varphi\,\vec{e}_1+\cos 2\varphi\cos\varphi\,\vec{e}_3\right] \\ \vec{\varepsilon}_2^{\,edge}=0 \\ \vec{\varepsilon}_3^{\,edge}=\dfrac{\vec{\Lambda}\vec{e}_3}{4\pi(1-v)}\dfrac{1}{r}\left[\cos 2\varphi\cos\varphi\,\vec{e}_1+(2v+\cos 2\varphi)\sin\varphi\,\vec{e}_3\right] \end{cases} \tag{12.93}$$

and the elastic shear strain $\vec{\alpha}_i^{\,edge}$

[3] *Hirth and Lothe, theory of dislocations, second edition, John Wiley and Sons, 1982, p.78*

$$\begin{cases} \vec{\alpha}_1^{edge} = \dfrac{\vec{\Lambda}\vec{e}_3}{4\pi(1-\nu)}\dfrac{1}{r}\left\{\left[-2(2-\nu)/3-\cos 2\varphi\right]\sin\varphi\,\vec{e}_1+\cos 2\varphi\cos\varphi\,\vec{e}_3\right\} \\ \vec{\alpha}_2^{edge} = \dfrac{\vec{\Lambda}\vec{e}_3}{4\pi(1-\nu)}\dfrac{1}{r}\left\{2(1-2\nu)\sin\varphi\,\vec{e}_2/3\right\} \\ \vec{\alpha}_3^{edge} = \dfrac{\vec{\Lambda}\vec{e}_3}{4\pi(1-\nu)}\dfrac{1}{r}\left\{\left[2(1+\nu)/3+\cos 2\varphi\right]\sin\varphi\,\vec{e}_3+\cos 2\varphi\cos\varphi\,\vec{e}_1\right\} \end{cases} \tag{12.94}$$

The edge dislocation in a usual medium which is prevented from bending

If the solid medium is subjected to boundary conditions, it is the adjustable parameter ς which will be chosen so as to satisfy said boundary conditions. The simplest case we can consider, for example, is to assume a rectangular solid in which an edge dislocation is introduced in the direction Ox_2 with a line charge $\vec{\Lambda}$ pointing in the direction Ox_3. In case the solid is completely free, we know that it bends in the direction Ox_1x_3 because of the term $\ln\left(x_1^2+x_3^2\right)$ of the component $u_3^{edge}(x_1,x_3)$ in the relations *(12.65)*. But if, due to the boundary conditions imposed on the solid, it is unable to bend, the term $\ln\left(x_1^2+x_3^2\right)$ in the relations *(12.65)* must be zero, which is possible since there is the free parameter ς to adjust $u_1^{edge}(x_1,x_3)$ and $u_3^{edge}(x_1,x_3)$. We then have

$$\frac{3-4\nu}{4(1-\nu)}-\frac{\varsigma}{2}=0 \quad\Rightarrow\quad \varsigma=\frac{3-4\nu}{2(1-\nu)} \tag{12.95}$$

The energies of the dislocation in this constrained lattice are deduced from *(12.81) and (12.88)*

$$\begin{cases} E_0^{edge}=\dfrac{nk_2\vec{\Lambda}^2}{4\pi}\left[\dfrac{16\nu^2-20\nu+8}{4(1-\nu)^2}\right]\ln\dfrac{R_\infty}{R}=\left[\dfrac{16\nu^2-20\nu+8}{16\nu^2-20\nu+7}\right]M_0^{edge}c_t^2 \\ E_{kin}^{edge}=\dfrac{mn\vec{\Lambda}^2}{8\pi}\left[\dfrac{16\nu^2-20\nu+7}{4(1-\nu)^2}\right]\left(\ln\dfrac{R_\infty}{R}\right)\mathbf{v}^2=\dfrac{1}{2}M_0^{edge}\mathbf{v}^2 \end{cases} \tag{12.96}$$

Since solid media usually have a Poisson modulus on the order of 1/3, the Einstein relation is modified to $E_0^{edge}=1.47\,M_0^{edge}c_t^2$.

The dislocation in an auxetic media, finite and free

In a more exotic environment where the modulus k_1 would be a negligible modulus vis-à-vis the k_2 modulus, so a medium for which the Poisson modulus would be close to -1, and which is called an *auxetic solid medium*, the energy of the edge dislocation is written

$$\begin{cases} E_0^{edge}=\dfrac{nk_2\vec{\Lambda}^2}{8\pi}\ln\dfrac{R_\infty}{R}=\left[\dfrac{8}{25}\right]M_0^{edge}c_t^2 \\ E_{kin}^{edge}=\dfrac{\rho\vec{\Lambda}^2}{8\pi}\left[\dfrac{25}{16}\right]\left(\ln\dfrac{R_\infty}{R}\right)\mathbf{v}^2=\dfrac{1}{2}M_0^{edge}\mathbf{v}^2 \end{cases} \tag{12.97}$$

In the auxetic medium, finite and free, the Einstein relation reads $E_0^{edge}\cong 0.32\,M_0^{edge}c_t^2$.

The edge dislocation in an auxetic media, which is prevented from bending

In the case of an auxetic medium prevented from bending, the energy of an edge dislocation reads

$$\begin{aligned} E_0^{edge} &= \frac{K_2\vec{\Lambda}^2}{4\pi}\left[\frac{11}{4}\right]\ln\frac{R_\infty}{R} = \left[\frac{44}{43}\right]M_0^{edge}c_t^2 \\ E_{cin}^{edge} &= \frac{mn\vec{\Lambda}^2}{8\pi}\left[\frac{43}{16}\right]\left(\ln\frac{R_\infty}{R}\right)\mathbf{v}^2 = \frac{1}{2}M_0^{edge}\mathbf{v} \end{aligned} \qquad (12.98)$$

In an auxetic media which is prevented from bending, something quite remarkable happens, namely that the Einstein equation is very close to an exact Einstein equation, since $E_0^{edge} = 1.02\ M_0^{edge}c_t^2$.

12.6 – Interactions between dislocations

Distortion fields associated with dislocations allow us to calculate the interaction forces acting between dislocations through the force of Peach and Koehler. In this section, we will present some cases of these interactions, limiting ourselves to the dislocations usually found in conventional, finite and free solids, that is to say, including edge dislocations satisfying the "traditional" relationships *(12.91) to (12.94)*.

The interactions between two dislocations

In the case of screw and edge dislocations, the force of Peach and Koehler acting per unit length of a dislocation *(2)* by the effect of a dislocation *(1)* is given by the expression *(11.12)*, namely

$$\vec{\boldsymbol{F}}_{PK(2)} = \sum_k\left(\vec{s}_{k(1)} \wedge \vec{\Lambda}_{k(2)}\right) + \Lambda_{(2)}\vec{m}_{(1)} + \frac{1}{2}\left(\vec{m}_{(1)} \wedge \vec{\Lambda}_{(2)}\right) + \vec{\Lambda}_{(2)}p_{(1)} \qquad (12.99)$$

in which the linear charges $\vec{\Lambda}_{k(2)}$, $\vec{\Lambda}_{(2)}$ and $\Lambda_{(2)}$ are obtained from the vector of Burgers $\vec{\boldsymbol{B}}_{(2)}$ of the dislocation *(2)* and the tangent vector $\vec{t}$ to the dislocation *(2)* by the following relationships

$$\begin{cases} \vec{\Lambda}_{k(2)} = -\vec{t}_{(2)}\boldsymbol{B}_{k(2)} \\ \vec{\Lambda}_{(2)} = \vec{\boldsymbol{B}}_{(2)} \wedge \vec{t}_{(2)} \\ \Lambda_{(2)} = -\vec{\boldsymbol{B}}_{(2)}\vec{t}_{(2)} / 2 \end{cases} \qquad (12.100)$$

With regards to the stress fields $\vec{m}_{(1)}$, $p_{(1)}$ and $\vec{s}_{k(1)}$ which appear in the force of Peach and Koehler, they are obtained from the elastic distortion fields generated by dislocation *(1)*, via relations *(12.2) and (12.14)*, and are written

$$\begin{cases} \vec{s}_{i(1)} = 2nk_2\vec{\alpha}_{i(1)}^{él} \\ \vec{m}_{(1)} = 4nk_2\vec{\omega}_{(1)}^{él} \\ p_{(1)} = n\left(k_0 - 2k_1\tau_{(1)}^{él}\right) \end{cases} \qquad (12.101)$$

For example, we will calculate in the following sections the interaction forces that occur between two parallel dislocations, for two screw strings, an edge and a screw string, and two edge strings.

The interaction between two parallel screw dislocations

A screw dislocation with charge $\Lambda_{(1)} = -\vec{B}_{screw(1)}\vec{e}_2 / 2$ in the direction Ox_2, with $\vec{t} = \vec{e}_2$, located at coordinates $x_1 = x_3 = 0$, is the source of a field of moments obtained from expression *(12.31)*, which we can recast in polar coordinate in the plane Ox_1x_3

$$\vec{m}_{(1)} = 4nk_2\vec{\omega}_{(1)} = \frac{2nk_2\Lambda_{(1)}}{\pi}\frac{\vec{r}}{r^2} = \frac{2nk_2\Lambda_{(1)}}{\pi}\frac{1}{r}\left(\cos\varphi\,\vec{e}_1 + \sin\varphi\,\vec{e}_3\right) \qquad (r > R) \tag{12.102}$$

If a second string parallel to the first one with a charge $\Lambda_{(2)} = -\vec{B}_{screw(2)}\vec{e}_2 / 2$, is situated at a distance r from it, with polar coordinates r and φ in the plane Ox_1x_3, it undergoes a force by unit length which is given by the force of Peach and Koehler $\vec{F}_{PK(2)}$

$$\vec{F}_{PK(2)} = \Lambda_{(2)}\vec{m}_{screw(1)} = \frac{2nk_2}{\pi}\frac{\Lambda_{(1)}\Lambda_{(2)}}{r}\frac{\vec{r}}{r} = \frac{nk_2}{2\pi}\frac{\boldsymbol{B}_{screw(1)}\boldsymbol{B}_{screw(2)}}{r}\left(\cos\varphi\,\vec{e}_1 + \sin\varphi\,\vec{e}_3\right) \tag{12.103}$$

The interaction force between two parallel screw dislocation is therefore repulsive if $\Lambda_{(1)}\Lambda_{(2)} > 0$ and attractive if $\Lambda_{(1)}\Lambda_{(2)} < 0$.

The interaction between an edge dislocation and a parallel screw dislocation

An edge dislocation in the direction Ox_2 ($\vec{t} = \vec{e}_2$), located at coordinates $x_1 = 0$ and $x_3 = 0$, of Burgers vector in the direction Ox_1, has a charge $\vec{\Lambda}_{(1)} = \boldsymbol{B}_{edge(1)}\vec{e}_3$. It is then the source of stress fields $\vec{m}_{(1)}$, $p_{(1)}$ et $\vec{s}_{k(1)}$ obtained from expressions *(12.62) and (12.94)*

$$\begin{cases} \vec{m}_{(1)} = 4nk_2\vec{\omega}_{(1)} = \dfrac{2nk_2\left(\vec{\Lambda}_{(1)}\vec{e}_3\right)}{\pi}\dfrac{1}{r}\cos\varphi\,\vec{e}_2 \\ p_{(1)} = -2nk_1\tau_{(1)} = \dfrac{nk_1\left(\vec{\Lambda}_{(1)}\vec{e}_3\right)}{\pi}\dfrac{1}{r}\left(\dfrac{1-2\nu}{1-\nu}\right)\sin\varphi \end{cases} \qquad (r > R) \tag{12.104}$$

$$\begin{cases} \vec{s}_{1(1)} = \dfrac{nk_2\left(\vec{\Lambda}_{(1)}\vec{e}_3\right)}{2\pi(1-\nu)}\dfrac{1}{r}\left\{\sin\varphi\left[-\dfrac{2}{3}(2-\nu) - \cos 2\varphi\right]\vec{e}_1 + \cos\varphi\cos 2\varphi\,\vec{e}_3\right\} \\ \vec{s}_{2(1)} = \dfrac{nk_2\left(\vec{\Lambda}_{(1)}\vec{e}_3\right)}{2\pi(1-\nu)}\dfrac{1}{r}\left\{\sin\varphi\left[\dfrac{2}{3}(1-2\nu)\right]\vec{e}_2\right\} \\ \vec{s}_{3(1)} = \dfrac{nk_2\left(\vec{\Lambda}_{(1)}\vec{e}_3\right)}{2\pi(1-\nu)}\dfrac{1}{r}\left\{\sin\varphi\left[\dfrac{2}{3}(1+\nu) + \cos 2\varphi\right]\vec{e}_3 + \cos\varphi\cos 2\varphi\,\vec{e}_1\right\} \end{cases} \tag{12.105}$$

If a second line parallel to the first one, but of screw type, is at a distance r, with charge $\Lambda_{(2)} = -\vec{B}_{screw(2)}\vec{e}_2 / 2$, it will undergo a force per unit length given by the force of Peach and Koehler $\vec{F}_{PK(2)}$

$$\vec{F}_{PK(2)} = \vec{s}_{2(1)} \wedge \vec{\Lambda}_{2(2)} = \frac{nk_2\left(\vec{\Lambda}_{(1)}\vec{e}_3\right)}{2\pi(1-\nu)}\frac{1}{r}\left[\left(\frac{2}{3} - \frac{4}{3}\nu\right)\sin\varphi\right]\boldsymbol{B}_{2(2)}\left(\vec{e}_2 \wedge \vec{e}_2\right) = 0 \tag{12.106}$$

We conclude that the interaction force between an edge dislocation and a parallel screw dislocation are always null, whatever their respective positions.

The interaction between two parallel edge dislocations

Given an edge dislocation in direction Ox_2 $(\vec{t}=\vec{e}_2)$, located at $x_1=0$ and $x_3=0$, with a Burgers vector in the direction Ox_1, with a charge $\vec{\Lambda}_{(1)}=\boldsymbol{B}_{edge(1)}\vec{e}_3$, which interacts with a parallel edge dislocation at polar coordinates $[r,\varphi]$ in the plane Ox_1x_3 (fig. 12.3), and with charges $\vec{\Lambda}_{i(2)}$ or $\vec{\Lambda}_{(2)}$ are given by

$$\begin{cases} \vec{\Lambda}_{1(2)}=-\vec{t}_{(2)}\,\boldsymbol{B}_{1(2)}=-\vec{e}_2\boldsymbol{B}_{edge(2)}\cos\theta \\ \vec{\Lambda}_{3(2)}=-\vec{t}_{(2)}\,\boldsymbol{B}_{3(2)}=-\vec{e}_2\boldsymbol{B}_{edge(2)}\sin\theta \\ \vec{\Lambda}_{(2)}=\vec{\boldsymbol{B}}_{coin(2)}\wedge\vec{t}_{(2)}=\boldsymbol{B}_{edge(2)}\cos\theta\vec{e}_3-\boldsymbol{B}_{edge(2)}\sin\theta\vec{e}_1 \end{cases} \tag{12.107}$$

Th first line *(1)* is the source of fields of stress $\vec{m}_{(1)}$, $p_{(1)}$ and $\vec{s}_{k(1)}$ given by *(12.104)* and *(12.105)*. In those fields, the line *(2)* undergoes the following force of Peach and Koehler by unit length

$$\vec{\boldsymbol{F}}_{PK(2)}=\sum_k\left(\vec{s}_{k(1)}\wedge\vec{\Lambda}_{k(2)}\right)+\vec{\Lambda}_{(2)}p_{(1)} \tag{12.108}$$

which is written

$$\vec{\boldsymbol{F}}_{PK(2)}=-\vec{\Lambda}_{1(2)}\wedge\vec{s}_{1(1)}-\vec{\Lambda}_{3(2)}\wedge\vec{s}_{3(1)}+\left(\boldsymbol{B}_{edge(2)}\cos\theta\vec{e}_3-\boldsymbol{B}_{edge(2)}\sin\theta\vec{e}_1\right)p_{(1)} \tag{12.109}$$

and finally

$$\begin{aligned} \vec{\boldsymbol{F}}_{PK(2)}=&\,\vec{e}_2 B_{edge(2)}\cos\theta\wedge\frac{nk_2\left(\vec{\Lambda}_{(1)}\vec{e}_3\right)}{2\pi(1-\nu)}\frac{1}{r}\left\{\sin\varphi\left[\frac{2}{3}\nu-\frac{4}{3}\nu-\cos 2\varphi\right]\vec{e}_1+\cos\varphi\cos 2\varphi\vec{e}_3\right\} \\ &+\vec{e}_2 B_{edge(2)}\sin\theta\wedge\frac{nk_2\left(\vec{\Lambda}_{(1)}\vec{e}_3\right)}{2\pi(1-\nu)}\frac{1}{r}\left\{\sin\varphi\left[\frac{2}{3}+\frac{2}{3}\nu+\cos 2\varphi\right]\vec{e}_3+\cos\varphi\cos 2\varphi\vec{e}_1\right\} \\ &+\left(B_{edge(2)}\cos\theta\vec{e}_3-B_{edge(2)}\sin\theta\vec{e}_1\right)\frac{nk_1\left(\vec{\Lambda}_{(1)}\vec{e}_3\right)}{\pi}\frac{1}{r}\left(\frac{1-2\nu}{1-\nu}\right)\sin\varphi \end{aligned} \tag{12.110}$$

By further transforming this expression, we obtain the classical expression of the interaction between two parallel edge dislocations

$$\begin{aligned} \vec{\boldsymbol{F}}_{PK(2)}=&\,\frac{nk_2\left(\vec{\Lambda}_{(1)}\vec{e}_3\right)}{2\pi(1-\nu)}\frac{1}{r}\left\{\left[B_{edge(2)}\cos\theta\right]\cos\varphi\cos 2\varphi+\left[B_{edge(2)}\sin\theta\right]\sin\varphi\cos 2\varphi\right\}\vec{e}_1 \\ &+\frac{nk_2\left(\vec{\Lambda}_{(1)}\vec{e}_3\right)}{2\pi(1-\nu)}\frac{1}{r}\left\{\left[B_{edge(2)}\cos\theta\right]\sin\varphi(2+\cos 2\varphi)-\left[B_{edge(2)}\sin\theta\right]\cos\varphi\cos 2\varphi\right\}\vec{e}_3 \end{aligned} \tag{12.111}$$

which can also be written by reintroducing the vector of charge $\vec{\Lambda}^{(2)}$ under the form

$$\begin{aligned} \vec{\boldsymbol{F}}_{PK(2)}=&\,\frac{nk_2\left(\vec{\Lambda}_{(1)}\vec{e}_3\right)}{2\pi(1-\nu)}\frac{1}{r}\left\{\left(\vec{\Lambda}_{(2)}\vec{e}_3\right)\cos\varphi\cos 2\varphi-\left(\vec{\Lambda}_{(2)}\vec{e}_1\right)\sin\varphi\cos 2\varphi\right\}\vec{e}_1 \\ &+\frac{nk_2\left(\vec{\Lambda}_{(1)}\vec{e}_3\right)}{2\pi(1-\nu)}\frac{1}{r}\left\{\left(\vec{\Lambda}_{(2)}\vec{e}_3\right)\sin\varphi(2+\cos 2\varphi)+\left(\vec{\Lambda}_{(2)}\vec{e}_1\right)\cos\varphi\cos 2\varphi\right\}\vec{e}_3 \end{aligned} \tag{12.112}$$

In figure 12.4, we have shown the direction of the components of the force of Peach and Koehler in the case where charge $\vec{\Lambda}_{(2)}$ of line *(2)* points along the axis Ox_3.

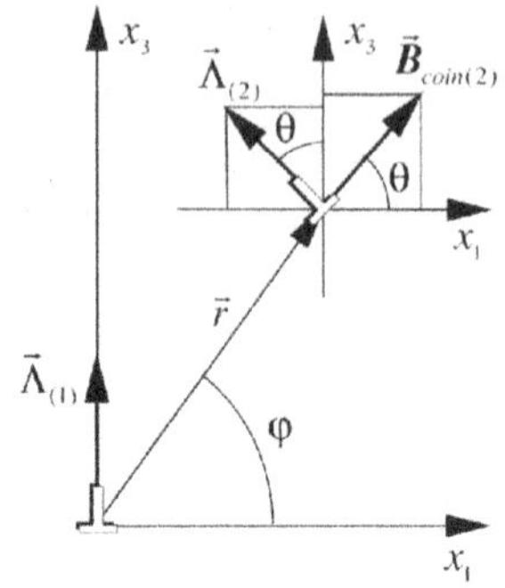

Figure 12.3 *- interaction between two parallel edge dislocations along* Ox_2 *with charges* $\vec{\Lambda}_{(1)}$ *and* $\vec{\Lambda}_{(2)}$

One interesting consequence of this figure is related to the fact that, as an edge dislocation can only easily move by sliding, it is the component of the force of Peach Koehler along the axis Ox_1 which is important in this particular case. The second interesting consequence of this figure is that the force along the axis Ox_1 tends to separate the two lines if the angle φ is less than 45 °, but it tends to superimpose them, if the angle φ is greater than 45 °. The latter case explains why there is formation of a stable ribbon of aligned edge dislocations, with their charges $\vec{\Lambda}$ more or less in the plane of the ribbon, and can thus form a dislocation membrane bounded by two lines of disclination, these are called *flexion joints* as were shown in fig. 9.17a.

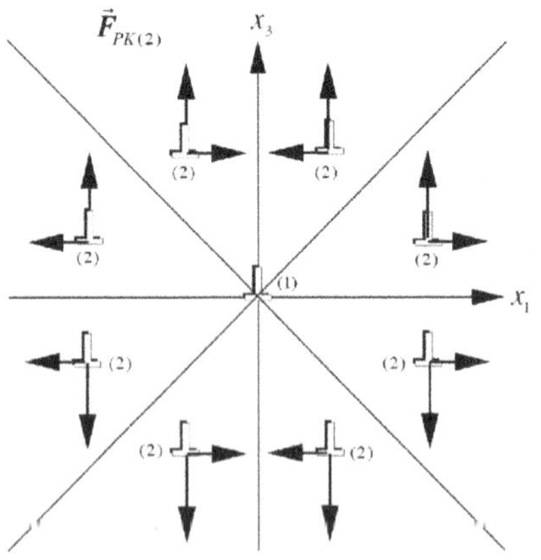

Figure 12.4 *- components of the force of Peach and Koehler due to the interaction between two edge dislocations parallel to axis* Ox_2 *and with charges* $\vec{\Lambda}_{(1)}$ *and* $\vec{\Lambda}_{(2)}$ *along the axis* Ox_3

12.7 – The string model of curved dislocation

The dynamic behavior of a dislocation that curves in a conventional solid, finite and free, can be addressed by introducing the concept of line tension of the dislocation, which allows us to write Newton's equation for an infinitesimal segment of string.

The line tension of a dislocation

Imagine a dislocation which passes through two fixed points $\boldsymbol{A}$ and $\boldsymbol{B}$ separated by a distance L, which is smoothly curved into an arc of radius of curvature r between the two points (fig. 12.5). At great distance from the string, it can be seen as a straight dislocation, but in the vicinity of the string, it has a length equal to $L+\Delta L$ because of its curvature. If we introduce the quantity l corresponding to the distance where we go from the short distance to the long distance, and which roughly corresponds to several times the length L, we can calculate the energy $\boldsymbol{E}_0$ of the dislocation segment of length L using the formulas *(12.43)* for a screw dislocation or *(12.99)* for an edge dislocation, the mathematical expressions are used to separate the energies of the string at short and long distance. We have

$$\boldsymbol{E}_0 = L\frac{\alpha n k_2 \vec{\boldsymbol{B}}^2}{4\pi}\ln\frac{R_\infty}{l} + (L+\Delta L)\frac{\alpha n k_2 \vec{\boldsymbol{B}}^2}{4\pi}\ln\frac{l}{R} = L\frac{\alpha n k_2 \vec{\boldsymbol{B}}^2}{4\pi}\ln\frac{R_\infty}{R} + \Delta L\frac{\alpha n k K_2 \vec{\boldsymbol{B}}^2}{4\pi}\ln\frac{l}{R} \quad (12.113)$$

where $1 \le \alpha \le 1/(1-\nu)$, since α is equal to 1 in the case of the screw dislocation and $1/(1-\nu)$ in the case of an edge dislocation.

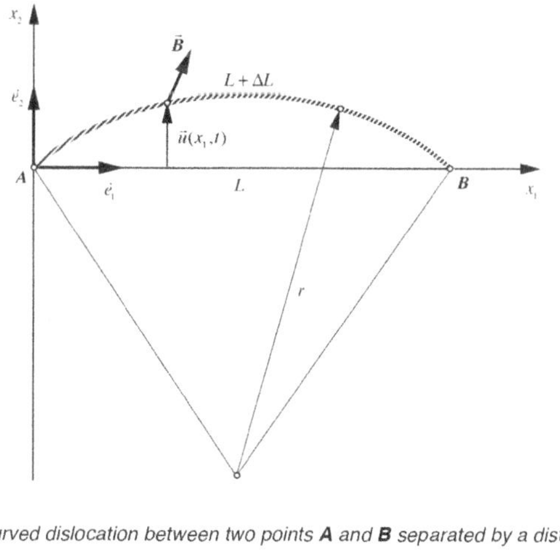

Figure 12.5 - *curved dislocation between two points **A** and **B** separated by a distance* L

The increase in energy ΔE_0 due to the curvature of the string reads

$$\Delta E_0 = \Delta L\frac{\alpha n k_2 \vec{\boldsymbol{B}}^2}{4\pi}\ln\frac{l}{R} \qquad \textit{with} \qquad l \approx L \quad (12.114)$$

The fact that the energy of a curved string is greater than the energy of the straight string implies that a curved string will try to become straight like a rubber band stretched between two points. In the case of an elastic, the elongation of a length equal to ΔL requires an energy equal to $\tau\Delta L$ where τ is the tension force of the elastic. By analogy with an elastic, we can introduce a concept of *line tension* τ *of the dislocation*, corresponding to a fictitious force tangential to the string, which is therefore

$$\Delta E_0 = \Delta L\tau = \Delta L\frac{\alpha n k_2 \vec{B}^2}{4\pi}\ln\frac{l}{R} \quad \Rightarrow \quad \tau = \frac{\alpha n k_2 \vec{B}^2}{4\pi}\ln\frac{l}{R} \tag{12.115}$$

In fact, the analogy between the line tension and an elastic is certainly not very rigorous and somewhat unclear[4]. But despite this, it plays a significant role in dislocation theory, for it is that representation that is used for solving dislocation dynamics problems.

The force of Peach and Koehler acting on a dislocation

Imagine a curved dislocation situated in its slip plane, that is to say in the plane containing both its Burgers vector $\vec{B}$ and $\vec{t}$, *the* vector tangential to the line. Suppose further that this dislocation is subjected to a shear stress tangential to the slip plane, which can be fully characterized by a torsor of moment $\vec{m}$ belonging to the slip plane and perpendicular to the plane of shear. The Peach and Koehler force acting on unit length of this dislocation can then be deduced from the relation *(11.13)*, taking only the terms containing the torsor moment

$$\vec{F}_{PK} = -\frac{1}{2}\left(\vec{B}\vec{t}\right)\vec{m} - \frac{1}{2}\left[\left(\vec{B}\wedge\vec{t}\right)\wedge\vec{m}\right] \tag{12.116}$$

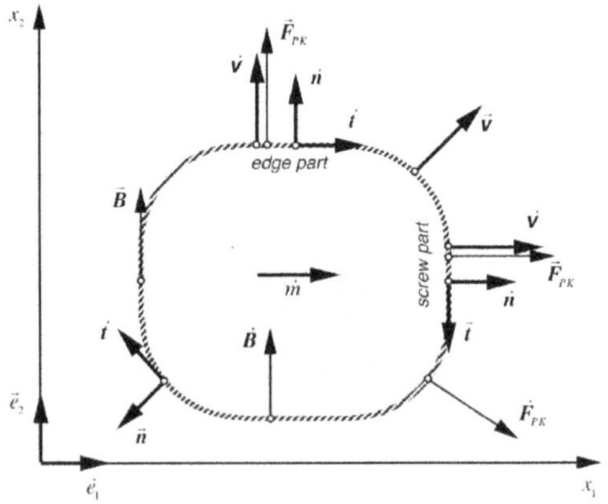

Figure 12.6 - *Close loop of dislocation in its slip plane*

We represent in figure 12.6 a closed loop dislocation in its slip plane Ox_1x_2, with a tangential vector $\vec{t}$ going clockwise and with Burgers vector $\vec{B} = B\vec{e}_2$ in the direction Ox_2. This dislocation is subjected to pure shear stress s_{23} applied in the plane Ox_2x_3. This shear stress can be perfectly represented by a torsor of moment $\vec{m}$ headed in the direction Ox_1, with $\vec{m} = -2s_{23}\vec{e}_1$, so that the strength of the Peach and Koehler force $\vec{F}_{PK}$ can be written as

[4] *J. P. Hirth, T. Jossang, J. Lothe, J. Appl. Phys., vol. 37, p.110, 1966*

$$\vec{F}_{PK} = \left[(\vec{e}_2 \vec{t}) \vec{e}_1 \right] B s_{23} + \left[(\vec{e}_2 \wedge \vec{t}) \wedge \vec{e}_1 \right] B s_{23} \tag{12.117}$$

We then check that the Peach and Koehler force is always directed along the normal vector $\vec{n}$ of the dislocation, contained in the slip plane Ox_1x_2, whatever the direction $\vec{t}$, so that

$$\vec{F}_{PK} = \vec{n}\, B s_{23} \tag{12.118}$$

The dynamic equation of a non-relativistic string

Now consider a similar dislocation to that described in the previous section, but with any shape in its slip plane Ox_1x_2, and subjected to pure shear stress $s_{23}(t)$ which may be time dependent (fig. 12.7). Describe its shape by a displacement vector $\vec{u}(x_1, t)$ in the slip plane, oriented in the direction Ox_2.

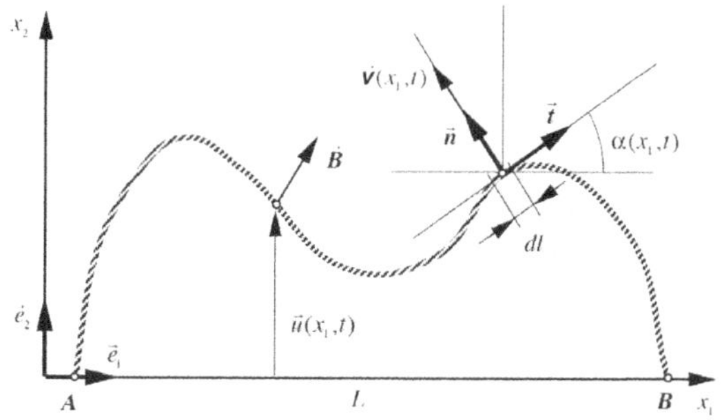

Figure 12.7 - *definition of displacement $\vec{u}(x_1, t)$ of a dislocation in its slip plane*

The goal is then to find the dynamic equation which manages the evolution of the shape $\vec{u}(x_1, t)$ of the dislocation subjected to shear stress. For this, we will try to write Newton's equation of an infinitesimal segment of string of length dl, which has an inclination angle α relative to the axis Ox_1, which is obviously defined by

$$\tan\alpha(x_1, t) = \frac{\partial u(x_1, t)}{\partial x_1} \tag{12.119}$$

If the dislocation moves in the sliding plane, the local velocity $\vec{\mathbf{v}}(x_1, t)$ can only be perpendicular to the string, and its projection on the axis Ox_2 must be equal to the time derivative $\dot{u}(x_1, t)$ of the displacement. The local velocity is thus written

$$\vec{\mathbf{v}}(x_1, t) = \vec{n}\; \mathbf{v}(x_1, t) = \vec{n}\; \frac{\dot{u}(x_1, t)}{\cos\alpha(x_1, t)} \tag{12.120}$$

As for the local acceleration $\vec{\mathbf{a}}(x_1, t)$ of the dislocation, it can also only be perpendicular to the string, and such that its projection onto the axis Ox_2 is equal to the second time derivative of the displacement $\ddot{u}(x_1, t)$. The local acceleration is written therefore

$$\vec{\mathbf{a}}(x_1, t) = \dot{\vec{\mathbf{v}}}(x_1, t) = \vec{n}\; \mathbf{a}(x_1, t) = \vec{n}\; \frac{\ddot{u}(x_1, t)}{\cos\alpha(x_1, t)} \tag{12.121}$$

Newton's equation of non-relativistic dislocation segment dl then involves the inertial mass at rest of the segment $M_0 dl$, and is written thus

$$M_0 dl\, \vec{\boldsymbol{a}} = \sum \vec{f}_i \tag{12.122}$$

where the $\vec{f}_i$ represent the diverse forces acting on the segment dl.

We can now find the set of forces that act on this infinitesimal segment (fig. 12.8), namely:

- *the force of Peach and Koehler* due to the shear stress $s_{23}(t)$

$$\vec{f}_{PK} = \vec{\boldsymbol{n}} \boldsymbol{B} s_{23} dl \tag{12.123}$$

- *the tension forces acting tangentially on the two extremities of the segment*

$$\vec{f}_\tau^{(1)} = -\tau \vec{\boldsymbol{t}}\Big|_{x_1+dx/2} \quad \text{and} \quad \vec{f}_\tau^{(2)} = \tau \vec{\boldsymbol{t}}\Big|_{x_1-dx/2} \tag{12.124}$$

- *a braking force* related to the possible existence of a viscous friction of the moving dislocation within the solid (this is for example the case of braking of dislocations by their interaction with phonons and electrons in metals). If it exists, the braking force is proportional at the first order to the speed of the segment, and can be written by introducing a *coefficient of viscous friction* B_f *by unit of length of the string*

$$\vec{f}_{freinage} = -B_f dl\, \vec{\boldsymbol{v}}(x_1, t) \tag{12.125}$$

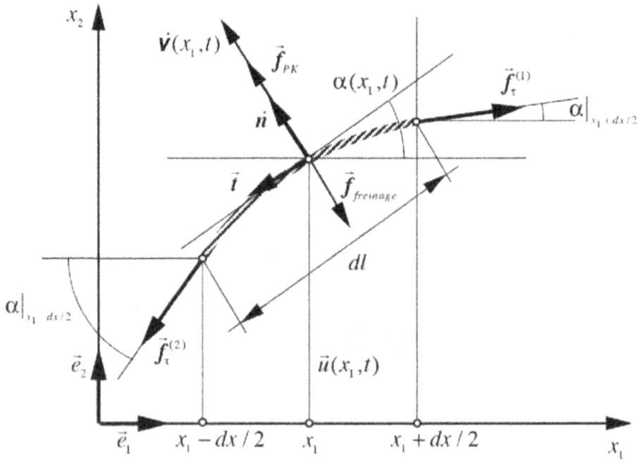

Figure 12.8 *- the forces acting on an infinitesimal segment of string in the sliding plane*

With these forces, the non-relativistic vectorial equation of Newton for the segment dl is written

$$M_0 dl\, \vec{\boldsymbol{a}} = \vec{\boldsymbol{n}} \boldsymbol{B} s_{23} dl - \tau \vec{\boldsymbol{t}}\Big|_{x_1+dx/2} + \tau \vec{\boldsymbol{t}}\Big|_{x_1-dx/2} - B_f dl\, \vec{\boldsymbol{v}} \tag{12.126}$$

which becomes, by introducing the values of $\vec{\boldsymbol{v}}(x_1, t)$ and $\vec{\boldsymbol{a}}(x_1, t)$

$$M_0 dl\, \frac{\ddot{u}}{\cos\alpha}\, \vec{\boldsymbol{n}} = \boldsymbol{B} s_{23} dl\, \vec{\boldsymbol{n}} - \tau \vec{\boldsymbol{t}}\Big|_{x_1+dx/2} + \tau \vec{\boldsymbol{t}}\Big|_{x_1-dx/2} - B_f dl\, \frac{\dot{u}}{\cos\alpha}\, \vec{\boldsymbol{n}} \tag{12.127}$$

And taking into account the following relations

$$\begin{cases} dl = dx / \cos\alpha \\ \textit{projection of } \vec{n} \textit{ on } Ox_2 = \cos\alpha \\ \textit{projection of } -\tau\vec{t}\big|_{x_1+dx/2} \textit{ on } Ox_2 = \tau\sin\alpha\big|_{x_1+dx/2} \\ \textit{projection of } +\tau\vec{t}\big|_{x_1-dx/2} \textit{ on } Ox_2 = -\tau\sin\alpha\big|_{x_1-dx/2} \end{cases} \qquad (12.128)$$

we can project equation *(12.135)* on axis Ox_2, and we have

$$M_0 \frac{\ddot{u}}{\cos\alpha} + B_f \frac{\dot{u}}{\cos\alpha} = Bs_{23} + \frac{\left(\tau\sin\alpha\big|_{x_1+dx/2} - \tau\sin\alpha\big|_{x_1-dx/2}\right)}{dx} \qquad (12.129)$$

The ratio in the last term is the infinitesimal expression of the partial derivative of $\tau\sin\alpha$ with respect to x_1, so that

$$M_0 \frac{\ddot{u}}{\cos\alpha} + B_f \frac{\dot{u}}{\cos\alpha} - \frac{\partial}{\partial x_1}(\tau\sin\alpha) = Bs_{23} \qquad (12.130)$$

All we have to do is to use *(12.127)* in order to express $\cos\alpha$ and $\sin\alpha$ as a function of $\vec{u}(x_1, t)$

$$\tan\alpha = \frac{\partial u}{\partial x_1} \;\Rightarrow\; \begin{cases} \sin\alpha = \dfrac{\tan\alpha}{\sqrt{1+\tan^2\alpha}} = \dfrac{\partial u}{\partial x_1}\Big/\sqrt{1+(\partial u/\partial x_1)^2} \\ \cos\alpha = \dfrac{1}{\sqrt{1+\tan^2\alpha}} = 1\Big/\sqrt{1+(\partial u/\partial x_1)^2} \end{cases} \qquad (12.131)$$

and we obtain the following differential equation, which is called the *"string model of dislocation"*

$$\left(M_0 \frac{\partial^2 u}{\partial t^2} + B_f \frac{\partial u}{\partial t}\right)\sqrt{1+(\partial u/\partial x_1)^2} - \frac{\partial}{\partial x_1}\left[\tau \frac{\partial u}{\partial x_1}\Big/\sqrt{1+(\partial u/\partial x_1)^2}\right] = Bs_{23} \qquad (12.132)$$

that calculates the dynamics $\vec{u}(x_1, t)$ of a non-straight dislocation on its slip plane, in a solid subjected to shear stress $s_{23}(t)$.

In this equation, the line tension τ may depend on the orientation of the string in the sliding surface since the line tension is different for a screw dislocation and an edge dislocation, in which case we must introduce τ as a function of the angle $\alpha(x_1, t)$ with the form $\tau = \tau\big(\alpha(x_1,t)\big) = \tau\big(\arctan(\partial u(x_1,t)/\partial x_1)\big)$. But, generally, we introduce an approximate constant value of τ, which already allows us to obtain satisfactory results.

12.8 – Applications of the string model

The string model is extremely useful and effective for treating plasticity and anelasticity problems due to dislocation motion and involving conventional solids, such as metals for example. But it is beyond the scope of this book to detail these phenomena, which can be read about in many books on this particular subject. However, we will outline how this problem can be addressed on the basis of the string model.

The plasticity and anelasticity due to the dynamics of the dislocations

The dislocation motion interacting with obstacles is one of the main phenomena responsible for the plasticity and the anelasticity of conventional materials[5]. Plasticity is then associated with irreversible long-distance movement of dislocations under the effect of an external stress, while anelasticity is associated with short, but recoverable, dissipative movements of dislocations strongly anchored in the solid and subject to a cyclical external constraint. If these microscopic movements at long or short distance can be calculated for a dislocation from the string model, then it is easy to deduce the speed of plastic $\dot{\varepsilon}_i^{pl}$ or anelastic $\dot{\varepsilon}_i^{an}$ macroscopic deformation of the solid due to a set of dislocation density Λ through the relationships of Orowan *(10.37)* *(10.40)* or *(10.41)*.

The control of mobility of dislocations by obstacles

In conventional materials, such as metals for example, the movement of dislocations is typically controlled by interactions at short or long range, with obstacles more or less localized, as phonons and electrons, other dislocations, point defects, precipitates, grain boundaries, crystal lattice itself, etc.

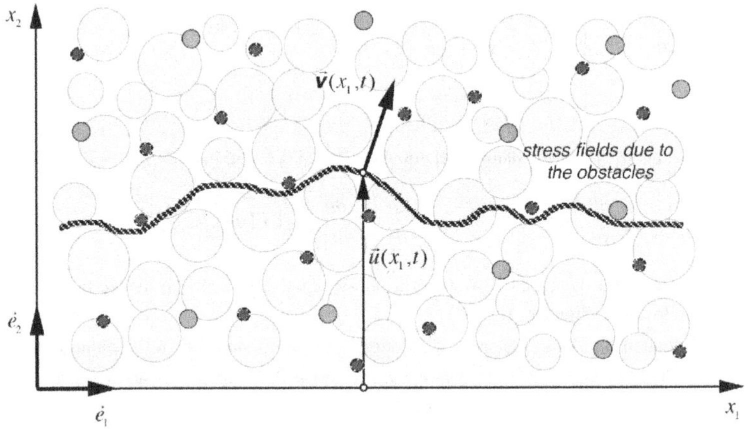

Figure 12.9 *- The stress fields due to obstacles represented in the slip plane*

To introduce this type of interaction in the string model, we must know the spatial distribution of the obstacles in the solid, and the internal stress fields generated by them. These stress fields due to obstacles can be expressed and displayed in the dislocation slip plane (fig. 12.9), where they become responsible for a Peach and Koehler force acting on the dislocation (relations of table 11.1).

In this way one can add in the equation of the string *(12.132)* all the forces $f_n\left(x_1, u(x_1, t)\right)$ due to the N obstacles surrounding a dislocation

[5] *J. P. Hirth, J. Lothe, «Theory of dislocations», second edition, John Wiley and Sons, New York, 1982; V. L. Indenbom, J. Lothe, «Elastic strain fields and dislocation mobility», Elsevier, North-Holland, 1992*

$$\left(M_0 \frac{\partial^2 u}{\partial t^2}+B_f \frac{\partial u}{\partial t}\right)\sqrt{1+(\partial u/\partial x_1)^2}-\frac{\partial}{\partial x_1}\left[\tau\frac{\partial u}{\partial x_1}\Big/\sqrt{1+(\partial u/\partial x_1)^2}\right] = \boldsymbol{B}s_{23}+\sum_{n=1}^{N}f_n\left(x_1,u(x_1,t)\right) \qquad (12.133)$$

But it is clear that such an approach to dislocation dynamics quickly proves very complex. In general, we address these problems of dislocation interactions with obstacles in a much more pragmatic way, by developing simple models based on the string model and judiciously adapted to the problem at hand. To further study this, we find many examples of dislocation interaction mechanisms with obstacles, illustrated by experimental results and theoretical models in many books on dislocations, or in review articles as "*dislocation point defect interactions*" [6] and "*dislocation-lattice interaction*" [7] .

The role of thermal activation during interactions with obstacles

The string equation *(12.133)* in the presence of interactions with obstacles is a purely mechanical equation, which cannot take into account the effects of temperature, such as migration of obstacles by diffusion or obstacle crossing by thermal activation. Introducing the effects of temperature in the string equation is theoretically possible by developing an image of "*Brownian*" dislocation, that is to say by introducing a thermal fluctuation of the local force $\boldsymbol{F}_{fluctuation}\left(x_1,u(x_1,t),t\right)$ due to local temperature in the equation of the string, modeled after the term of thermal fluctuations in the model of the *Langevin equation* [8]

$$\left(M_0 \frac{\partial^2 u}{\partial t^2}+B_f \frac{\partial u}{\partial t}\right)\sqrt{1+(\partial u/\partial x_1)^2}-\frac{\partial}{\partial x_1}\left[\tau\frac{\partial u}{\partial x_1}\Big/\sqrt{1+(\partial u/\partial x_1)^2}\right] = \boldsymbol{B}s_{23}+\sum_{n=1}^{N}f_n\left(x_1,u(x_1,t)\right)+\boldsymbol{F}_{fluctuation}\left(x_1,u(x_1,t),t\right) \qquad (12.134)$$

Such an approach of thermal fluctuation phenomenon proves again very complex, so that, in general, we also address the more pragmatic problems by developing models of the thermal activation suitably adapted to the problem to be treated.

For interested readers, typical examples of this type of thermal activation approach is developed by the author in the articles *"overview on dislocation-point defect interaction: the brownian picture of dislocation motion"* [9] and *"theory of plasticity and anelasticity due to dislocation creep through a multi-scale hierarchy of obstacles"* [10]. In the latter, it is shown among other things that the thermally activated motion of a dislocation moving in a hierarchy of different barriers is of a fractal appearance (fig. 12.10), with interesting effects on the plasticity and anelasticity of the

---

[6] *G. Gremaud, chapter 3.3, in «Mechanical spectroscopy», Trans Tech Publications, Zürich, 2001, p. 178-246*

[7] *W. Benoit, chapter 3.2, in «Mechanical spectroscopy», Trans Tech Publications, Zürich, 2001, p.158-177*

[8] *Isihara, statistical physics, Academic Press, N.-Y., 1971, chapitre 7*

[9] *G. Gremaud, Materials Science and Engineering, A 370, p. 191-198, 2004*

[10] *G. Gremaud, Materials Science and Engineering, A 521-522, p. 12-17, 2009*

macroscopic solid.

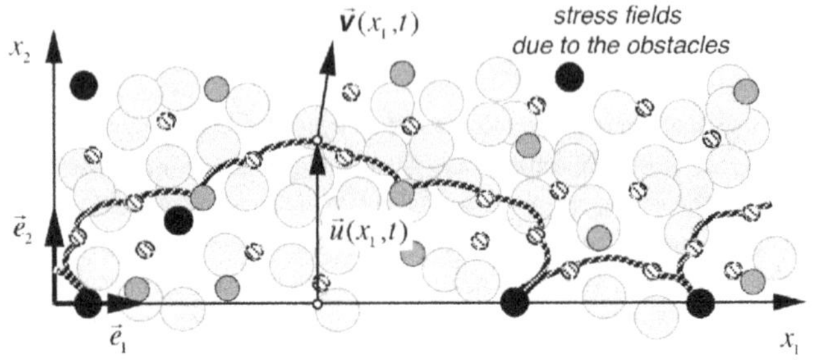

***Figure 12.10** - the fractal behavior of a dislocation moving amid many types of obstacles*

12.9 – The effects of a strong static stress on a pinned dislocation

Suppose a dislocation segment of length L anchored on its sliding surface at two pinning points ***A*** and ***B*** located on the axis Ox_1, $x_1 = 0$ and $x_1 = L$ as shown in figure 12.7. Such pinning of the dislocation in the medium may be due to the existence of strong localized interactions between the dislocation and obstacles (other dislocations, precipitates, etc.).

The static deformation of a pinned dislocation

If this segment is subjected to a constant shear stress s_{23}, it will take a curved static form which we will try to determine. We make the simplifying assumption that one can use an approximately constant line tension, which is independent of the orientation of the segment in the sliding plane. The static equation of the string model *(12.140)* can then be written

$$\frac{\partial}{\partial x_1}\left[\frac{\partial u}{\partial x_1}\Big/\sqrt{1+(\partial u/\partial x_1)^2}\right] = -\frac{\mathbf{B}s_{23}}{\tau} \qquad (12.135)$$

This equation can easily be integrated a first time with respect to x_1. We have, by introducing an integration constant C, a new differential equation

$$\frac{\partial u}{\partial x_1}\Big/\sqrt{1+(\partial u/\partial x_1)^2} = -\frac{\mathbf{B}s_{23}}{\tau}x_1 + C \qquad (12.136)$$

This equation can also be modified to extract a value of the derivative $\partial u/\partial x_1$, and we obtain the following differential equation of first order

$$\frac{\partial u}{\partial x_1} = \frac{C - x_1/r}{\sqrt{1-(C-x_1/r)^2}} \quad \text{with} \quad r = \frac{\tau}{\mathbf{B}s_{23}} \qquad (12.137)$$

The solution is simply

$$u(x_1) = r\sqrt{1-(C-x_1/r)^2} + K \qquad (12.138)$$

where K is a second integration constant. The two integration constants C and K are obtained from the boundary conditions of the string, namely the pinning points for which $u(0)=0$ and $u(L)=0$

$$C=\frac{L}{2r} \quad et \quad K=-r\sqrt{1-C^2}=-\sqrt{r^2-(L/2)^2} \qquad (12.139)$$

With these values, the deformation of the string segment is described by

$$u(x_1)=\sqrt{r^2-(L/2-x_1)^2}-\sqrt{r^2-(L/2)^2} \qquad (12.140)$$

This deformation is an arc of circle which goes through the pinning points ***A*** and ***B***, whose *radius of curvature* corresponds to $r=\tau/(\boldsymbol{B}s_{23})$, as illustrated in figure 12.5.

The limit of the static deformation and the source of dislocations of Frank-Read

The radius of curvature of the segment $r=\tau/(\boldsymbol{B}s_{23})$ is inversely proportional to the static shear stress s_{23}, which means that it decreases when s_{23} increases. However, it is clear that there is a minimum limit for the radius of curvature, which occurs for a critical stress such that the radius of curvature $s_{23}\big|_{cr}$ becomes equal to $L/2$, therefore

$$s_{23}\big|_{cr}=\frac{2\tau}{\boldsymbol{B}L} \qquad (12.141)$$

For any value of s_{23} greater than $s_{23}\big|_{cr}$, there cannot be static solutions for the deformation of the string. What appears instead is a complex dynamic solution of equation *(12.132)*, which corresponds to a *mechanism of source of dislocations of Frank-Read* [11] (fig. 12.11).

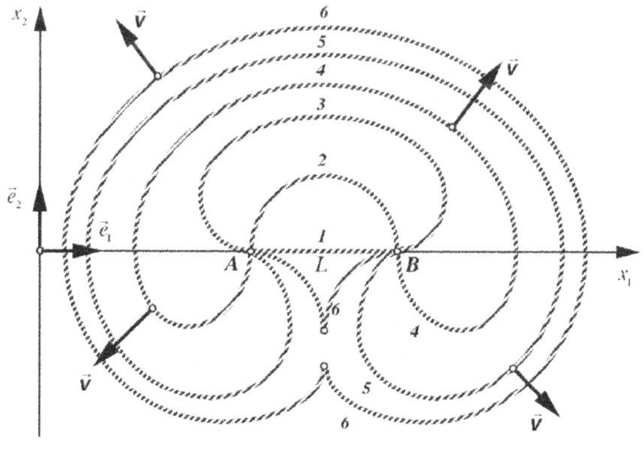

Figure 12.11 *- the mechanism of source of dislocations of Frank-Read*

The initial rectilinear segment represented by *(1)* in figure 25.7 bends between the two pinning points to form a half-circle *(2)*. Then it continues to extend beyond the pinning points, steps *(3)*, *(4)* and *(5)*, until the strand segment leaving ***A*** join the segment strand starting from ***B*** *(6)*. Here, as the two strands have the same Burgers vector, they bind to each other forming, first of

[11] *F. C. Frank, W. T. Read, Phys. Rev., vol.79, p.722, 1950*

all a new segment *(1)* growing between the pinning points **A** and **B**, and secondly a closed loop *(6)* that will not stop growing. This mechanism is equivalent to an uninterrupted source of dislocation loops. It is essentially this phenomenon, which is very well observed by electron microscopy, which makes it possible to perform large plastic deformations of certain solids such as metals. And it is this phenomenon that may be responsible for the existence of a non-zero source of dislocation charges in the continuity equation *(10.9)* of the density of dislocation charges.

12.10 – The effects of a dynamical stress on a pinned dislocation

Assume again a dislocation segment pinned in its sliding plane at two points **A** and **B** located on the axis Ox_1, at $x_1 = 0$ and $x_1 = L$, as shown in figure 12.7. And suppose further that this segment is subject to an external dynamical stress $s_{23}(t)$ such that $s_{23}(t) << s_{23}\big|_{cr}$. To solve the problem of this moving pinned dislocation, it is clear that the string model *(12.134)* is far too sophisticated. In the case of low stress, it is useful to develop simplified models.

The classical model of the string for weak constraints [12]

In the case where $s_{23}(t) << s_{23}\big|_{cr}$, the angle α that the string forms with respect to axis Ox_1 is always weak so that

$$\tan\alpha = \partial u / \partial x_1 << 1 \tag{12.142}$$

We rewrite the string model by supposing that $\partial u / \partial x_1 \cong 0$, and one obtains the well-known *'classic model'* of the string

$$M_0 \frac{\partial^2 u}{\partial t^2} + B_f \frac{\partial u}{\partial t} - \tau \frac{\partial^2 u}{\partial x_1^{\,2}} = \boldsymbol{B} s_{23} \tag{12.143}$$

In this model we show that the static deformation obtained in the case of a stress s_{23} is approximated by a parabola of equation

$$u(x_1) = \frac{\boldsymbol{B} s_{23}}{2\tau}(L - x_1)\, x_1 \tag{12.144}$$

which maximum is worth

$$\max(u) = u(x_1 = L/2) = \frac{\boldsymbol{B} L^2}{8\tau} s_{23} \tag{12.145}$$

The «rigid rod model» of a pinned dislocation submitted to weak stresses

The model of the classic string can further be simplified by looking for an approximate equation for the mean displacement $\bar{u}$ of the dislocation, defined by relation

$$\bar{u} = \frac{1}{L}\int_0^L u(x_1)dx_1 = \frac{1}{L}\int_0^L \frac{\boldsymbol{B} s_{23}}{2\tau}(L - x_1)x_1 dx_1 = \frac{\boldsymbol{B} L^2}{12\tau} s_{23} \tag{12.146}$$

12 *A. Granato and K. Lücke, J. Appl. Phys., vol. 27, p. 583, 1956; A. Granato and K. Lücke, J. Appl. Phys., vol. 27, p. 789, 1956*

For a constant stress s_{23}, the average statical displacement $\bar{u}$ satisfies the following equation, by introducing a *restoring coefficient* K_u *due to the line tension* τ of the string

$$K_u \bar{u} = \boldsymbol{B} s_{23} \quad \text{with} \quad K_u = \frac{12\tau}{L^2} \tag{12.147}$$

We can then find a dynamical model which is approximative for the pinned dislocation, by writing that the average displacement $\bar{u}(t)$ satisfies the following equation

$$M_0\, \ddot{\bar{u}}(t) + B_f\, \dot{\bar{u}}(t) + K_u\, \bar{u}(t) = \boldsymbol{B} s_{23} \tag{12.148}$$

This equation is called the *"rigid rod model"* because it describes the average displacement of the dislocation, as if it were a rigid bar subjected to a restoring force. It is in fact nothing but the equation of a *damped harmonic oscillator*, and is obviously a rough approximation of reality, since it presupposes that the dynamic deformation of the dislocation remains a parabola which is clearly not the case if the speed of variation of s_{23} becomes very large!

The «phonon relaxation» of dislocations in metals

In a metal, there is always a substantial density Λ of dislocations. They generally form a three-dimensional network known as the *Frank network* [13], consisting of nodes to which dislocations converge. These nodes are to be considered as anchor points for lattice dislocations, so that we can consider that, for low applied stresses, dislocations are represented by segments, of average length $\bar{L}$ that bend under the effect of applied stresses. If such a metal is subjected to an ultrasonic dynamic stress $s_{23}(t) = s_0 \sin \omega t$ of low amplitude but high frequency, these segments will begin to vibrate.

In the equation of the rigid rod model *(12.148)*, we use the rest mass M_0 per unit length of dislocation, and the viscous friction coefficient B_f per unit length of string. In the absence of other interactions, the viscous friction coefficient B_f of a metal is essentially due to the interaction of mobile dislocations with phonons of the lattice [14]. The friction coefficient is controlled by two mechanisms: the *"phonon wind"* which appears with a linear temperature dependence of B_f, and the *"slow phonon relaxation"*, which appears with a parabolic behavior of B_f while temperature is below the Debye temperature of the metal, and with a constant behavior of B_f when the temperature is higher than the Debye temperature of the metal. The behavior of the dislocation segments under the effect of ultrasound then depends primarily on the relative values of M_0 and B_f.

To study this behavior, we can show that the short distance microscopic movements of the dislocation segments will be perceived as a macroscopic anelastic shear deformation α_{23}^{an}. Indeed, applying the relationships of Orowan *(10.40)* to the case of the string shown in figure 12.6, assuming that the dislocation loop extends at a constant velocity $\boldsymbol{V}$ in all directions, we can verify that we always have the following relationships, throughout the string portions

---

[13] *F. C. Frank, in Report of the Conference on Defects in Crystalline Solids, Phys. Soc., London, p. 159, 1954*

[14] *V. I. Alshits, the Phonon-dislocation interaction and its rôle in dislocation dragging and thermal resistivity, chapter 11 in «Elastic strain fields and dislocation mobility», Elsevier, North-Holland, 1992*

$$\begin{cases} \dot{\vec{\alpha}}_1^{pl} = 0 \\ \dot{\vec{\alpha}}_2^{pl} = -\Lambda B \mathbf{v} \vec{e}_3 / 2 \\ \dot{\vec{\alpha}}_3^{pl} = -\Lambda B \mathbf{v} \vec{e}_2 / 2 \end{cases} \tag{12.149}$$

which imply that

$$\dot{\alpha}_{23}^{pl} = \dot{\alpha}_{32}^{pl} = \frac{1}{2} \Lambda B \mathbf{v} = \frac{1}{2} \Lambda B \dot{\vec{u}} \tag{12.150}$$

Thus, at the macroscopic scale of the metal, the microscopic motion represented by equation *(12.148)* can be put in the form of an equation of anelasticity

$$\vec{u} = \frac{2}{\Lambda B} \alpha_{23}^{an} \quad \Rightarrow \quad \frac{2M_0}{\Lambda B^2} \ddot{\alpha}_{23}^{an} + \frac{2B_f}{\Lambda B^2} \dot{\alpha}_{23}^{an} + \frac{2K_u}{\Lambda B^2} \alpha_{23}^{an} = s_{23} \tag{12.151}$$

which corresponds exactly to the phenomenological equation of anelasticity *(7.89)* which we introduced in chapter 7, namely

$$M_{an} \ddot{\alpha}_{23}^{an} + B_{an} \dot{\alpha}_{23}^{an} + K_{an} \alpha_{23}^{an} = s_{23} \tag{12.152}$$

In the presence of such an equation, we saw in chapter 7 that the response to a solicitation $s_{23}(t) = s_0 \sin \omega t$ might be a resonance at frequency $\omega_0 = \sqrt{K_{an} / M_{an}} = \sqrt{K_u / M_0}$, or a relaxation with a relaxation time $\tau_{an} = B_{an} / K_{an} = B_f / K_u$, depending on whether the product $\omega_0 \tau_{an}$ is greater or lesser than 1, so whether $B_f > \sqrt{M_0 K_u}$ or $B_f < \sqrt{M_0 K_u}$.
Experimentally, there is a relaxation around 20 to 150 MHz in many metals, which could probably be attributed to the interaction dislocation-phonon [15]. Therefore, it was concluded that in general the coefficient of friction B_f due to phonon outweighs the inertial mass M_0 of dislocations, and therefore we can neglect the term in M_0 in the string equation when dealing with dynamic problems of dislocations in metals.
There are several interesting consequences to this experimental observation:

- the fact that the dislocation segments have a relaxation in the field of MHz in the metal implies that the dislocations become perfectly still for frequencies well above the relaxation frequency, so that this field of very high frequency is ideal for measuring pure elastic properties of the lattice without interference due to dislocations,

- this same fact implies that if we study dislocations by mechanical spectroscopy in a frequency range well below the frequency of the phonon relaxation, it is perfectly possible to neglect the term B_f, and consequently the term M_0 in the string model, meaning that the pinned dislocation segments are always on the mechanical balance with the low frequency applied stress,

- as the coefficient of friction B_f due to the phonons prevails over the inertial mass M_0 of the dislocations in the metals, the dislocations are over-amortized, so that pinned dislocation segments cannot present vibrations like we could observe on a tight rope. This means among other things that, under the effect of thermal disturbances of the network, a dislocation segment has to behave as a "Brownian segment" which would deform continuously under the effect of local thermal perturbations of stress.

[15] *A. Granato, K. Lücke, J. Appl. Phys., vol. 27, p. 789, 1956*

It is on the basis of these very important observations that we can quite easily develop many models of interactions between dislocations and obstacles, such as the reader may consult in numerous books on dislocations, and in the articles [2,3,5,6] cited above.

APPENDIX

Mathematical formulas

Appendix A

Mathematical formulas

The mathematical concepts needed for the comprehension of this book are summed up in this appendix, written up as a review of differential calculus. We establish in this annex the formulas for integral derivatives on mobile domains.

A.1 - Vectorial calculus

The scalar product

The *scalar product of* two vectors $\vec{u}$ and $\vec{v}$ gives us a scalar a. In a orthonormal framework, the scalar product is calculated as the sum of the products of the components

$$a = \vec{u}\,\vec{v} = u_1 v_1 + u_2 v_2 + u_3 v_3 = \sum_i u_i v_i \qquad (A.1)$$

The *intrinsic properties* of the scalar product are the following

$$\begin{cases} a = \vec{u}\,\vec{v} = |\vec{u}||\vec{v}|\cos\alpha \\ a = \vec{u}\,\vec{v} = \vec{v}\,\vec{u} \\ \vec{u} \perp \vec{v} \quad \Rightarrow \quad \vec{u}\,\vec{v} = 0 \\ |\vec{u}| = \sqrt{\vec{u}\,\vec{u}} = \sqrt{\vec{u}^2} = L_u \quad (norm) \end{cases} \qquad (A.2)$$

The vectorial product

The *vectorial product* of two vectors $\vec{u}$ and $\vec{v}$ gives us a vector $\vec{w}$. In a orthonormal framework, the *vectorial product* of two vectors can be calculated by using the formalism of determinant calculation in the following way

$$\vec{w} = \vec{u} \wedge \vec{v} = \begin{vmatrix} \vec{e}_1 & \vec{e}_2 & \vec{e}_3 \\ u_1 & u_2 & u_3 \\ v_1 & v_2 & v_3 \end{vmatrix} = \sum_i \vec{e}_i \left(u_j v_k - u_k v_j \right) \qquad (A.3)$$

The *intrinsic properties* of the vectorial product are the following

$$\begin{cases} |\vec{w}| = |\vec{u} \wedge \vec{v}| = |\vec{u}||\vec{v}|\sin\alpha = S \quad (area) \\ \vec{w} = \vec{u} \wedge \vec{v} = -\vec{v} \wedge \vec{u} \\ \vec{w} \perp \vec{u} \quad et \quad \vec{w} \perp \vec{v} \\ \vec{u} \parallel \vec{v} \quad \Rightarrow \quad \vec{u} \wedge \vec{v} = 0 \end{cases} \qquad (A.4)$$

The mixed product

The *mixed product* of three vectors $\vec{u}$, $\vec{v}$ and $\vec{w}$ is composed of the scalar product and the vectorial product which gives us a scalar b.

In a orthonormal framework, the *mixed product* of three vectors is equal to the determinant of the components of the three vectors

$$b = [\vec{u}, \vec{v}, \vec{w}] = \vec{u}(\vec{v} \wedge \vec{w}) = \begin{vmatrix} u_1 & u_2 & u_3 \\ v_1 & v_2 & v_3 \\ w_1 & w_2 & w_3 \end{vmatrix} = \sum_i u_i (v_j w_k - v_k w_j) \tag{A.5}$$

The *intrinsic properties* of the mixed product are the following:

$$\begin{cases} b = [\vec{u}, \vec{v}, \vec{w}] = \vec{u}(\vec{v} \wedge \vec{w}) = V \quad (volume) \\ b = \vec{u}(\vec{v} \wedge \vec{w}) = \vec{v}(\vec{w} \wedge \vec{u}) = \vec{w}(\vec{u} \wedge \vec{v}) \end{cases} \tag{A.6}$$

Compositions of products

The three products of vectors, the scalar product, the vectorial product and the mixed product can be composed together. These compositions verify the following equations

$$\begin{cases} \vec{u} \wedge (\vec{v} \wedge \vec{w}) = (\vec{u}\,\vec{w})\vec{v} - (\vec{u}\,\vec{v})\vec{w} \\ (\vec{u} \wedge \vec{t})(\vec{v} \wedge \vec{w}) = (\vec{u}\,\vec{v})(\vec{t}\,\vec{w}) - (\vec{u}\,\vec{w})(\vec{t}\,\vec{v}) \\ (\vec{u} \wedge \vec{t}) \wedge (\vec{v} \wedge \vec{w}) = [\vec{u}, \vec{t}, \vec{w}]\vec{v} - [\vec{u}, \vec{t}, \vec{v}]\vec{w} \\ [\vec{u}, \vec{v}, \vec{w}]\vec{t} = [\vec{t}, \vec{v}, \vec{w}]\vec{u} + [\vec{u}, \vec{t}, \vec{w}]\vec{v} + [\vec{u}, \vec{v}, \vec{t}]\vec{w} \end{cases} \tag{A.7}$$

A.2 - Vectorial Analysis

The gradient operator

The *gradient of a scalar field* f is a vectorial field $\vec{u}$ defined in each point $\boldsymbol{A}$ of space by the following limit taken on the surface $\boldsymbol{S}$ limiting a volume $\boldsymbol{V}$ taken around point $\boldsymbol{A}$ (figure A.1)

$$\vec{u}\big|_A = \overrightarrow{\text{grad}}\, f\big|_A = \lim_{V \to 0} \frac{1}{\boldsymbol{V}} \oiint_S f\, d\vec{S} \tag{A.8}$$

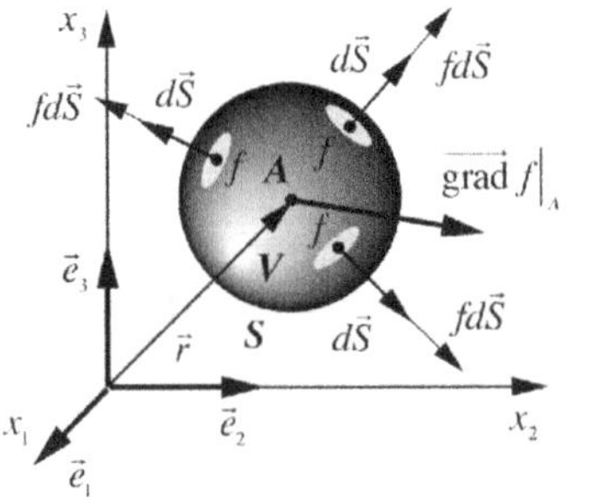

Figure A.1 *- gradient operator*

The direction of vector $\vec{u}$ is consequently perpendicular to the level surfaces of the function f in space and its norm is proportional to the velocity of variation of the function f in that direction. This intrinsic *definition* shows that the gradient is an *invariant* of field f, meaning that it does not depend on the system of coordinates that one chooses. In a orthonormal framework, the gradient can be calculated thanks to the following relation

$$\vec{u}(\vec{r}) = \overrightarrow{\text{grad}}\, f = \frac{\partial f}{\partial x_1}\vec{e}_1 + \frac{\partial f}{\partial x_2}\vec{e}_2 + \frac{\partial f}{\partial x_3}\vec{e}_3 = \sum_i \frac{\partial f}{\partial x_i}\vec{e}_i \qquad (A.9)$$

The *intrinsic properties* of the gradient operator are the following:

- the derivative of f in a given direction $\vec{s}$ in space is given by the scalar product

$$\frac{df}{ds} = \vec{s}\,\overrightarrow{\text{grad}} f \qquad (A.10)$$

- if $\overrightarrow{\text{grad}}\, f = \overrightarrow{\text{grad}}\, g$, then $f = g + constant$ (A.11)

- if $\vec{u} = \overrightarrow{\text{grad}}\, f$, $\vec{u}$ is called a *conservative vectorial field* and it can be derived from the *scalar potential* f. In this case, if the field $\vec{u}$ is known, the field f can be easily found, by resolution of $df = u_1 dx_1 + u_2 dx_2 + u_3 dx_3$, modulo an additive constant. If $\vec{u} = \overrightarrow{\text{grad}}\, f$, we also have

$$\int_{P_1}^{P_2} \vec{u}\, d\vec{r} = f(\boldsymbol{P}_2) - f(\boldsymbol{P}_1) \qquad (A.12)$$

$$\oint_C \vec{u}\, d\vec{r} = 0 \quad ; \quad \forall\ closed\ contour\ \boldsymbol{C} \qquad (A.13)$$

- the *gradient theorems*

$$\oiint_S f\, d\vec{S} = \iiint_V \overrightarrow{\text{grad}}\, f\, dV \qquad (A.14)$$

$$\oint_C f\, d\vec{r} = -\iint_S \overrightarrow{\text{grad}}\, f \wedge d\vec{S} \qquad (A.15)$$

The curl (rotational) operator

The *curl (rotational) of a vectorial field* $\vec{u}$ is another vectorial field $\vec{v}$ which satisfies in all point $\boldsymbol{A}$ of space the following limit taken on a contour C around point $\boldsymbol{A}$ and perpendicular to any direction $\vec{n}$ in space (figure A.2)

$$\vec{n}\vec{v}\big|_A = \vec{n}\,\overrightarrow{\text{rot}}\,\vec{u}\big|_A = \lim_{S\to 0} \frac{1}{S} \oint_C \vec{u}\, d\vec{r} \qquad (A.16)$$

The direction of vector $\vec{v}$ is consequently perpendicular to the surface of maximum circulation of a vector $\vec{u}$ around $\boldsymbol{A}$ and its norm is proportional to the velocity of circulation of $\vec{u}$ around $\boldsymbol{A}$ in that direction.

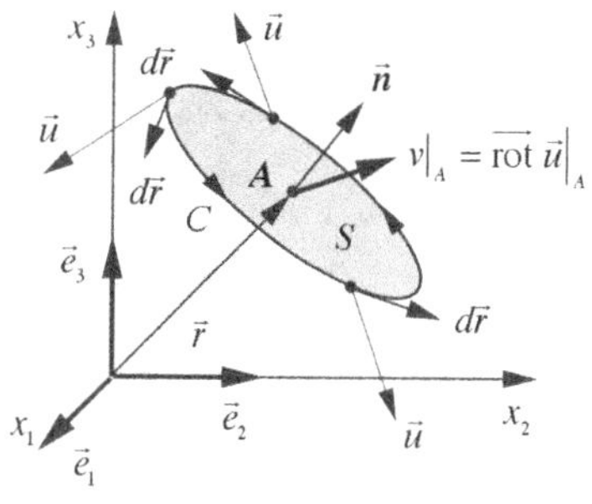

Figure A.2 *- rotational operator*

This *intrinsic definition* shows that rotational is an *invariant* of field $\vec{u}$, meaning that it does not depend on the system of coordinates we have chosen. In a orthonormal framework, the rotational can be calculated thanks to the formalism of the determinant, in the following fashion

$$\vec{v}(\vec{r}) = \overrightarrow{\mathrm{rot}}\,\vec{u} = \begin{vmatrix} \vec{e}_1 & \vec{e}_2 & \vec{e}_3 \\ \partial/\partial x_1 & \partial/\partial x_2 & \partial/\partial x_3 \\ u_1 & u_2 & u_3 \end{vmatrix} \qquad (A.17)$$

The *intrinsic properties* of the rotational operator are the following:

- $\overrightarrow{\mathrm{rot}}\,\overrightarrow{\mathrm{grad}}\, f \equiv 0$ (A.18)
- if $\overrightarrow{\mathrm{rot}}\,\vec{u} \equiv 0$, $\vec{u}$ is said to be *irrotational vectorial field.* In this case, the previous property implies that $\vec{u} = \overrightarrow{\mathrm{grad}}\, f$
- if $\overrightarrow{\mathrm{rot}}\,\vec{u} = \overrightarrow{\mathrm{rot}}\,\vec{v}$, then $\vec{u} = \vec{v} + \overrightarrow{\mathrm{grad}}\, f$ (A.19)
- if $\vec{v} = \overrightarrow{\mathrm{rot}}\,\vec{u}$, $\vec{v}$ is said to be a *rotational vectorial field* and it derives from the *vector potential* $\vec{u}$. In this case, if the vectorial field $\vec{v}$ is known, the vectorial field $\vec{u}$ can be found up to the gradient of a scalar field
- rotational theorems

$$\oiint_S d\vec{S} \wedge \vec{v} = \iiint_V \overrightarrow{\mathrm{rot}}\,\vec{v}\, dV \qquad (A.20)$$

$$\oint_C \vec{v}\, d\vec{r} = \iint_S \overrightarrow{\mathrm{rot}}\,\vec{v}\, d\vec{S} \qquad (A.21)$$

The divergence operator

The *divergence of a vectorial field* $\vec{u}$ is a scalar field g satisfying in every point A of space the following limit taken on surface S limiting volume V taken around point A (figure A.3)

$$g\big|_A = \mathrm{div}\,\vec{u}\big|_A = \lim_{V \to 0} \frac{1}{V} \oiint_S \vec{u}\, d\vec{S} \qquad (A.22)$$

The scalar g represents as a consequence the limit of the field flux $\vec{u}$ through the surface S and is only different from zero if the field $\vec{u}$ diverges locally around a point A . This *intrinsic definition* shows that the divergence is an *invariant* of the field $\vec{u}$, meaning that it does not depend on the system of coordinates that was chosen.

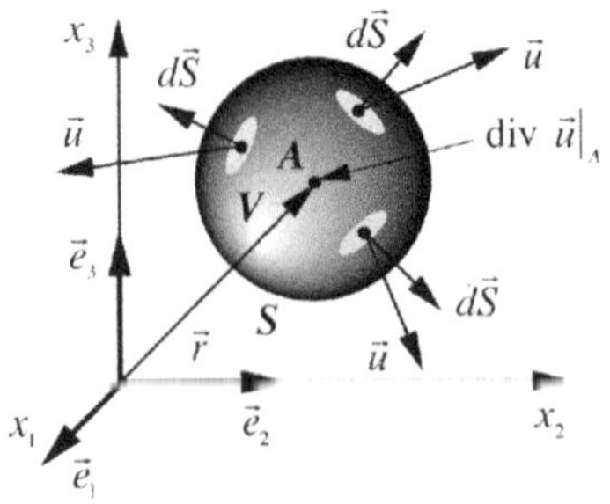

Figure A.3 - *divergence operator*

In an orthonormal framework, the divergence can be calculated thanks to the following relation

$$g(\vec{r}) = \operatorname{div} \vec{u} = \frac{\partial u_1}{\partial x_1} + \frac{\partial u_2}{\partial x_2} + \frac{\partial u_3}{\partial x_3} = \sum_i \frac{\partial u_i}{\partial x_i} \qquad (A.23)$$

The *intrinsic properties* of the divergence operator are the following:

- $\operatorname{div} \overrightarrow{\operatorname{rot}} \vec{u} \equiv 0$ (A.24)

- if $\operatorname{div} \vec{v} \equiv 0$, $\vec{v}$ is said to be a *conservative vectorial field or solenoidal field* and the previous property implies that $\vec{v} = \overrightarrow{\operatorname{rot}} \vec{u}$

- if $\operatorname{div} \vec{u} = \operatorname{div} \vec{v}$, then $\vec{u} = \vec{v} + \overrightarrow{\operatorname{rot}} \vec{w}$ (A.25)

- if $g = \operatorname{div} \vec{u}$, $\vec{u}$ is said to be a *divergent vectorial field.* In this case, if the scalar g is known, the vectorial field $\vec{u}$ can be calculated up to a rotational vectorial field

- the *divergence theorem*

$$\oiint_S \vec{v}\, d\vec{S} = \iiint_V \operatorname{div} \vec{v}\, dV \qquad (A.26)$$

The Laplacian operator

The *laplacian of a scalar field* f, or the *laplacian of a vectorial field* $\vec{u}$, is a scalar field g, or a vectorial field $\vec{v}$, defined intrinsically by applying twice the vectorial operators div, $\overrightarrow{\operatorname{grad}}$ or $\overrightarrow{\operatorname{rot}}$ by the following relations

$$\begin{cases} g(\vec{\xi}) = \Delta f = \operatorname{div} \overrightarrow{\operatorname{grad}} f \\ \vec{v}(\vec{\xi}) = \Delta \vec{u} = \overrightarrow{\operatorname{grad}} \operatorname{div} \vec{u} - \overrightarrow{\operatorname{rot}}\, \overrightarrow{\operatorname{rot}} \vec{u} \end{cases} \qquad (A.27)$$

These *intrinsic definitions* show that the Laplacian is an *invariant* of field f or $\vec{u}$, meaning that it does not depend on the system of coordinates that were chosen, since it is defined in terms of the operators div, $\overrightarrow{\operatorname{grad}}$ or $\overrightarrow{\operatorname{rot}}$. In an orthonormal framework, the Laplacian can be calculated thanks to the following relations

$$\begin{cases} g(\vec{r}) = \Delta f = \operatorname{div} \overrightarrow{\operatorname{grad}} f = \dfrac{\partial^2 f}{\partial x_1^2} + \dfrac{\partial^2 f}{\partial x_2^2} + \dfrac{\partial^2 f}{\partial x_3^2} = \sum_i \dfrac{\partial^2 f}{\partial x_i^2} \\ \vec{v}(\vec{r}) = \Delta \vec{u} = \overrightarrow{\operatorname{grad}} \operatorname{div} \vec{u} - \overrightarrow{\operatorname{rot}}\, \overrightarrow{\operatorname{rot}} \vec{u} = \dfrac{\partial^2 \vec{u}}{\partial x_1^2} + \dfrac{\partial^2 \vec{u}}{\partial x_2^2} + \dfrac{\partial^2 \vec{u}}{\partial x_3^2} = \sum_i \dfrac{\partial^2 \vec{u}}{\partial x_i^2} \end{cases} \qquad (A.28)$$

The *intrinsic properties* of the laplacian operator are the following:

- $\begin{cases} \Delta \vec{u} = \overrightarrow{\operatorname{grad}} \operatorname{div} \vec{u} \quad \textit{if} \quad \overrightarrow{\operatorname{rot}} \vec{u} = 0 \\ \Delta \vec{u} = -\overrightarrow{\operatorname{rot}}\, \overrightarrow{\operatorname{rot}} \vec{u} \quad \textit{if} \quad \operatorname{div} \vec{u} = 0 \end{cases}$ (A.29)

- $\Delta \vec{u} = \vec{e}_1 \Delta u_1 + \vec{e}_2 \Delta u_2 + \vec{e}_3 \Delta u_3$ (A.30)

- The *Green formulas*

$$\iiint_V f \Delta g\, dV = \oiint_S f\, \overrightarrow{\operatorname{grad}}\, g\, d\vec{S} - \iiint_V \overrightarrow{\operatorname{grad}} f\, \overrightarrow{\operatorname{grad}}\, g\, dV \qquad (A.31)$$

$$\iiint_V (f \Delta g - g \Delta f)\, dV = \oiint_S \left(f\, \overrightarrow{\operatorname{grad}}\, g - g\, \overrightarrow{\operatorname{grad}} f \right) d\vec{S} \qquad (A.32)$$

The « del » operator

The formal operator *"del"* $\vec{\nabla}$ is an operator defined by the relation

$$\vec{\nabla} = \vec{e}_1 \frac{\partial}{\partial x_1} + \vec{e}_2 \frac{\partial}{\partial x_2} + \vec{e}_3 \frac{\partial}{\partial x_3} = \sum_i \vec{e}_i \frac{\partial}{\partial x_i} \tag{A.33}$$

To use this operator, some simple rules are observed including the following:

- *"del"* behaves as a vector,

- if *"del"* works on a product, the result is the sum of expressions obtained by considering successively the derivation of each of the terms (derivation rule of a product),

- *"del"* must come before the terms on which it acts. If, after transformation, *"del"* acts on a term on which it did not before, we consider this term as a constant as far as *"del"* is concerned,

- The classic operators of vectorial analysis are then expressed in the following form:

$$\begin{cases} \overrightarrow{\text{grad}}\, f = \vec{\nabla} f \\ \text{div}\,\vec{u} = \vec{\nabla}\vec{u} \\ \overrightarrow{\text{rot}}\,\vec{u} = \vec{\nabla} \wedge \vec{u} \end{cases} \quad \text{and} \quad \begin{cases} \Delta f = (\vec{\nabla}\vec{\nabla}) f = \vec{\nabla}^2 f \\ \Delta \vec{u} = (\vec{\nabla}\vec{\nabla})\vec{u} = \vec{\nabla}^2 \vec{u} \end{cases} \tag{A.34}$$

On the linearity, distributivity and iteration of the vectorial operators

The vectorial operators satisfy a set of rules that are very convenient, linked to *linearity, distributivity* and *iteration* of these diverse operators as shown by the following formulas, where α et β are constants

$$\begin{cases} \overrightarrow{\text{grad}}(\alpha f + \beta g) = \alpha\, \overrightarrow{\text{grad}}\, f + \beta\, \overrightarrow{\text{grad}}\, g \\ \text{div}(\alpha\vec{u} + \beta\vec{v}) = \alpha\, \text{div}\,\vec{u} + \beta\, \text{div}\,\vec{v} \\ \overrightarrow{\text{rot}}(\alpha\vec{u} + \beta\vec{v}) = \alpha\, \overrightarrow{\text{rot}}\,\vec{u} + \beta\, \overrightarrow{\text{rot}}\,\vec{v} \\ \Delta(\alpha f + \beta g) = \alpha \Delta f + \beta \Delta g \\ \Delta(\alpha\vec{u} + \beta\vec{v}) = \alpha \Delta\vec{u} + \beta \Delta\vec{v} \end{cases} \tag{A.35}$$

$$\begin{cases} \overrightarrow{\text{grad}}(fg) = f\, \overrightarrow{\text{grad}}\, g + g\, \overrightarrow{\text{grad}}\, f \\ \overrightarrow{\text{grad}}\,(\vec{u}\vec{v}) = \vec{u} \wedge \overrightarrow{\text{rot}}\,\vec{v} + \vec{v} \wedge \overrightarrow{\text{rot}}\,\vec{u} + (\vec{u}\vec{\nabla})\vec{v} + (\vec{v}\vec{\nabla})\vec{u} \\ \text{div}(f\vec{v}) = f\, \text{div}\,\vec{v} + \vec{v}\, \overrightarrow{\text{grad}}\, f \\ \text{div}(\vec{u} \wedge \vec{v}) = \vec{v}\, \overrightarrow{\text{rot}}\,\vec{u} - \vec{u}\, \overrightarrow{\text{rot}}\,\vec{v} \\ \overrightarrow{\text{rot}}(f\vec{v}) = f\, \overrightarrow{\text{rot}}\,\vec{v} + \overrightarrow{\text{grad}}\, f \wedge \vec{v} \\ \overrightarrow{\text{rot}}(\vec{u} \wedge \vec{v}) = \vec{u}\, \text{div}\,\vec{v} - \vec{v}\, \text{div}\,\vec{u} + (\vec{v}\vec{\nabla})\vec{u} - (\vec{u}\vec{\nabla})\vec{v} \\ \Delta(fg) = f \Delta g + g \Delta f + 2\, \overrightarrow{\text{grad}}\, f\, \overrightarrow{\text{grad}}\, g \end{cases} \tag{A.36}$$

$$\begin{cases} \overrightarrow{\text{rot}}\ \overrightarrow{\text{grad}}\, f \equiv 0 \\ \text{div}\ \overrightarrow{\text{rot}}\,\vec{v} \equiv 0 \\ \text{div}\ \overrightarrow{\text{grad}}\, f = \Delta f \\ \overrightarrow{\text{rot}}\, \overrightarrow{\text{rot}}\ \vec{v} = \overrightarrow{\text{grad}}\, \text{div}\,\vec{v} - \Delta\vec{v} \end{cases} \tag{A.37}$$

A.3 - Derivatives of integrals on volumes, surfaces and mobile contours

The derivative of an integral on a mobile volume

Given the integral of a function $f(\vec{r},t)$ on a volume V_m mobile with velocity $\vec{\phi}(\vec{r},t)$ in space and the problem of calculation of temporal variation of this integral. To carry out this calculation, we must consider the volume V_m at instant t, which becomes volume V'_m at instant $t+\Delta t$ (figure A.4). By definition the temporal derivative of the integral is equal to the following limit, calculated for Δt going to zero

$$\begin{aligned}\frac{d}{dt}\iiint_{V_m} f(\vec{r},t)dV &= \lim_{\Delta t\to 0}\frac{1}{\Delta t}\left(\iiint_{V'_m} f(\vec{r},t+\Delta t)dV - \iiint_{V_m} f(\vec{r},t)dV\right)\\ &= \lim_{\Delta t\to 0}\frac{1}{\Delta t}\left(\iiint_{V_m}\left[f(\vec{r},t+\Delta t)-f(\vec{r},t)\right]dV + \iiint_{V'_m-V_m} f(\vec{r},t+\Delta t)dV\right)\end{aligned} \qquad (A.38)$$

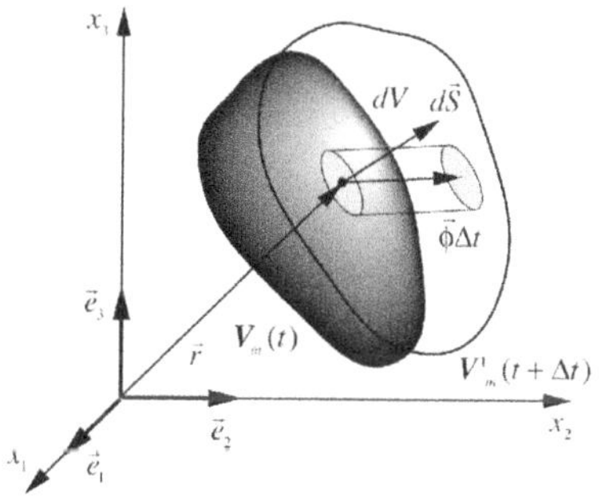

Figure A.4 - *derivative of an integral on a mobile volume*

The two limits of this last expression can be calculated by introducing the volume $dV = \vec{\phi}\Delta t d\vec{S}$ defined in the figure with a "cylindrical" shape, and we have

$$\lim_{\Delta t\to 0}\frac{1}{\Delta t}\iiint_{V_m}\left[f(\vec{r},t+\Delta t)-f(\vec{r},t)\right]dV = \iiint_{V_m}\frac{\partial f(\vec{r},t)}{\partial t}dV \qquad (A.39)$$

$$\lim_{\Delta t\to 0}\frac{1}{\Delta t}\iiint_{V'_m-V_m} f(\vec{r},t+\Delta t)dV = \lim_{\Delta t\to 0}\frac{1}{\Delta t}\oiint_{S_m} f(\vec{r},t+\Delta t)\vec{\phi}\Delta t d\vec{S} = \oiint_{S_m}\vec{\phi}f(\vec{r},t)d\vec{S} = \iiint_{V_m}\operatorname{div}\left(f(\vec{r},t)\vec{\phi}\right)dV \qquad (A.40)$$

so that the formula we seek is written under the form

$$\frac{d}{dt}\iiint_{V_m} f\,dV = \iiint_{V_m}\left[\frac{\partial f}{\partial t}+\operatorname{div}\left(f\vec{\phi}\right)\right]dV \qquad (A.41)$$

Just as the argument of the integral can be written

$$\frac{\partial f}{\partial t}+\operatorname{div}\left(f\vec{\phi}\right) = \frac{\partial f}{\partial t}+(\vec{\phi}\vec{\nabla})f + f\operatorname{div}\vec{\phi} \qquad (A.42)$$

the formula for the calculation of the derivative of the integral on a field $\vec{g}$ is deduced from *(A.*

68)

$$\frac{d}{dt}\iiint_{V_m}\vec{g}\,dV=\iiint_{V_m}\left[\frac{\partial\vec{g}}{\partial t}+(\vec{\phi}\vec{\nabla})\vec{g}+\vec{g}\,\mathrm{div}\vec{\phi}\right]dV \qquad (A.43)$$

The derivative of an integral on a mobile surface

Given the field integral $\vec{g}(\vec{r},t)$ on a surface S_m mobile with velocity $\vec{\phi}(\vec{r},t)$ and the problem of the calculation of the temporal variations of this integral. To carry out this calculation, we must consider the surface S_m at instant t, which becomes the surface S'_m at instant $t+\Delta t$ (figure A.5). By definition, the temporal derivative of the integral is equal to the following limit, calculated for Δt tending to zero

$$\frac{d}{dt}\iint_{S_m}\vec{g}(\vec{r},t)d\vec{S}=\lim_{\Delta t\to 0}\frac{1}{\Delta t}\left(\iint_{S'_m}\vec{g}(\vec{r},t+\Delta t)d\vec{S}'-\iint_{S_m}\vec{g}(\vec{r},t)d\vec{S}\right) \qquad (A.44)$$

$$\frac{d}{dt}\iint_{S_m}\vec{g}(\vec{r},t)d\vec{S}=\lim_{\Delta t\to 0}\frac{1}{\Delta t}\left(\iint_{S_m}[\vec{g}(\vec{r},t+\Delta t)-\vec{g}(\vec{r},t)]d\vec{S}\right)+\lim_{\Delta t\to 0}\frac{1}{\Delta t}\left(\iint_{S'_m}\vec{g}(\vec{r},t+\Delta t)d\vec{S}'-\iint_{S_m}\vec{g}(\vec{r},t+\Delta t)d\vec{S}\right) \qquad (A.45)$$

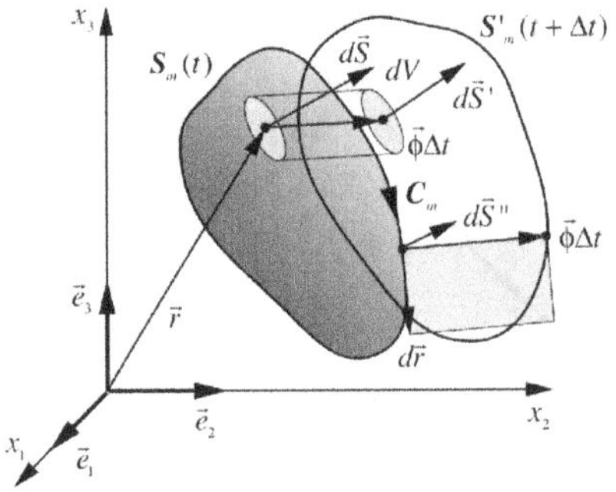

***Figure A.5** - derivative of an integral on a mobile surface*

The first limit is deduced immediately

$$\lim_{\Delta t\to 0}\frac{1}{\Delta t}\left(\iint_{S_m}[\vec{g}(\vec{r},t+\Delta t)-\vec{g}(\vec{r},t)]d\vec{S}\right)=\iint_{S_m}\frac{\partial\vec{g}(\vec{r},t)}{\partial t}d\vec{S} \qquad (A.46)$$

To calculate the second limit, we must define S''_m as a lateral surface linking the contours of surfaces S_m and S'_m, S_t as the closed surface made up of the surfaces S_m, S'_m and S''_m, V_m as the volume comprised inside the closed surfaces S_t and C_m as the contour around the surface S_m, then we effect the following transformations, by using the divergence and rotational theorems

$$\iint_{S'_m}\vec{g}\,d\vec{S}'-\iint_{S_m}\vec{g}\,d\vec{S}=\iint_{S_t}\vec{g}\,d\vec{S}-\iint_{S''_m}\vec{g}\,d\vec{S}''=\iiint_{V_m}\mathrm{div}\,\vec{g}\,dV-\oint_{C_m}\vec{g}\left(d\vec{r}\wedge\vec{\phi}\Delta t\right)$$

$$=\iint_{S_m}\mathrm{div}\,\vec{g}\left(\vec{\phi}\Delta t\,d\vec{S}\right)-\Delta t\oint_{C_m}d\vec{r}\left(\vec{\phi}\wedge\vec{g}\right)=\Delta t\iint_{S_m}\vec{\phi}\,\mathrm{div}\,\vec{g}\,d\vec{S}-\Delta t\iint_{S_m}\overrightarrow{\mathrm{rot}}\left(\vec{\phi}\wedge\vec{g}\right)d\vec{S} \qquad (A.47)$$

We have for the second limit

$$\lim_{\Delta t\to 0}\frac{1}{\Delta t}\left(\iint_{S'_m}\vec{g}(\vec{r},t+\Delta t)d\vec{S}\,'-\iint_{S_m}\vec{g}(\vec{r},t+\Delta t)d\vec{S}\right)=\iint_{S_m}\left[\vec{\phi}\operatorname{div}\vec{g}-\overrightarrow{\operatorname{rot}}\left(\vec{\phi}\wedge\vec{g}\right)\right]d\vec{S} \tag{A.48}$$

so that the formula we seek is written

$$\frac{d}{dt}\iint_{S_m}\vec{g}\,d\vec{S}=\iint_{S_m}\left[\frac{\partial\vec{g}}{\partial t}+\vec{\phi}\operatorname{div}\vec{g}-\overrightarrow{\operatorname{rot}}\left(\vec{\phi}\wedge\vec{g}\right)\right]d\vec{S} \tag{A.49}$$

If the integral is calculated on a closed surface, the previous relationship is written

$$\oiint_{S_m}\overrightarrow{\operatorname{rot}}\left(\vec{\phi}\wedge\vec{g}\right)d\vec{S}=\iiint_{V_m}\operatorname{div}\left[\overrightarrow{\operatorname{rot}}\left(\vec{\phi}\wedge\vec{g}\right)\right]dV\equiv 0 \tag{A.50}$$

so that the derivation formula becomes

$$\frac{d}{dt}\oiint_{S_m}\vec{g}\,d\vec{S}=\oiint_{S_m}\left(\frac{\partial\vec{g}}{\partial t}+\vec{\phi}\operatorname{div}\vec{g}\right)d\vec{S} \tag{A.51}$$

This last relation could have been derived directly from formula *(A.43)* by using the divergence theorem

$$\begin{aligned}\frac{d}{dt}\oiint_{S_m}\vec{g}\,d\vec{S}&=\frac{d}{dt}\iiint_{V_m}\operatorname{div}\vec{g}\,dV=\iiint_{V_m}\left[\frac{\partial\operatorname{div}\vec{g}}{\partial t}+\operatorname{div}\left(\vec{\phi}\operatorname{div}\vec{g}\right)\right]dV\\&=\iiint_{V_m}\operatorname{div}\left(\frac{\partial\vec{g}}{\partial t}+\vec{\phi}\operatorname{div}\vec{g}\right)dV=\oiint_{S_m}\left(\frac{\partial\vec{g}}{\partial t}+\vec{\phi}\operatorname{div}\vec{g}\right)d\vec{S}\end{aligned} \tag{A.52}$$

The derivative of an integral on a mobile contour

Given the integral of a field $\vec{g}(\vec{r},t)$ on a contour C_m mobile with velocity $\vec{\phi}(\vec{r},t)$ and the problem of calculating the temporal variation of that integral. To carry that calculation, we must consider a contour C_m at instant t, which becomes contour C'_m at instant $t+\Delta t$ (figure A.6). By definition, the temporal derivative of the integral is equal to the following limit, calculated for Δt tending to zero

$$\begin{aligned}\frac{d}{dt}\int_{C_m}\vec{g}(\vec{r},t)d\vec{r}&=\lim_{\Delta t\to 0}\frac{1}{\Delta t}\left(\int_{C'_m}\vec{g}(\vec{r},t+\Delta t)d\vec{r}\,'-\int_{C_m}\vec{g}(\vec{r},t)d\vec{r}\right)\\&=\lim_{\Delta t\to 0}\frac{1}{\Delta t}\left\{\int_{C_m}\left[\vec{g}(\vec{r},t+\Delta t)-\vec{g}(\vec{r},t)\right]d\vec{r}\right\}\\&+\lim_{\Delta t\to 0}\frac{1}{\Delta t}\left[\int_{C'_m}\vec{g}(\vec{r},t+\Delta t)d\vec{r}\,'-\int_{C_m}\vec{g}(\vec{r},t+\Delta t)\,d\vec{r}\right]\end{aligned} \tag{A.53}$$

The first term of the limit transforms by giving a partial derivative

$$\lim_{\Delta t\to 0}\frac{1}{\Delta t}\left\{\int_{C_m}\left[\vec{g}(\vec{r},t+\Delta t)-\vec{g}(\vec{r},t)\right]d\vec{r}\right\}=\int_{C_m}\frac{\partial\vec{g}(\vec{r},t)}{\partial t}\,d\vec{r} \tag{A.54}$$

To calculate the second term of the limit, we have to define C_t like a closed contour

$A_1A_2B_2B_1A_1$ and S as a surface surrounded by the contour C_t, and then to effect the following transformations, by applying the Stokes theorem of the rotational

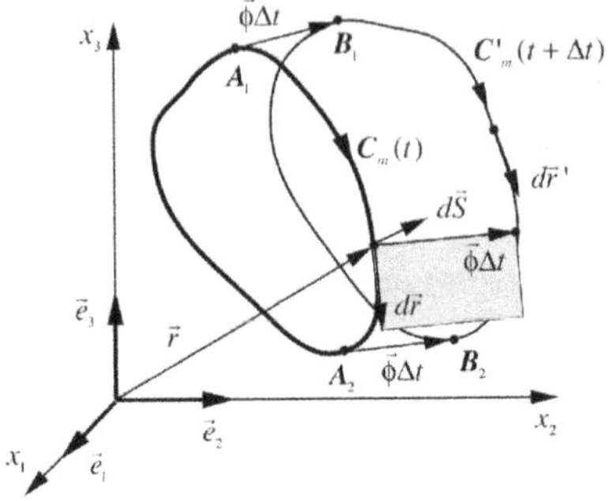

Figure A.6 *- derivative of an integral on a mobile contour*

$$\int_{C'_m} \vec{g}\,d\vec{r}\,' - \int_{C_m} \vec{g}\,d\vec{r} = -\oint_{C_t} \vec{g}\,d\vec{r} - \int_{A_1}^{B_1} \vec{g}\,d\vec{r} + \int_{A_2}^{B_2} \vec{g}\,d\vec{r}$$
$$= -\iint_S \overrightarrow{\text{rot}}\,\vec{g}\,d\vec{s} - \left[\vec{g}\vec{\phi}\right]_{A_1}\Delta t + \left[\vec{g}\vec{\phi}\right]_{A_2}\Delta t \qquad (A.55)$$

$$\int_{C'_m} \vec{g}\,d\vec{r}\,' - \int_{C_m} \vec{g}\,d\vec{r} = -\int_{C_m} \overrightarrow{\text{rot}}\,\vec{g}\left(d\vec{r} \wedge \vec{\phi}\Delta t\right) - \left[\vec{g}\vec{\phi}\right]_{A_1}\Delta t + \left[\vec{g}\vec{\phi}\right]_{A_2}\Delta t$$
$$= -\Delta t \int_{C_m} \left(\vec{\phi} \wedge \overrightarrow{\text{rot}}\,\vec{g}\right)d\vec{r} - \left[\vec{g}\vec{\phi}\right]_{A_1}\Delta t + \left[\vec{g}\vec{\phi}\right]_{A_2}\Delta t \qquad (A.56)$$

The second term of the limit is then

$$\lim_{\Delta t \to 0} \frac{1}{\Delta t}\left[\int_{C'_m} \vec{g}\,d\vec{r}\,' - \int_{C_m} \vec{g}\,d\vec{r}\right] = \left[\vec{g}\vec{\phi}\right]_{A_2} - \left[\vec{g}\vec{\phi}\right]_{A_1} - \int_{C_m} \left(\vec{\phi} \wedge \overrightarrow{\text{rot}}\,\vec{g}\right)d\vec{r} \qquad (A.57)$$

so that the derivation formula sought after is written

$$\frac{d}{dt}\int_{C_m} \vec{g}\,d\vec{r} = \left[\vec{g}\vec{\phi}\right]_{A_2} - \left[\vec{g}\vec{\phi}\right]_{A_1} + \int_{C_m} \left(\frac{\partial \vec{g}}{\partial t} - \vec{\phi} \wedge \overrightarrow{\text{rot}}\,\vec{g}\right)d\vec{r} \qquad (A.58)$$

On a closed contour, the points A_1 and A_2 are equal, so that it becomes in this case

$$\frac{d}{dt}\oint_{C_m} \vec{g}\,d\vec{r} = \oint_{C_m} \left(\frac{\partial \vec{g}}{\partial t} - \vec{\phi} \wedge \overrightarrow{\text{rot}}\,\vec{g}\right)d\vec{r} \qquad (A.59)$$

Furthermore, this last formula could have been directly established by relation *(A.52)*. Indeed

$$\frac{d}{dt}\oint_{C_m} \vec{g}\,d\vec{r} = \frac{d}{dt}\iint_S \overrightarrow{\text{rot}}\,\vec{g}\,d\vec{S} = \iint_{S_m}\left[\frac{\partial\,\overrightarrow{\text{rot}}\,\vec{g}}{\partial t} + \vec{\phi}\,\text{div}\,\overrightarrow{\text{rot}}\,\vec{g} + \overrightarrow{\text{rot}}\left(\overrightarrow{\text{rot}}\,\vec{g} \wedge \vec{\phi}\right)\right]d\vec{S}$$
$$= \iint_{S_m} \text{rot}\left[\frac{\partial \vec{g}}{\partial t} + \text{rot}\,\vec{g} \wedge \vec{\phi}\right]d\vec{S} = \oint_{C_m}\left(\frac{\partial \vec{g}}{\partial t} - \vec{\phi} \wedge \text{rot}\,\vec{g}\right)d\vec{r} \qquad (A.60)$$

Made in the USA
Monee, IL
04 November 2026

67380105R00174